# BOAS OF THE WEST INDIES

# BOAS OF THE WEST INDIES

## *Evolution, Natural History, and Conservation*

R. Graham Reynolds, Robert W. Henderson, Luis M. Díaz,
Tomás M. Rodríguez-Cabrera, and Alberto R. Puente-Rolón

With a Foreword by Jonathan B. Losos

Comstock Publishing Associates
an imprint of
Cornell University Press
Ithaca and London

Funding to support production of this book was provided by the Silver Boa Trust

First published 2023 by Cornell University Press

Design and composition by Julie Allred, BW&A Books
Printed in China

Library of Congress Cataloging-in-Publication Data
Names: Reynolds, R. Graham, author. | Henderson, Robert W., 1945– author. | Rodríguez Cabrera, Tomás M., 1983– author. | Díaz, Luis Manuel, 1972– author. | Puente-Rolón, Alberto R., 1970– author. | Losos, Jonathan B., writer of foreword.
Title: Boas of the West Indies : evolution, natural history, and conservation / R. Graham Reynolds, Robert W. Henderson, Luis Manuel Díaz, Tomás M. Rodríguez-Cabrera, and Alberto R. Puente-Rolón; with a foreword by Jonathan B. Losos.
Description: Ithaca [New York] : Comstock Publishing Associates, an imprint of Cornell University Press, 2023. | Includes bibliographical references and index.
Identifiers: LCCN 2021056179 | ISBN 9781501765452 (hardcover)
Subjects: LCSH: Boidae—West Indies. | Boidae—Classification. | Boidae—Conservation. | Boidae—Geographical distribution.
Classification: LCC QL666.O63 R495 2023 | DDC 597.96/7—dc23/eng/20211209
LC record available at https://lccn.loc.gov/2021056179

I dedicate my work in this volume to the future students of West Indian Herpetology.

R. Graham Reynolds

Thirty years ago, I dedicated my contribution to the first *Boas of the West Indies* to Rose, Ky, and Karen. It seems apropos to dedicate my contribution to this updated *Boas of the West Indies* to Ky Henderson, to Karen and Brian Hannemann, and to the memory of many happy adventures shared with Rose Henderson. And to Mary S. Kohler who, for more than twenty years, enthusiastically and generously supported my *Corallus* fieldwork.

Robert W. Henderson

I dedicate this book to the new generations of herpetologists and to all people who have a passion for snakes. Also, to four very special people: my mother, Miriam, for giving me not only life but also love, guidance, and education; my wife, Ariatna, for her ever-present love and support in everything I do; Sofia, my daughter, for bringing great joy and inspiration to our lives; and Leo, for being such a great son.

Luis M. Díaz

I dedicate my work in this volume to my son and pride, Alejandro Michel Rodríguez González, who also shares my passion for snakes.

Tomás M. Rodríguez-Cabrera

I dedicate my work in this book to Fernando J. Bird-Picó and Peter J. Tolson. They allowed me to have my first experiences with boas, herpetology, and field work.

Alberto R. Puente-Rolón

# CONTENTS

Chapter 5

## SPECIES ACCOUNTS, 59



Chapter 6

## A BRIEF HISTORY OF THE STUDY OF WEST INDIAN BOAS, 225

## EPILOGUE, 231

# FOREWORD

West Indian reptiles and amphibians have played an important role in the development of our understanding of evolutionary diversity. For example, on page eight of their classic book, *The Theory of Island Biogeography*, MacArthur and Wilson used the herpetological species diversity on West Indian islands as a key example that motivated their pathbreaking work on the determinants of island species richness.

At about the same time in the late 1960s, Ernest Williams was initiating his investigations on the evolutionary ecology of *Anolis* lizards, developing a research program that launched the career of many now-leading ecologists, evolutionary biologists, behaviorists, and others, as well as making *Anolis* a household name in biodiversity circles.

Although anoles have hogged the limelight, studies on other of the thousand or so West Indian reptiles and amphibians have made important and general contributions as well. Frogs in the genus *Eleutherodactylus*, their diversity rivalling that of anoles, come immediately to mind, as well as research on many other taxa including curly tailed lizards, whiptail lizards, dwarf geckoes, iguanas, tree frogs, toads, vine snakes, and racers.

One group notably absent from this list is the boas. Despite being represented by eighteen extant species and occurring on nearly one hundred islands, relatively little is known about the basic natural history of these snakes; much less have they importantly affected our understanding of broader issues in ecology, evolution, and conservation.

Despite the scores of scientists who have devoted their careers to studying the zoology of this region for well over a century, we know so little about these snakes that for many species among the best sources of information we have today are the personal observations of the authors of this book, most of whom are young scientists still in the early stages of their careers.

Part of the reason, undoubtedly, is that these snakes aren't often abundant or particularly easy to find. I have spent three decades studying anoles in the Caribbean, and I can count the number of boas I've come across on the fingers of my hands (well, maybe throw in a couple of toes). Now, admittedly, I've been looking up more than down, so maybe that says more about me than the snakes, but still, they're not easy to study in most places.

Recent discoveries are demonstrating that there is a lot we've missed about these creatures. Not only cryptic species waiting to be recognized by modern species delimitation methods but also entirely new creatures living in places we never knew harbored boas. For most of these species, the basics of their biology—their natural history—remain almost totally unknown. Who knows what amazing new discoveries remain to be made?

As an anole enthusiast, I didn't know whether to be proud or dismayed to learn that anoles have been recorded in the diet of most West Indian boas, in many cases being a primary part of the diet of juveniles. And, as the pages of this volume attest, anoles are likely to be on the menu of most of the rest of the species, once we learn more about those species' natural history.

Despite their predator–prey relationship, anoles and boas show many similar patterns of diversity. For example, independent evolutionary radiation producing the same, convergent set of ecomorphs on multiple islands is the hallmark of anole evolution. Similarly, at least three island groups harbor or have harbored multiple boa species that are ecologically and morphologically different. Comparisons across islands show that the same ecological types are represented on many islands, and molecular phylogenetic studies show that this is the result of convergent evolution, similar species on different islands not being closely related.

Many island boa and anole species have adapted, and in some cases thrive, in human-disturbed landscapes. Just as with anoles, the population density of island boas is substantially greater than for their mainland counterparts, and probably due to the same reason: paucity of competitors and predators. For boas, there may be an additional reason for insular exuberance: they have an abundant prey base composed of species—such as anoles—that themselves are extraordinarily abundant.

And, finally, the West Indian anole and boa evolutionary radiations are similar in producing some spectacular species unlike anything you'd expect from an anole or a boa. For anoles, the most sensational are the species formerly in the genus *Chamaeleolis* that at first glance could be mistaken for a chameleon (hence their common name, "false chameleons"). For the boas, the honors go to the slender *Chilabothrus gracilis*, a Hispaniolan species that looks much more like the colubrid vine snakes one sees in the Costa Rican jungle than a member of the boa clan. The natural history of the aptly named Hispaniolan gracile boa is little known, sadly a common statement for Haiti's imperiled wildlife. Hopefully one of the readers of this volume will be inspired to learn more about this remarkable species.

My prediction is that as we learn more about West Indian boa diversity and natural history, there will be many more parallels, as well as exciting differences, from anoles and other better-studied taxa. And that's why this volume is such a valuable contribution. By pulling together all that we know about these fascinating creatures, and by including many unpublished observations previously residing only in the heads and notebooks of the authors, *Boas of the West Indies: Evolution, Natural History, and Conservation* has not only compiled all that is known about these snakes but hopefully will set the stage for a continued renaissance in their study.

Jonathan B. Losos, Ph.D.
Director, Living Earth Collaborative
William H. Danforth Distinguished
University Professor
Washington University in St. Louis

# ACKNOWLEDGMENTS

There are many, many individuals to thank for assistance on this project as well as supporting our research on boas over the decades.

For loaning samples and specimens related to this project we thank Jeff Murray; the Florida Museum of Natural History; Kevin de Queiroz and Addison Wynn of the Smithsonian National Museum of Natural History; Jim Hanken, Jonathan Losos, and José Rosado of the Harvard Museum of Comparative Zoology; Rich Glor and Rafe Brown of the University of Kansas; Bryan Stuart of the North Carolina Museum of Natural Sciences; and Jimmy McGuire and Carol Spencer of the Museum of Vertebrate Zoology at Berkeley.

For providing additional and useful information on these species we thank the following individuals: Nancy Albury, Nicole Angeli, Fernando Bird-Picó, Jeanelle Brisbane, Joseph Burgess, Jenny Daltry, Glenn Gerber, Sean Giery, Alejandro Hernández Gómez, Blair Hedges, Sebastian Hoefer, Sixto Incháustegui, Scott Johnson, Miguel A. Landestoy T., Skip Lazell, Ricky Lockett, Isadore Monah, Bryan Naqqi Manco, Jeff Murray, Brent Newman, Willow Outhwaite, Renata Platenberg, Kim Roberts, Marie Rush, the late Albert Schwartz, Mark de Silva, Dustin Smith, Fausto Starace, Richard Thomas, Roger Thorpe, Peter J. Tolson, Sondra Vega-Castillo, Nicolas Vidal, Isabel Vique, Lázaro W. Viñola, and J. P. Zegarra. We also consulted field notes of the late Albert Schwartz and James D. ("Skip") Lazell. We further thank J. P. Zegarra for providing access to scanned images of extensive literature he has accumulated for the species *Chilabothrus granti* and Brent Newman for providing early access to his in-press manuscripts.

We are extremely grateful to the following individuals who contributed their excellent photographs, which have greatly enhanced this volume: Kraig Adler, Father Alejandro Sanchez Muñoz, Aliesky del Río, José R. Almodovar, Baptiste Angin, Joseph Burgess, Jenny Daltry, Day's Edge Productions, Rosario Domínguez, René Durocher, Wayne Fidler, Anthony Geneva, Sean Giery, Katie Grudecki, S. Blair Hedges, Matthijs Kuijpers, Miguel A. Landestoy T., Jeff Lemm, Ricky Lockett, Raimundo López-Silvero, Bryan Naqqi-Manco, Brent Newman, Matthew Niemiller, Esau Pierre, Robert Powell, Pedro Genero Rodríguez, Marie Rush, Richard Sajdak, Dustin Smith, Jon Suh, Rolando Teruel, Mike Treglia, Rhea Warren, Shannan Yates, and J. P. Zegarra. The following individuals graciously read drafts of sections of this book and provided valuable feedback: Marinus Hoogmoed, William Lamar, Robert Powell, Dustin Smith, Samuel Turvey, and Tomás Waller. Aryeh H. Miller read much of the draft and provided substantial feedback. We also thank Peter J. Tolson and two anonymous reviewers for excellent comments on previous versions of this manuscript.

We are grateful to Kitty Liu, Jacqulyn Teoh, and Allegra Martschenko of Comstock Publishing Associates. To Kitty for her enthusiasm in seeing our vision and her wise editorial insight to guide the structure of the project, and to Jacqulyn and Allegra for editorial assistance with the many, many details necessary to produce this work. Thanks as well to the deft copyediting work done by Eva Silverfine.

## ADDITIONAL ACKNOWLEDGMENTS AND THANKS

Graham wrote most of his portions of this volume in between boa field work while on Ambergris Cay, Turks and Caicos Islands (TCI), as well as in the Loker Reading Room in Widener Library at Harvard University and in the Herpetology Lab at the Museum of Comparative Zoology. He is grateful to his family for allowing him time away from home on these writing retreats. Graham also thanks Duke University, the University of Tennessee Knoxville, the University of Massachusetts Boston, Harvard University and the Museum of Comparative Zoology, and the University of North Carolina Asheville for supporting and funding his work in the Caribbean over the last two decades. Graham has been extremely fortunate to have had incredible mentors over the years and wishes to thank the following people for training and supporting him: Joanne Bartsch (while at Carolina Day School); Susan Alberts (while at Duke); Ben Fitzpatrick, Brian O'Meara, Jim Fordyce, Dan Simberloff, Graciela Cabana, and Sandy Echternacht (while at University of Tennessee Knoxville); Liam Revell (while at University of Massachusetts Boston); and Jonathan Losos (while at Harvard).

Thanks to the following friends and colleagues for support, training, assistance, and fun while studying boas: Nicole Angeli, Kevin Aviles-Rodriguez, Brian Baker, Diane Barber, Sandy Buckner, Joe Burgess, Shelley Cant-Woodside, Giuliano Colisimo, Colin Donihue, Jason Fredette, Anthony Geneva, Glenn Gerber, Nick Herrmann, Wendy Jesse, Scott Johnson, Oriol Lapiedra, Inbar Maayan, Bryan Naqqi Manco, Aryeh Miller, J. B. Minter, Pavitra Muralidhar, Matthew Niemiller, Stesha Pasachnik, Keeley Peek, Andrea Persson, Quyhn Quach, Molly Reger, Liam Revell, Dustin Smith, Alyssa Vanerelli, George Waters, Kristin Winchell, J. P. Zegarra, and many others.

Thanks as well to tremendous assistance and support from Jim Hanken, Jonathan Losos, José Rosado, Joe Martinez, Tsuyoshi Takahashi, and Jared Hughes of Harvard's Museum of Comparative Zoology.

Graham has received an immense amount of help over many years studying boas in the Caribbean and would like to thank the following people and institutions for their tremendous assistance in the field, advice, and information related to boa work in each of these regions: *Bahamas*: Kevin Aviles-Rodriguez, Brian Baker, Joe Burgess, Sandy Buckner, Eric Carey, Omar Daly, Giselle Dean, Lynn Gape, Anthony Geneva, Sean Giery, Alexis Harrison, Nick Herrmann, Wendy Jesse, Scott Johnson, Mark Keasler, Charles Knapp, Craig Layman, Jonathan Losos, Pavitra Muralidhar, Matt Niemiller, Quyhn Quach, George Rathgeber, Liam Revell, Kristin Winchell, Shelley Cant-Woodside, Shannan Yates, Bonnie Young, Everyone at the Bahamas National Trust, the Bahamas BEST Commission, the Ministry of Agriculture, Ardastra Gardens, Stella Maris Resort, and Bimini Shark Lab. *Turks and Caicos*: Joe Burgess, Giuliano Colisimo, Glenn Gerber, Paul Mahoney, Bryan Naqqi Manco, Lindsay Mensen, Aryeh Miller, Keeley Peek, Bonnie Raphael, Molly Reger, Eric Salamanca, George Waters, Mark Welch, Mark Woodring, Alyssa Vanerelli, the staff at Ambergris Cay, especially the kitchen staff (Verona, Mimi, and Jody); the TCI National Trust, and the School for Field Studies. *Jamaica*: Joe Burgess, Susan Koenig, and Brent Newman. *US Virgin Islands*: Nicole Angeli, Diane Barber, Ezra Ellis, Aryeh Miller, J. B. Minter, Andrea Persson, Renata Platenberg, Dustin Smith, Peter Tolson, and J. P. Zegarra. *British Virgin Islands*: Skip Lazell, Lianna Jarecki, and the staff at Guana Island. *Dominican Republic*: Stesha Pasachnik, Miguel A. Landestoy T., Rich Glor, Luke Mahler, and Groupo Jaragua. *Puerto Rico*: Fernando Bird-Picó, Marylin Colon, Jose Luis Herrera, Robert Reed, Liam Revell, Richard Thomas, Sondra Vega-Castillo, Kristin Winchell, Joseph Wunderle, and J. P. Zegarra, University of Puerto Rico Río Piedras (and the El Verde Field Station), Universidad Interamericana de Puerto Rico Bayamon Mata de Plátano Field Station, and the U.S. Fish and Wildlife (USFWS) Southeast Region. A big thank you to Jeff Murray and Ryan Potts for information and support. There are many others as well.

Graham's work on boas in the Caribbean has been funded and supported by the following sources: Harvard University, Harvard Museum of Comparative Zoology, University of Massachusetts Boston, University of North Carolina Asheville, University of Tennessee Knoxville, Mohammed bin Zayed Species Conservation Fund, USFWS, National Science Foundation, American Philosophical Society, San Diego Zoo Global, Brownstone Family Foundation, the Reger Family, North Carolina Zoo, Ft. Worth Zoo, Division of Fish and Wildlife Government of the Virgin Islands, Darwin Plus: Overseas Territories Environment and Climate Fund, American Genetics Association, the American Museum of Natural History, National Geographic Society, donors to the Reynolds Lab at University of North Carolina Asheville, and the Silver Boa Trust.

—R. Graham Reynolds

Henderson is grateful to colleagues and family who spent many happy nights in pursuit of species of *Corallus* in the West Indies. They are Craig Berg, Mark de Silva, Joel Friesch, Gary Haas, Billie Harrison, Eric Hileman, Shawn Miller, Isidore Monah, John Murphy, Bob Powell, Sylvia Powell, Marie Rush, Rich and Linn Sajdak, Mike Treglia, Alan Winstel, and Derek Yorks. Rose Henderson shared the initial forays into treeboa work on Grenada and St. Vincent; it was wonderfully exciting, great fun, and remains a cherished memory. Similarly, having his son, Ky, accompany him on several trips to Grenada was a dream come true.

Bob is appreciative of the cooperation he has received from forestry and wildlife officials in St. Vincent and the Grenadines and in Grenada. On Grenada: Anthony (Jerry) Jeremiah, Aden Forteau, Alan Joseph, the late Rolax Frederick, Robert Dunn, and George Vincent; on St. Vincent and the Grenadines: Glenroy Gaymes, Amos Glasgow, Brian Johnson, Calvin Nicholls, FitzGerald Providence, and Cornelius Richards. On Grenada and Carriacou, our "headquarters" changed over the years, but personnel at Maffikin Apartments, Lazy Lagoon, Siesta Hotel, and Grandview Hotel always tolerated our unusual pursuits and strange hours. We are thankful for their tolerance and hospitality.

Over the past 30 years, funding for fieldwork with species of *Corallus* has come from a variety of sources. For nearly 20 years, Mary S. Kohler, by way of the Windway Foundation, the Mary and Terry Kohler Special Charitable Account, and out-of-pocket, has generously funded and expressed genuine interest in Bob's *Corallus* fieldwork; he can't thank her enough. The Milwaukee Public Museum has been on board from the beginning, and he appreciates the museum's continued support. Additionally, the late Albert Schwartz, the late Jack A. Puelicher, the late Robert W. Bourgeois, the Robert W. Bourgeois Memorial Fund, the National Science Foundation (through grants to Robert Powell), the Central Florida Herpetological Society, and Allen M. Young have all contributed to funding fieldwork.

—Robert W. Henderson

I wish to express my gratitude to the colleagues, friends, and authorities who have accompanied and supported my herpetological work in different ways over three decades (in Cuba and Dominican Republic), particularly Jose A. Álvarez Lemus, Miguel Ángel Abad, Eduardo Abreu, Giraldo Alayón, Emilio Alfaro, Ansel Fong, Ángel Arias, Yvonne Arias, Gerardo Begué, Yonder Berdecia, Antonio Cádiz, Barchile Calzada, Angel Céspedes, Ramón Cueto, Fraser Durie, Alberto R. Estrada, Wolfgang Feichtinger, Rolando Fernández de Arcila, Evelyn Gabot, Xiomara Gálvez, Orlando H. Garrido, Julio A. Genaro, Amnerys González, Esteban Gutiérrez, Blair Hedges, María Elena Ibarra, Jesús Imbert, Eduardo Iñigo, Sixto J. Incháustegui, Yamilka Joubert, Masakado Kawata, Arturo Kirkconnell, Cristian Marte, Rosendo Martínez, Karina Massieu, Celeste Mir, José Morales (Fefo), Luis V. Moreno, Robert Murphy, Nils Navarro, Ernesto Palacio, Abel Pérez, Esther Pérez, Chris Raxworthy, Marcos Rodríguez, Ana Sanz, Paul Sosa, Claus Steinlein, Rolando Teruel, Carlos Suriel, Pavel Valdés, and Nicasio Viña. Special gratitude to Jeff Lemm, Cristian Marte, Rene Durocher, and Matthijs Kuijpers for contributing photographs. Peter Tolson, Veronika Zahradníčková, Ivan Rehak, Gerardo García, and Craig Adler have contributed literature and advice. Thanks to Alexander Arango for giving me access to his living collection of snakes at the National Zoo in Havana. Alejandro Hernandez Gómez kindly provided information presented in this book and allowed me to photograph his boas.

—Luis M. Díaz

Many people have assisted Tomás and provided accommodation over the past 12 years of field work with *Chilabothrus angulifer* across Cuba: Adalberto Vallejo, Alejandro Abella, Alejandro Hernández Gómez, Alejandro M. Rodríguez, Aliesky del Río, Armando R. Longueira, Cándida Sampedro, Carlos Hernández, Ernesto Morell, Francisca ("Pancha") Amador, Hansel Caballero, Humberto Vela, Javier ("Fili") Pérez, Javier

Torres, Josué H. Pérez, Julio D. León, Leosveli Vasallo and family, Luis B. ("Cuco") Pérez, Martín Núñez, Raimundo López-Silvero, Roberto Martínez and family ("Mil Cumbres," Managed Resource Protected Area [MCPA]), Rolando Teruel, Rosalina ("Nina") Montes, Rosario ("Charo") Domínguez, Ruben Marrero, Sheila Rodríguez, Tomás García and family ("Santa Cruz" river canyon), Yaira López, and Yoana Lisca.

Some of the abovementioned people and others also provided valuable information on the natural history and other aspects related to the Cuban boa: Alberto Clark, Alexander Arango, Angel Arias, Ansel Fong, Armando Falcón, Arturo Hernández, C. Sampedro, Dennis Denis, David Ortiz, Elier Fonseca, Elsa Rodríguez, E. Morell, Gerardo Begué, Haydée González, Horacio Grillo, H. Vela, José L. Linares, J. D. León, Lázaro W. Viñola, Leopoldo M. Vasallo, Leosvany Vasallo, Leosveli Vasallo, L.B. Pérez, Luis F. de Armas, L. Yusnaviel García, Maikel Cañizares, Mario V. Muñoz, Maydiel Cañizares, Nelson Capote, Nils Navarro, Orestes Hernández, Osvaldo Fariña, Peter J. Tolson, Pedro N. Otero, the late Rafael A. Fuentes, Rafael A. Pérez, Rafael Roche, R. López-Silvero, R. Martínez (MCPA), Roberto Martínez (Cienfuegos Botanical Garden), Roberto Varela, Rolando Cardoso, R. Teruel, Rosa C. Rodríguez, Rubén Chamizo, T. Menelio Rodríguez Águila, and Yudaimy de la C. Pérez.

The Mohamed bin Zayed Species Conservation Fund provided funding for field expeditions and equipment to work with threatened endemic snakes in central Cuba, which included the Cuban boa. The Rufford Foundation also provided funding for some equipment that has been useful while working with Cuban boas.

—Tomás M. Rodríguez-Cabrera

Puente-Rolón thanks his family (Sondra I. Vega Castillo, Alejandro D. Puente, and Andira Del Alba Puente) for their support, help, and understanding. Also, thanks to Richard Thomas and Harold Heatwole for their advice and guidance. Thanks to the Department of Natural and Environmental Resources and the U.S. Fish and Wildlife, Caribbean Field Office for all their support and efforts to protect our boa species.

—Alberto R. Puente-Rolón

# ABBREVIATIONS

ARPR Alberto R. Puente-Rolón
asl above sea level
BP Before Present
BVI British Virgin Islands
C Celsius
cal calibrated
CE Common Era
CITES Convention on International Trade in Endangered Species of Wild Fauna and Flora
cm centimeter
d day
elev. elevation
FAD first appearance date
g gram
h hour
ha hectare
in litt. *in litteris*; communicated in writing
IUCN International Union for the Conservation of Nature
kg kilogram
km kilometer
KUH University of Kansas Biodiversity Institute Herpetology Collection
LAD last appearance date
LMD Luis M. Díaz
LSUMZ Louisiana State University Museum of Natural Science (also LSU MNS)
max. maximum
m meter
MCZ Museum of Comparative Zoology, Harvard University
µm micrometer
mm millimeter
mo month
MY millions of years ago
pers. comm. personal communication
pers. obs. personal observation
p-h person-hour
RGR R. Graham Reynolds
RWH Robert W. Henderson
SVL snout–vent length
TCI Turks and Caicos Islands
TL total length
TMRC Tomás M. Rodríguez-Cabrera
UMMZ University of Michigan Museum of Zoology
unpubl. unpublished data
USFWS U.S. Fish and Wildlife Service
USVI U.S. Virgin Islands
yr year(s)

# BOAS OF THE WEST INDIES

**Figure 0.1**. The greater Caribbean Region, showing some major islands and island groups discussed in the text.

# INTRODUCTION

This volume is intended to celebrate and extend the now classic and out-of-print work by Tolson and Henderson (1993)—the last focused volume on the West Indian boas (see end of this section regarding "West Indian"). This fascinating group of largely insular snakes consists of three genera: *Boa*, *Chilabothrus*, and *Corallus*, which are represented by eighteen extant species and at least two extinct species in the Caribbean region. Thus, this work focuses only on the islands of the Bahamas (Lucayan) Archipelago, the Greater Antilles, and the Lesser Antilles and does not include natural populations of boas found on continental islands in the greater Caribbean region, such as *Boa constrictor*, *Boa imperator*, *Corallus ruschenbergerii*, and *Epicrates maurus*, as these would best be treated in a volume that was inclusive of the mainland Neotropics (although we do discuss introductions of these species to the Caribbean). Further, we do not include the genus *Tropidophis*, commonly known as dwarf boas, pygmy boas, or wood snakes, as they belong to a highly evolutionarily divergent and non-booid lineage. We include short accounts for two recently described fossil species—one which did not occur in the West Indies but is a member of the otherwise exclusively West Indian genus *Chilabothrus*, and one recently characterized extinct *Boa* species from Marie-Galante Island in the Guadeloupean Archipelago. Three additional fossil *Boa*, the Antigua boa, the Guadeloupe boa, and the Martinique boa, have yet to be fully characterized, and each is included here as *Boa* sp.

Numerous taxonomic arrangements have flowed through the boa literature over the years, with additions of new species as well as the elevation and sinking of specific, generic, and familial epithets, resulting in a somewhat confusing web in which to navigate for anyone who is not a boa systematist. Such alterations are generally a positive outcome because they follow the process of science, that is, the accretionary growth of knowledge. Happily, boa taxonomy was recently organized in Pyron et al. (2014) and in a monographic treatment by Reynolds and Henderson (2018). Here we follow the taxonomy of that latter work, noting that since its publication one new species has been described (*Chilabothrus ampelophis*; Landestoy T. et al. 2021b). We further follow the use of common names recommended in Hedges et al. (2019) in an attempt to insert stability into the use of common names. We further note that we distinguish between the words "boa" and "boid" and "booid." The first term, boa, refers generally to any member of the superfamily Booidea; the second term, boid, refers specifically to a member of the family Boidae; and the last term, booid, refers specifically to members of the superfamily Booidea. We use the term boa throughout the text unless specificity is required.

We provide a comprehensive review of the literature on West Indian boas, attempting to incorporate as much published material as possible to provide an extensive reference list. We cite references parenthetically in the text to facilitate tracing ideas and data to published papers and authors. As we have collectively conducted over a century of work on these boas, we also include substantial amounts of our own data and observations, some of which have been published and some of which have not. We occasionally provide annotations for this, such as the designation "personal observation," when such distinction is useful, but for the most part provide such information without citation to add additional depth and detail to the text. Thus, many accounts contain data and information drawn directly from our field notes, and statements of fact without citation should be regarded as coming directly from our own data and experiences and communicated to the literature via this volume.

To the extent that we are able, we have tried to include relevant and comprehensive distributional data for each species and subspecies treated here. Nevertheless, in some cases we feel that adding precise locality information would be detrimental to the species—particularly for critically endangered species or threatened populations. In those cases, we intentionally obfuscate some locality data. While it brings us no joy to do so, in all cases we try to be as forthcoming as possible regarding distribution, striking a balance between providing as much information as possible while also protecting especially sensitive populations. Maps are intended to serve as approximations of species' ranges, and for the above reasons do not include point locality data (although this is available in other volumes). Several species of West Indian boas have poorly characterized ranges (e.g., *Chilabothrus fordii*, *C. gracilis*), hence we attempt to use distributional data as well as topographical and environmental data to suggest likely areas of occurrence. For species with recognized subspecies, we include those ranges in alternate colors.

We do not include much anecdotal information related to captivity for the species represented here. Although a tremendous amount of information has been accumulated by the dedicated and passionate keepers of these species, much of this information is not available in the primary literature, and we feel it is best left to a different type of work than the present one.

We have further added a wide variety of images to illustrate the range of phenotypes and behaviors of these incredible snakes—especially as several are cryptic, endangered, or otherwise unlikely to be seen by anyone but specialists. We are indebted to the photographers, credited in the text and thanked in the acknowledgements, who have helped tremendously in this regard.

Why "West Indian"? The term West Indian refers to the collection of islands between North and South America, lying within what is also called the Caribbean Basin. The term itself is one imparted to the region by Europeans, and it gained particular use in the English language to refer to colonial territories and possessions in the western Atlantic. The word Caribbean is another term that refers to the indigenous group named the Caribs by Europeans, the name being a corrupted version of the name of the group who referred to themselves as "Kalinago." Thus, both words referencing this geographic region in English come to us from European colonialists. Some people of the region prefer the term Caribbean, sensing that West Indian has more of a colonial connotation, whereas Caribbean is at least an attempt to use an indigenous name (even if incorrectly). However, Caribbean tends to refer to all islands in the region as well as the coasts of the mainland (e.g., the Caribbean coast of Panama). West Indian connotes the islands that form the core of the region itself, exclusive of the continental islands. Further, West Indian is the term most widely used by current and former herpetological works in the region. For these reasons, we have chosen to use West Indian, with the use of Caribbean interspersed, but with the acknowledgement that both terms were born out of colonialism.

## EVOLUTION AND BIOGEOGRAPHY OF THE BOAS

Although the full story of the arrival, dispersal, and evolutionary change in the West Indian boas is shrouded by time, we have nevertheless learned a significant amount about how these boas came to be. Through centuries of observation, dedicated fieldwork, and the advent and application of modern molecular and statistical techniques, we are now able to tell a remarkable story about the origins of the boas in the region, as well as a fascinating story about the evolutionary process.

The boas—snakes that we collectively recognize as members of the superfamily Booidea—are currently represented by sixty-seven species and thirty-three subspecies distributed across a huge swath of the globe (Reynolds and Henderson 2018). Boas are found in Africa, southern Europe and the eastern Mediterranean, the Middle East, south and central Asia, Madagascar, the South Pacific, and nearly across the depth of the Western Hemisphere from Canada to Argentina (Reynolds and Henderson 2018). These species represent a tremendous diversity of shapes and sizes, habitat preferences, climactic tolerances, and natural histories. For example, boas occupy tiny tropical islands in the Caribbean, dry deserts in Africa, cold montane conifer forests in North America, seasonally flooded llanos in South America, and dense rainforests in Central America. The boas date back to a divergence from other snakes, such as the ancestors of pythons and shield-tailed snakes, around the mid-Cretaceous (Burbrink et al. 2020). All extant lineages of booid snakes (members of the superfamily Booidea) began diverging near the boundary of the Cretaceous and the Paleogene (Burbrink et al. 2020), or about 75 million years ago (MY), when the present-day continents were much closer together. As these lineages evolved, they further diverged into the six boa families we recognize today (Pyron et al. 2014; Reynolds and Henderson 2018; Burbrink et al. 2020). The family Candoiidae is restricted to Melanesia and small portions of Micronesia and Polynesia. Sanziniidae is found only on Madagascar and satellite islands. Calabariidae is restricted to west Africa, and Erycidae is found across northern Africa through southern Europe to central Asia. Other lineages of snakes commonly called boas, such as members of the genera *Tropidophis*, *Casarea*, and *Bolyeria*, are not members of the superfamily Booidea. The Western Hemisphere is where boa diversification has accelerated since the Miocene, with 65.6% of the sixty-seven booid species we recognize today present in the region. The family Charinidae is currently represented in North and Central America, while the diverse family Boidae currently occupies much of the area south of the U.S.-Mexico border, through the Caribbean, and down to the northern reaches of Patagonia. This latter family, Boidae, contains the three genera that are present in the West Indies: *Boa*, *Chilabothrus*, and *Corallus*.

### *Boa*

The genus *Boa* is represented in the West Indies by two extant species (and likely four or more extinct species) occurring in the Windward Islands of the Lesser Antilles. *Boa nebulosa*, or the Dominica Boa, occurs exclusively on the island of Dominica. Farther down the Windward Chain (but giving Martinique a miss), the species *Boa orophias* is endemic to the island of St. Lucia. These two boa species are, like other terrestrial squamates of the Windward Antilles, likely a product of ancient dispersal from South America via the Orinoco River. These species are thought to be sister lineages to each other, suggesting a single colonization from the South American coast to the Windward Antilles. No genetic data have yet been published from *B. orophias*, and *B. nebulosa* was only recently studied from a molecular phylogenetic perspective (Bezerra de Lima 2016; RGR, unpubl. data).

The absence of *Boa* on Martinique, located between Dominica and St. Lucia, is a bit of a mystery, although it does certainly seem likely that boas occurred there and might have since gone extinct. In Breuil (2002; translated from French by RGR), the author argues that "the ancient presence of a boa constrictor (*Boa constrictor*) on Martinique is well-established owing to descriptions by two historical writers." Collectively, Breuil (2002, 2009, 2011) and Lorvelec et al. (2007) suggest that the evidence weighs in favor of considering *B. constrictor* previously having inhabited Martinique. Breuil (2002) and Lorvelec et al. (2007) suggest that descriptions in L'Anonyme [Anonymous] de Carpentras (1618–1620) describe the presence of a boa, although Breuil (2009, 2011) states that it remains unknown whether this author was specifically referring to Martinique. The assertion of Breuil (2002, 2009, 2011) and Lorvelec et al. (2007) is thus hinging upon a series of reports by Pere Labat (1724), who reliably recorded some observations of snakes on Martinique. Labat noted that an animal that appeared to be 10 feet (3 m) long (presumably literal foot lengths) chased frogs and was an enemy of *Bothrops*—the venomous viper on Martinique. Putting aside the length estimate, which are notoriously exaggerated, this description sounds more like a racer (chasing a frog), possibly *Erythrolamprus cursor*, or a clelia (an enemy of *Bothrops*), possibly *Clelia* sp., which is not otherwise recorded from Martinique but occurred on St. Lucia until the nineteenth century when they went extinct from human activity. Notably, Breuil (2002, 2009, 2011) mentions that no naturalists subsequently observed large snakes on Martinique despite their extensive collections of other taxa. Thus, the evidence suggesting the presence of a boa is reduced to just the estimate of the animal's size. *Clelia* can reach lengths of 1.6 m, and the description from Labat (1724) is "l'animal paraissait avoir 10 pieds de long" (translation: the animal appeared to be 10 feet in length), which suggests some uncertainty in the estimate (i.e., Pere Labat did not actually measure the animal directly). Taken together, these observations suggest that we did not, in fact, have solid evidence that boas occurred on Martinique—contrary to what is suggested by Breuil (2002, 2009, 2011) and Lorvelec et al. (2007). Nevertheless, Dewynter et al. (2019) recently insisted on listing *Boa* from Martinique ("La présence passée d'un Boa en Martinique est confirmée"; translation: the historical presence of a *Boa* in Martinique is confirmed), and Bochaton (2020) described the presence of *Boa* vertebra in middens from Dizac Beach, possibly modified into decorative items like beads. Given the biogeographic likelihood that *Boa* existed there, as well as the recent finding of vertebrae on the island, we choose to include *Boa* as part of the historical fauna of Martinique.

Subfossil *B. constrictor* are known from Antigua (Table 0.1), where they have been found at two sites. Material from Indian Creek dates to 1915–845 years

**Table 0.1.** Fossil and subfossil material attributed to boas in the West Indies. See the *C. angulifer* account for extensive fossil material for Cuba.

Radiometrically calibrated ages are given as mean years ± SD before present [BP].

| Island | Location | Age (years BP) | Species | Material | Reference |
|---|---|---|---|---|---|
| **Bahamas/Florida** | | | | | |
| Grand Turk | Coralie | 1280 ± 60–900 ± 50 | *Chilabothrus chrysogaster* | Vertebrae | Carlson 1999; Newsom and Wing 2004 |
| Florida | Thomas Farm, Gilchrist County | 18,000,000 | *Chilabothrus stanolseni* | Vertebrae | Onary and Hsiou 2018 |
| Abaco | Sawmill Sink | "Late Pleistocene" | *Chilabothrus exsul* | Vertebrae | Mead and Steadman 2017 |
| **Greater Antilles** | | | | | |
| Puerto Rico | Northern Karst Region | "Late Pleistocene" | *Chilabothrus inornatus* | Skull bones and vertebrae | Pregill 1981 |
| **Lesser Antilles** | | | | | |
| Antigua | Indian Creek | 1915 ± 80–845 ± 80 | *Boa constrictor* | Unknown | Steadman et al. 1984; Pregill et al. 1988 |
| Antigua | Burma Quarry | 4300 ± 150–2560 ± 70 | *Boa* sp. | Vertebra | Steadman et al. 1984; Pregill et al. 1988 |
| Basse-Terre | Cathedral Site | Saladoid | *Boa* sp. | Vertebra fragment | Bochaton 2020 |
| La Désirade | Pointe Gros Rempart 6 | 577–369 | *Boa* sp. | Vertebra | Bochaton 2020 |
| Marie-Galante | Cadet 2 Cave, Cadet 3 Rock Shelter, and Blanchard Cave | 34,000–15,000 | *Boa blanchardensis* | Skull bones and vertebrae | Bochaton and Bailon 2018 |
| Martinique | Dizac Beach | 1605–1330 | *Boa* sp. | Vertebrae | Bochaton 2020 |

before present (BP) and has been identified as "*Boa constrictor*" (Steadman et al. 1984; Pregill et al. 1988). Material from Burma Quarry is older, dating to 4300–2560 years BP, and consists of a single vertebra from a boid snake, but without conclusive features that would allow identification to genus (Steadman et al. 1984; Pregill et al. 1988). Thus, the evidence appears to weigh in favor of a *Boa* species present on Antigua into the Holocene; although owing to the lingering uncertainty regarding these specimens, and a lack of recent reexamination, we hesitate to consider the Antiguan boas at the specific level, preferring instead to continue to treat them as *Boa* sp.

Much speculation existed regarding whether boas colonized the Guadeloupean Archipelago. Breuil (2002) noted that some authors have posited the certainty of the existence of a boa on Guadeloupe prior to the introduction of mongoose, while others have rejected this hypothesis. Happily, the historical presence of *Boa* in the Guadeloupe islands appears to have been confirmed. Recently, fossil material from Marie-Galante has been described as *Boa blanchardensis*, an extinct species that did not survive to the Holocene (Bochaton and Bailon 2018). Fragments of skull and jaw bones, as well as trunk and caudal vertebrae, were well known and collected from three different sites on the island (Table 0.1; Bochaton et al. 2015). Additional vertebrae from a recent (577–369 BP) owl deposit on La Désirade might represent a Guadeloupe boa species that survived into the pre-Columbian Holocene (Bochaton 2020). Further, ~1500-year-old material from Basse-Terre fashioned into beads suggests that the Guadeloupe boa was present on all three main islands (Bochaton et al. 2021).

A further interesting anecdote is that a mid-nineteenth century reference in a contemporaneously well-publicized naturalist's account of the West Indies suggests the presence of a large snake on Nevis and St. Kitts (Lynch 1856). The author described it as "yellow striped with black," 10 feet long, and readily consuming domestic fowl. Other exaggerations by the author elsewhere in this text lend support to the likelihood that, while suggesting a boa based on some of these characteristics, almost certainly this is an account of the red-bellied racer, *Alsophis rufiventris*, which is decidedly yellowish ventrally and with dark striping anteriorly (although a mere 1 m long).

Thus, some fossil and subfossil specimens suggest that boas could very well have occupied most of the Windward Antilles at some point and that they might have dispersed as far north as Antigua prior to going extinct on some islands. Alternatively, it is possible that boas happened to establish on some islands and not others or that some Holocene material, such as that from Guadeloupe, Antigua, or Martinique, was the result of animals or animal material brought to the islands by people.

### *Chilabothrus*

The genus *Chilabothrus* was the name originally given to the first boa in the group described (*Chilabothrus inornatus* Duméril and Bibron 1844) in the Caribbean, although they were subsequently placed in the genus *Epicrates* (Boulenger 1893). Reynolds et al. (2013a) used molecular phylogenetic data to show that the insular *Epicrates* represented a distinct clade and should therefore be named *Chilabothrus*. This genus is represented by fourteen extant species distributed in the Greater Antilles (Cuba, Jamaica, Puerto Rico Bank, and Hispaniola) as well as Isla de Mona and the Lucayan Archipelago (Little Bahamas Bank, Great Bahamas Bank, Conception Bank, Crooked-Acklins Bank, Inagua Bank, Caicos Bank, and Turks Bank). Most of these major islands or island banks are single-species islands, in that they contain only one boa species. On Cuba, the large *C. angulifer* is found island-wide, although it is presently restricted to certain regions by habitat modification (see *C. angulifer* account). On Jamaica, the large *C. subflavus* has been reduced to a few pockets of protected areas from a former large range across the island. On Isla de Mona, the single species *C. monensis* is found island-wide, although it is a somewhat more recently divergent lineage from boas (*C. granti*) on Puerto Rico to the east. In the Lucayan Archipelago—a series of geologically similar island banks comprising the Bahamas and Turks and Caicos Islands—boas can be found on some, but not all, of the island banks. Where boas occur, there is only one species of boa, never more, although this likely was not the case in the past (more on this below). Hispaniola and Puerto Rico, on the other hand, have four and two species present, respectively, resulting in some remarkable evolutionary inferences (more below).

Our fossil record of boas in the West Indies is poor

(Table 0.1), being mostly represented by subfossils of extant species found in human middens in places like Grand Turk (Carlson 1999), in sinkholes on Abaco (Mead and Steadman 2017), or in deposits in the Lesser Antilles (e.g., Bochaton 2020). Thus, we are somewhat limited in our ability to consider what species might have occurred on these islands at various points in the past. For example—did Cuba or Jamaica ever have a small species of boa? It is certainly plausible on Cuba, as the present island existed as a series of separate paleo islands throughout the Miocene and Pliocene and thus might have housed additional boa species that have since gone extinct. Nevertheless, all fossil and subfossil material so far reported from this archipelago belong to the extant species *C. angulifer.* Alternatively, the fantastic radiation of *Tropidophis* species on Cuba (seventeen species currently described with more to come; Díaz and Cádiz 2020) might have provided such strong competition that boas evolved large body size to prey on hutias and solenodons instead of competing with the smaller tropes (Rodríguez-Cabrera et al. 2016a). On Jamaica it is possible that a species of boa existed prior to the near-complete inundation of the island around 8 MY. Such questions remain in the realm of speculation unless additional fossil material is uncovered.

Recently, a fascinating discovery has been made—the description of fossil *Chilabothrus* material from peninsular Florida dating to the Miocene (Onary and Hsiou 2018). Estimated to be 18 million years old, this extinct species (*Chilabothrus stanolseni*) suggests that boas colonized Florida from the Caribbean less than 5 million years after they colonized the West Indies. There are few examples of squamate reptiles crossing the Straits of Florida, although the well-known *Anolis* lizards are perhaps the best example (*Anolis carolinensis*). These fossils of *C. stanolseni* are further suggestive that we might soon discover more about the evolutionary history of *Chilabothrus*, as the fossils were repeatedly identified as different genera before a recent reevaluation determined that they most resembled *Chilabothrus*.

*Corallus*

The genus *Corallus*, like the genus *Boa*, is represented by two species that likely arrived relatively recently (~1 MY) in the Windward Antilles from South America. On St. Vincent, *Co. cookii* occurs island-wide, although its abundance is highly variable depending on habitat type. *Corallus grenadensis* occurs on Grenada as well as on at least ten islands in the Grenadines. These two species are sister lineages nested in the widespread species *Co. hortulana*, suggesting that an ancestor of the species *Co. hortulana* colonized the Windward Antilles from South America and gave rise to these two lineages that we today recognize as separate species (Colston et al. 2013; Reynolds and Henderson 2018).

## PHYLOGENETICS OF WEST INDIAN BOAS

A major building block in understanding the origins of West Indian boa diversity has been the ability to infer the evolutionary relationships among modern boas accurately. Like most studies of phylogeny, or evolutionary relationships, our understanding has changed as we continually generate new data and deploy new methods to reconstruct the origins of boa biodiversity. Early data sets used morphological data, or measurements and observations regarding the shape, size, and habits of boas, to attempt to infer evolutionary relatedness. This shifted toward quantifying anatomical features—meristic measurements and morphometric analysis of increasingly finer-scale features such as scale counts or skeletal elements. These early studies provided an important backbone to our understanding of boa relationships, but they also set up testable hypotheses to be investigated by increasingly fine-scale and more powerful data sets and statistical analyses. In particular, the application of genetic data gave us tremendous ability to reconstruct the evolutionary history of boas, particularly when paired with advanced statistical and computational techniques such as maximum likelihood and Bayesian phylogenetic inference. Genetic data initially took the form of examining whole proteins extracted from tissues, called allozymes, that allowed us to estimate relationships by examining whether species shared similar protein sizes (and thus, underlying gene variants). But it was our ability to access the genome directly, at the level of the base pairs (*A*s and *C*s and *G*s and *T*s) and thus to read the code for *producing* proteins that has given us the most accurate view of the origins of boa diversity. Indeed, molecular phylogenies have mostly resolved, or reconstructed, the evolutionary relationships among these boas. Beginning with analyses of a few hundred base pairs of mitochondrial DNA, researchers began to reconstruct some of the relationships that are still recognized today. More and more data have been added at a dizzying pace—just within the last two decades we have gone from a few hundred base pairs of mitochondrial data (Campbell 1997; Burbrink 2004), to twelve-gene analyses with 4000 base pairs of mitochondrial and nuclear data (Reynolds et al. 2014a) to the present day, when we are using data sets with tens of thousands of genetic loci and millions of base pairs to reconstruct the evolution of West Indian boas (RGR, unpubl. data) or even sequencing entire boa genomes (Bradnam et al. 2013). These genome-scale data sets are so large that they require specialized computers with lots of memory to analyze them—and are providing us with our finest-scale resolution of the relationships among these boas we have ever had. What a time to be an evolutionary geneticist!

Figure 0.2 is a phylogeny of West Indian boas based on a mitochondrial gene (cytochrome *b*, >1100 base pairs long; after Reynolds et al. 2014a). This phylogenetic tree shows evolutionary relationships among species more or less as they are currently understood, although some of the nodes are not well supported, and it is likely that *C. angulifer* diverges earlier than indicated here (RGR, unpubl. data). But it should be noted that even with tremendous data sets and statistical analyses, relationships represented on phylogenies are still hypotheses. What we mean is that they are testable hypotheses representing our best idea of these relationships but that they are subject to continual revision as new data come along. Before too long, we expect that we will have entire genomes for these species, which might yield additional clues about the evolutionary origins and relationships of boas; indeed, we are starting this process now (RGR, in preparation). Nevertheless, this phylogenetic hypothesis offers a robust view of the likely relationships among these lineages based on information from their mitochondrial genomes.

## A PRIMER ON THE STUDY OF THE EVOLUTION OF BOAS

Evolutionary biology is a historical science in that in order to make sense of current patterns of biodiversity we must reconstruct past events. Nevertheless, it is a robust science and one that we understand quite well. Consider this: we exist in a rarefied age when we have a remarkably solid understanding of how evolution works. What wonders we could have told the naturalists and scientists of the past—what questions we could have answered! A mere two centuries ago we had only an inkling that species might change through time. The much-misunderstood Jean Baptiste Lamarck provided us with the radical idea that species change through time in response to their environment. This was a revolutionary explication of the existence of a process that was not necessarily teleological and one that bucked the prevailing understanding that species were immutable or did not change. That Lamarck is mostly credited as the guy who thought giraffes stretched their necks is a misunderstanding of the fundamental shift in our scientific understanding that he foretold. Lamark, Charles Lyell, Thomas Malthus, Alfred Russell Wallace, Erasmus Darwin, Gregor Mendel, and Charles Darwin—these individuals whether intentionally or by proxy laid the foundations for us to apply the scientific method to understand the origins of biodiversity.

The "evolutionary process" is the name that we give when we refer to a collection of natural forces that work to shape the breadth and depth of biodiversity. At the smallest scale evolution is a population-genetic process. That is, evolution occurs in populations, not in individuals and, fundamentally, is the change in the frequency of genetic variants, or alleles, in populations through time. Consider this: a population of boas contains a certain level of genetic diversity at any given time point. This diversity is manifest by the presence of different variants of different genetic regions, colloquially (but not technically) called "genes." All individual boas in the population have the same genetic regions (genes), but some individuals have slightly different variations of these genes that may or may not make them different in some anatomical, physiological, or behavioral way. This variation is a good thing—it is what allows populations to respond (evolutionarily) to

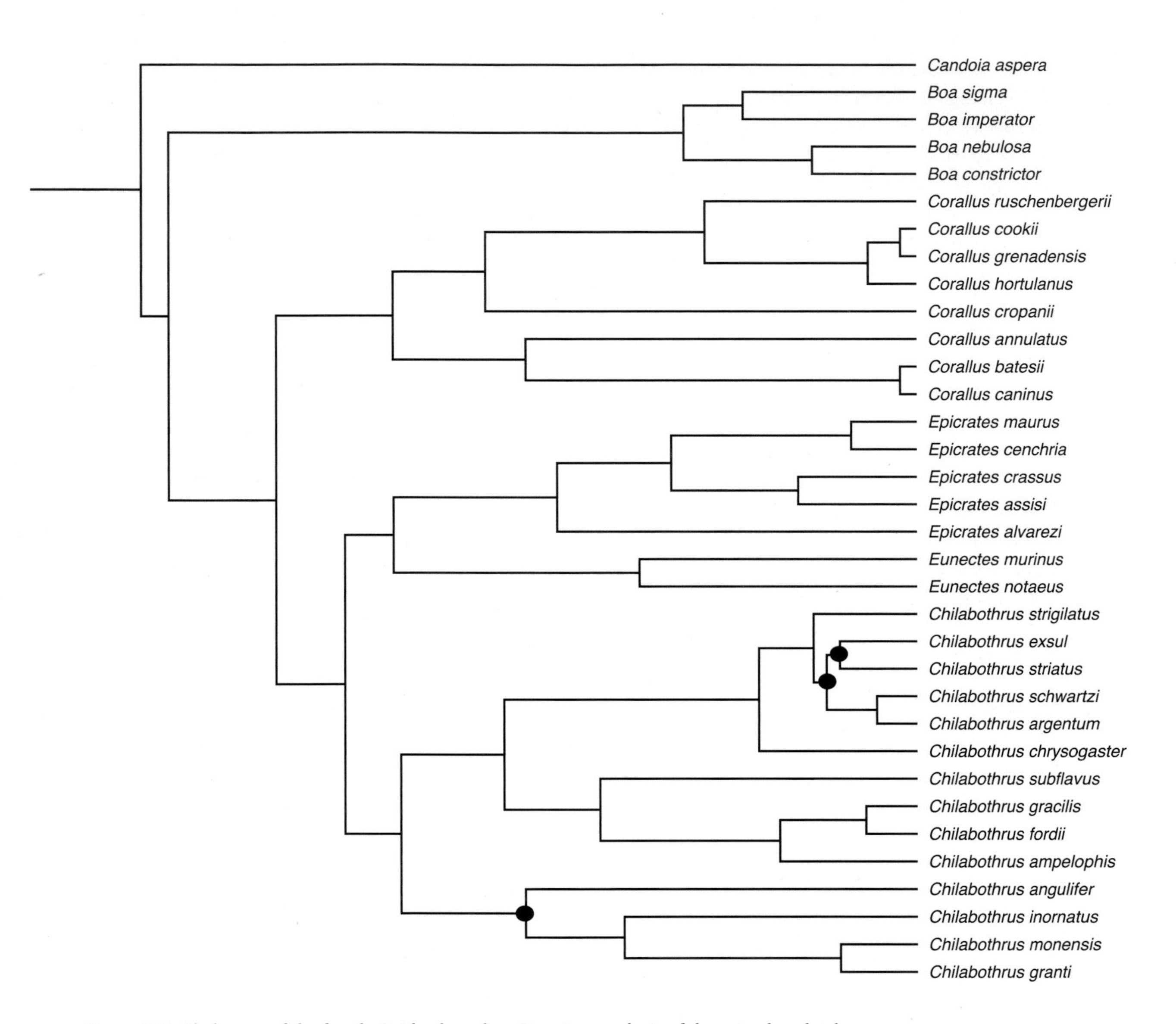


**Figure 0.2.** Phylogeny of the family Boidae based on Bayesian analysis of the mitochondrial locus cytochrome *b* (1100 base pairs; after Reynolds et al. 2013a, 2014a). The phylogeny is missing one West Indian taxon (*Boa orophias*) and three mainland taxa, for which no genetic data were available to us. Note that some relationships are subject to change depending on the type of data and the type of analysis conducted (e.g., black circles are areas with low nodal support). Upcoming studies using genome-scale data are resolving these relationships (RGR, unpubl. data).

the constantly changing environment. An environment might be roughly similar over a period of generations—but over tens to hundreds of generations the environment might change drastically from the perspective of any given species. The frequency of these genetic variants in the population is what we term "allelic diversity," or simply genetic diversity. Thus, we can quantify the frequency of certain alleles in the population, in the same way that we might count the frequency of colors of jellybeans in a jar. Where this really becomes interesting is when we see that with each generation of boas, we get different frequencies of these alleles. Each time the population reproduces, the allele frequency in the population changes. This is both predictable in a theoretical framework (as an expected deviation from Hardy–Weinberg Equilibrium, in case you'd like to pursue this idea further) as well as readily observable using genetic analysis techniques. One can visualize this as the jar of jellybeans changing composition each generation, or each time someone grabs a handful to eat and then dumps replacements into the jar from the bag—some generations the frequency of the red jellybean goes up, some generations the frequency of green goes up. Why does this happen? It is a combination of both predictable (deterministic) and random (stochastic) processes operating in the population (some people intentionally "select" red jellybeans and pick them out, others just grab handfuls). We might predict that the frequency of an allele would increase if, perhaps, it imparted an advantage in the current environment, such as the well-known fact (wink) that no one likes licorice jellybeans and thus those might "survive." Alternatively, the same allele, even if advantageous, could decrease in frequency or even disappear altogether purely owing to random chance—the individual that bore that allele might die before reproducing (someone accidentally ate a licorice jellybean—yuck!). Life is unpredictable, and thus stochasticity plays a major role in how allele frequencies change. But it is our ability to understand the forces that generate these stochastic and deterministic patterns that play out each generation, and to extrapolate that to longer time periods, that allows us to formulate the body of knowledge that describes the evolutionary process. Determinism and stochasticity have played a major role in the evolution of West Indian boas (as they have in all species) and in a way that has left a remarkable signature on the animals we see today.

For example, we know that boa populations are capable of evolving smaller body size on islands. Indeed, the frequent observation that island populations of living things are often differently sized from mainland populations is referred to as island gigantism or island dwarfism. We see it across so many different groups that it has been proposed as a general rule—that when a species colonizes an island it will evolve to be smaller if it is a "big" species, like a mammoth or a hippopotamus, or it will evolve to be bigger if it is a smaller species, like a tortoise or an iguana. Such patterns, while common, are not necessarily the rule, but they do provide a remarkable and dramatic example of how the evolutionary process plays out on islands. *Boa imperator*, the Central American boa, is an excellent example. This species has colonized dozens of islands off the coast of Central America in the last 6000–10,000 years. The Caribbean coastal islands are remarkable in that the boas on these islands are dwarfed, or obtain a much smaller body size, than relatives on the mainland. This shift in mean body size in these populations has likely happened rather quickly on an evolutionary timescale—over thousands, rather than millions, of years. A thousand years is only about two hundred boa generations, so to witness such a dramatic reduction in body size over that time period is quite remarkable. But even more astounding is that this has happened over, and over, and over again (Card et al. 2016, 2019). Boas have independently colonized these islands and in each case evolved smaller body size. If this had happened once, we might call it stochastic, or random, that the individuals surviving on the island happened to be smaller. Alternatively, if we see this happening many, many times, then we suspect that it is a deterministic, or predictable, process. Indeed, it is likely that boas on these islands are shifting to using smaller prey, such as birds, and thus natural selection is strongly favoring individuals that are smaller and require smaller meals or are more capable of stalking and capturing the available prey. Selection must be strong on these islands to effect such a rapid change, and, amazingly, selection is consistent across islands!

So how does this inform our understanding of West Indian boa evolution? Well, it posits that boas on these islands might have undergone similar evolutionary processes owing to similar selective forces. Could it be that the ancestor of *Chilabothrus* colonized the

West Indies from South America and then got bigger or smaller? This seems very likely to have happened. Consider an early diverging lineage of West Indian boas—the Cuban boa (*Chilabothrus angulifer*). Attaining a size of over 5 m, this qualifies as a truly giant snake among any Neotropical boas! Alternatively, consider the newly discovered Hispaniolan vineboa (*C. ampelophis*), known to achieve lengths of only 68 cm (Landestoy T. et al. 2021b). Nevertheless, the story of West Indian boa evolution gets more complicated and fascinating.

## BIG BOAS, SMALL BOAS: DETERMINISM IN WEST INDIAN BOA EVOLUTION

On islands of the Greater Antilles and Bahamas Banks, the genus *Chilabothrus* is represented by fourteen species—some really big and some quite small. Some islands or island groups, such as Hispaniola, Puerto Rico, and the Bahamas, have more than one species. Where multiple species occur, there is one large species and one to three smaller species. Concomitantly, the large species are ecological generalists, in that they occupy both terrestrial and arboreal habitats and consume a wide variety of prey. On the other hand, the small species tend to be ecological specialists, occupying either terrestrial or arboreal habitats and feeding on a smaller range of prey. For example, on Hispaniola the large species *C. striatus* is found in nearly every habitat type—from virgin dry forest to moist montane forest to urban areas in Santo Domingo to monoculture sugarcane fields. They feed on birds, rats, domestic poultry, hutias, and basically anything they can catch. The three small species—*C. gracilis*, *C. fordii*, and *C. ampelophis*—are much more specialized. *Chilabothrus gracilis* lives in mesic forested environments, most commonly around streams, mangrove swamps, and rivers, as well as open areas adjacent to moist forest. They are entirely arboreal with a long and skinny shape that allows them to navigate easily among thin branches of trees and even large clumps of tall grasses. Here they hunt one type of prey—lizards, particularly sleeping *Anolis* lizards. This is ecological specialization in both habitat and diet. *Chilabothrus gracilis* will never compete for resources with adult *C. striatus* (although there might be some overlap with juvenile *C. striatus*, which are also arboreal lizard eaters; Henderson et al. 1987). The small species *C. fordii* lives in xeric environments, such as the Plaine du Cul-de-Sac/Valle de Neiba on the border between Haiti and the Dominican Republic. This species is largely terrestrial, spending the day under rocks and foraging on the ground or in low bushes during the night. The diet of *C. fordii* is less well known (see the species account) but includes lizards and introduced rodents (a "new" food source in an evolutionary context) and likely occasionally birds. *Chilabothrus ampelophis*, first discovered in 2020, is known from only five specimens but likely represents the smallest boa on Hispaniola (Landestoy T. et al. 2021b) and is likely the smallest species in the family Boidae. They are slender snakes, found in a small area of dry forest; the diet is unknown, although presumed to include mostly *Anolis* lizards (Landestoy T. et al. 2021b). Thus, the island of Hispaniola supports a large generalist species and three small specialists adapted to specific habitat types (mesic forest, xeric scrub, and dry forest; respectively). We see a similar situation on Puerto Rico, as well as in the Bahamas, where small specialist species occur alongside or near to large generalist species. Thus, this begs the question of whether a small ancestral species evolved and then spread to other islands or whether selection has repeatedly led to the evolution of small specialists on these islands. This question has been addressed by various authors over the last few decades (Tolson 1987; Kluge 1989; Rodríguez-Robles and Greene 1996), but a recent study has shed full light on these observations (Reynolds et al. 2016a), and the explanation follows.

Remarkably, the small boa species are not each other's closest relatives—instead, molecular phylogenetic reconstruction has shown that large and small body size (and concomitant specialization) have evolved independently and repeatedly on each of these islands. Thus, a deterministic process is operating in some West Indian islands to produce a similar faunistic outcome—colonization of an island by an ancestral boa, followed by speciation and the evolution of large generalists and small specialists. One hypothesis for what might be driving these divergences is, of course, diet, as was found in the case of island dwarfism in *Boa imperator*. Predictions regarding diet in snakes are mostly driven by two anatomical features—body size and head shape. The latter, also technically known as trophic morphology, is directly related to the types and sizes of prey that are consumed—prey must fit in the confines of the head and be capable of being subdued and manipulated by the snake. Not unlike the famous beaks of Darwin's finches (*Geospiza* spp.) of the Galapagos, whose beak sizes and shapes are directly related to the types of seeds that they consume (Grant and Grant 2011), boa heads might tell a story about how the small species evolved to be specialists. In other words, is the repeated evolution of small body size directly related to consistent changes in head shape? Did body size and trophic morphology evolve in tandem in a similar fashion across these islands? Just examining the heads of some of these species, it seems that not only might this be the case but also that the heads of the small species greatly resemble the heads of juveniles of the large species. Could small boas just be neotenic large boas? Such a hypothesis is intriguing, and undoubtably some form of arrested development is likely contributing to small size in these species—although a detailed developmental study would be needed to confirm this. Interestingly, however, the heads of different species of small boas are not similar in shape when measured in multiple dimensions and analyzed statistically (Reynolds et al. 2016a). So, yes, boas repeatedly evolved smaller body size and feed on similar specialized prey resources (like *Anolis* lizards), but they did not change head shape in a consistent fashion over the evolution of these lineages. This is an excellent example of the interplay between determinism (repeated evolution of small specialist boas) and stochasticity (head shape changes randomly, not consistently, in small boas).

The evolutionary process is indeed complex, but increased attention to these boas from evolutionary biologists is shaping our understanding of the origins of boa biodiversity and giving us a remarkably complete view of their evolutionary history.

# Chapter 1

# THE SHARED HISTORY OF BOAS AND HUMANS IN THE WEST INDIES

Boas and humans have a long-shared history in the West Indies, and, not surprisingly, humans have had a profound impact on the natural history of every species of boa on the islands where boas occur. Humans directly or indirectly made dramatic alterations to, or virtually eliminated, habitats that once supported boa populations. They caused the extinction or extirpation of mammalian species that almost certainly were part of the boas' prey base; they introduced species that were capable of exploiting boas as food; and they most likely killed boas when encountered. In this chapter we present an abbreviated history of human colonizations in the West Indies and what impact humans might have had or certainly did have regarding habitats, prey bases, and introduced predators. Many of the dates that appear in the text and tables are in flux. As new archaeological and paleontological data become available, especially for first appearance dates (FAD) for humans (Table 1.1) and last appearance dates (LAD) for mammals, cited dates likely will change.

Four colonization waves were associated with increasingly complex material culture (Keegan 1994): Lithic, Archaic, Ceramic, and historical. The earliest known archaeological sites were found on Cuba and Hispaniola, but little is known about the impact of Lithic Age (~7000–4000 years BP) populations on the environment. However, "human transformation of island ecosystems began at initial colonization and often accelerated through time as populations grew and human activities intensified" (Rick et al. 2013). In terms of habitat alterations, perhaps we can assume that the impact of Lithic Age colonizers on species of *Chilabothrus* was negligible. No animal remains have been reported from Lithic sites (Cooke et al. 2017).

Likewise, little is known about population size and distribution of Archaic Age (~5000–2000 BP) peoples who likely originated somewhere along the northern Colombian, Venezuelan, or Guyanan coasts (Callaghan 2003), although a diverse material culture is known from Cuba, Hispaniola, Puerto Rico, and several islands in the northern Lesser Antilles, most notably Antigua (Fitzpatrick 2015). Virtually no Archaic sites occur in the southern Caribbean (south of the Guadeloupe Passage). Whether that absence can be attributed to oceanographic conditions, seafaring deficiencies, volcanism, or an element of fear or belief systems that discouraged settlement has not been affirmed with any certainty (Callaghan 2010; Fitzpatrick 2015), although many archaic coastal sites are probably now underwater. Archaic Age sites usually are located in coastal situations, and evidence suggests that these groups of people were more sedentary than once thought. Archaic peoples were fisher–foragers; most of their protein came from the sea (mollusks, fish, turtles), and archaeobotanical remains indicate the presence of cultivated plants, including species not native to the Caribbean (implying some movement and management; Fitzpatrick 2015). Mounting evidence suggests that Archaic peoples lived in permanent villages, that their impacts on the landscape (including burning) were substantial, and "that island ecologies were degraded before Saladoid [Ceramic Age] peoples arrived" (Fitzpatrick and Keegan 2007). Records of mammals in Archaic sites are limited but do include

**Table 1.1.** Occurrence dates (years BP) for human remains and island area for West Indian islands that currently or at one time harbored boa populations.

Occurrence dates given as calibrated radiometric age, earliest colonization (single asterisk), or Ceramic Age colonization (double asterisk). Most of the information taken from Cooke et al. (2017), who gleaned data from various sources; asterisked data are from Fitzpatrick 2015.

| Island | Occurrence dates (years BP) | Area ($km^2$) |
|---|---|---|
| **Bahamas / Turks and Caicos (TCI)** | | |
| Great Abaco | 923 | 1224 |
| Crooked Island | 692–358 | 148 |
| Eleuthera | 1594–690 | 457 |
| Grand Turk (TCI) | 1211–721 | 18 |
| Long Cay (TCI) | 950* | 1 |
| Middle Caicos (TCI) | 854–733 | 430 |
| Middleton Cay (TCI) | 800* | 0.1 |
| New Providence | 1150* | 228 |
| San Salvador | 1278–931 | 162 |
| **Greater Antilles** | | |
| Cuba | 5904–4942; 7250* | 105,806 |
| Isla de la Juventud | 1015* | 2237 |
| Hispaniola | 6024–5640; 6350* | 74,546 |
| Jamaica | 1208–729; 1185* | 11,026 |
| Puerto Rico | 4654–4180; 6855* | 8761 |
| Vieques | 3731–2790; 4050* | 148 |
| Isla Mona | 4955* | 58 |
| St. Thomas | 3923–2570; 3035* | 70 |
| Virgin Gorda | 1068 | 21 |
| **Lesser Antilles** | | |
| Marie-Galante | 4889–1274; 1485** | 171 |
| Dominica | 2404–1341; ~2400** | 787 |
| Martinique | 3709–2552; 3710* | 1167 |
| St. Lucia | 1370–792; 1395* | 640 |
| St. Vincent | 1451–848; 1650* | 352 |
| Grenada | 1460–1265; 1850* | 323 |
| Baliceaux | 566–168; 565* | 2 |
| Mustique | 1705* | 7 |
| Union | 1027; 1025* | 9 |
| Carriacou | 1472–1330; 1460* | 35 |

*Heteropsomys insulans*, an echimyid rodent in Puerto Rico; apparently mammals were not prominent in the diet of Archaic peoples (Cooke et al. 2017). That *Chilabothrus angulifer*, *C. inornatus*, and *C. striatus* were already being impacted by the activities of humans seems likely, and it is known that some vertebrate species were moved around the Caribbean by early inhabitants (Kemp et al. 2020).

The Ceramic Age (~2500 calibrated [cal] years BP [i.e., the radiocarbon date has been corrected using current methodologies] in Puerto Rico and the Virgin Islands and later elsewhere) "sites span nearly the entire eastern Caribbean, from Puerto Rico in the north, down through the Lesser Antilles to Grenada in the south, across to Trinidad and Tobago, and into the Middle and Lower Orinoco River basin in Venezuela. This population dispersal was relatively rapid and involved a cultural complex known as Saladoid" (Fitzpatrick 2015). Interestingly, the southern Lesser Antilles were settled centuries later than more northerly sites in Puerto Rico and the Leeward Islands. Although some of the Saladoid dispersals led to islands that had not previously been colonized (i.e., St. Lucia, St. Vincent, the Grenadines, and Grenada), others were already occupied by Archaic groups (Fitzpatrick 2015). Like Archaic Age sites, the majority of early Ceramic Age sites are located along coasts. On St. Vincent, for example, Callaghan (2007) found evidence of settlements on low-lying coastal areas in cactus scrub or secondary rainforest, habitats frequented by *Corallus cookii*. Elsewhere in the West Indies, a wide variety of habitats were exploited, including inland areas, especially those with access to freshwater (Fitzpatrick 2015). Evidence suggests a strong focus on horticulture and the introduction of nonnative plants. In addition to marine-based protein, early Ceramic Age peoples exploited native land crabs, frogs, lizards, snakes, sea birds, and mammals; midden records include hutias (Echimyidae: Capromyinae) in the Greater Antilles and oryzomyine rice rats (Cricetidae: Sigmodontinae) in the Lesser Antilles (Newsom and Wing 2004; Turvey et al. 2010); both hutias and rice rats were/are likely important components of boa diets.

During the late Ceramic Age (1500–500 years BP) the number of colonized sites on various islands steadily increased, reflecting population growth, territoriality, and regionalization. The first archaeological evidence of colonization of the Bahamas and Jamaica occurred ~1600–1500 cal years BP (Berman and Gnivecki 1995; Fitzpatrick 2015). Post-1500 cal years BP, mammals native to the Neotropical mainland were transported to the southern Lesser Antilles, with direct evidence of this practice reported from archaeological sites on Grenada and Carriacou. These included agouti (*Dasyprocta leporina*), opossum (*Didelphis marsupialis*), armadillo (*Dasypus novemcinctus*), peccary (*Tayassu* or *Pecari*), deer (*Odocoileus* and *Mazama*), and dogs (*Canis lupus familiaris*; Giovas et al. 2012). Similarly, the rodent *Isolobodon portoricensis* was transported from its native Hispaniola to Puerto Rico and the Virgin Islands (Newsom and Wing 2004). In addition to these translocations, land crabs, amphibians, lizards, snakes, birds, and rice rats continued to be exploited for food.

Amerindians during this time probably practiced shifting cultivation ("swidden agriculture") to create agricultural plots for manioc and other crops, and this "would have resulted in the first major anthropogenic disturbance and disruption of the natural forests and vegetation" (Newsom and Wing 2004). The "Drake Manuscript" (Drake and Pierpont 1996) depicts hunting of the "COVLLEVVRE NOIRE (black snake, plate 122)" for food, while other images from that manuscript suggest that Amerindians hunted a huge variety of wildlife and cleared land for cultivation. On the Caicos Bank, midden sites on Middle Caicos hold remains of reptiles and their prey, dating to about 1000 years BP (Carlson 1994). Pearls, the most thoroughly studied Ceramic Age site in Grenada, was a large, 60.7-ha site (Newsom and Wing 2004) on rich agricultural land (Bullen 1964) and a center of Saladoid culture. A large area would have been cleared for the village and surrounding gardens. It would have been planted for 4 or 5 years and then allowed to return to secondary forest before being cleared again (perhaps 20 years later) using a slash-and-burn technique (W. Keegan, in litt. to RWH, 6 November 2008). The arboreal boa *Corallus grenadensis* was surely encountered there. Apparently, a waxing and waning of potential treeboa habitat would have been associated with settlements similar to Pearls. Even today, pottery sherds can be found scattered on the surface at Pearls anywhere recent agricultural activity has occurred, and *Co. grenadensis* remains relatively common.

The fourth colonization wave was the historical (post-1492 Common Era; ~500 years BP). The Taíno groups in the Bahamas, Cuba, and on Hispaniola were the first to encounter Europeans. By then, they had developed "complex forms of social organization suitable to transforming landscapes for productive use. Eighty species of plants were cultivated on land, and intensification included terracing on hillsides and earth mounds replacing slash-and-burn agriculture in some places" (Cooke et al. 2017). The indigenous population of the Antilles at the time of European contact was estimated to be anywhere from 100,000 to 6–8 million (Keegan 1996). European arrival spelled "the collapse of native Taíno societies and rapid depopulation, largely through the transmission of European zoonotic diseases" (Cooke et al. 2017). Soon after arrival, a wide range of commercial field crops (e.g., sugarcane and cotton) were introduced from Europe, Africa, Asia, and mainland South America, as well as noncommercial invasive plants and the establishment of feral ungulates (e.g., pigs, cattle, goats), thereby further altering plant communities and ecosystem structure (Watts 1987). Shortly after European arrival, invasive black rats (*Rattus rattus*), house mice (*Mus musculus*), and feral cats (*Felis catus*) and dogs became established (Watts 1987), all of which have had a profound impact on native herpetofauna.

The most impactful changes to West Indian landscapes, and those that would have the most direct negative impact on boa populations, occurred during the eighteenth and nineteenth centuries. That was a time of widespread forest clearance for sea cotton, sugarcane, sisal, and coffee plantations and the export of native hardwoods. In the Lesser Antilles, most or all forests were eliminated, excepting only those on the steepest portions of volcanic islands (e.g., St. Vincent and Grenada). During Beard's (1949) forest assessments in 1943 and 1944, however, he found expanses of primary (virgin) forest on St. Lucia but not on Dominica. Forests on the low-lying Bahamas and Grenadines often were completely destroyed. Larger islands were similarly impacted; by 1899 Puerto Rico, for example, had lost more than 99% of its primary forest (Brash 1987), and by the mid-twentieth century about 90% of Cuba's primary forests were gone (del Risco Rodríguez 1995; Funes Monzote 2004). At the same time, competitive (brown rat, *Rattus norvegicus*) and predatory (small Indian mongoose, *Urva auropunctata*) invasive mammals were introduced either accidentally or deliberately.

## THE IMPACT OF HUMANS ON THE BOA PREY BASE

Despite the astounding and relatively recent impact following the colonization of the region by Europeans, the impact of humans on their environments in the pre-Columbian West Indies should not be underestimated (Newsom and Wing 2004; Fitzpatrick and Keegan 2007). These peoples dramatically altered habitats and caused observable and significant declines in the numbers of many faunal populations. Species of *Boa*, *Chilabothrus*, and *Corallus* inhabited West Indian islands long before the arrival of humans, which likely began about 7000 years BP. Boas hunted and reproduced in habitats that were altered only by natural events (e.g., hurricanes, volcanic eruptions) and were hunted only by native predators (e.g., raptors, herons). Once humans reached the islands, landscapes were altered, vast expanses of native trees were eliminated, and vegetation alien to the islands was introduced. Humans also introduced additional predators that included rats, cats, dogs, and mongooses. But the predators also included humans, who likely consumed boas as evidenced by some material in middens (e.g., Newsom and Wing 2004) and recorded observations (e.g., Drake and Pierpont 1996). The boas' diets consisted of native lizards (anoles, iguanas), birds, and mammals (insectivores, rodents) (Table 1.2). At times, however, based on the excavation of prehistoric middens, boas had to compete with humans for trophic resources, specifically hutias and rice rats (although the extent to which this might have impacted boa populations is unclear).

West Indian boas prey on frogs, lizards, birds, bats, and nonvolant mammals (primarily rodents; Tables 1.2, 1.3, 1.4, 1.6, 5.3). Frogs are rarely taken and have been recorded in the diets of only five species. Bats have also been recorded in the diets of only eight species, but we suspect that predation on bats is more widespread among the boas than is currently documented; they can be an important, although localized (i.e., in caves), prey. Many caves that formerly harbored

**Table 1.2.** Prey groups exploited by West Indian boas.

X = known to be included in diet; — = not documented in diet;
? = presumed to be in diet.

| Species | Frogs | Lizards | Birds | Bats | Rodents |
|---|---|---|---|---|---|
| *Boa nebulosa* | — | X | X | — | X |
| *Boa orophias* | — | X | X | X | X |
| *Chilabothrus ampelophis* | — | ? | — | — | — |
| *Chilabothrus angulifer* | X | X | X | X | X |
| *Chilabothrus argentum* | — | X | X | — | — |
| *Chilabothrus chrysogaster* | — | X | X | — | X |
| *Chilabothrus exsul* | X | X | X | — | X |
| *Chilabothrus fordii* | — | X | — | — | X |
| *Chilabothrus gracilis* | — | X | — | — | — |
| *Chilabothrus granti* | — | X | — | — | X |
| *Chilabothrus inornatus* | — | X | X | X | X |
| *Chilabothrus monensis* | ? | X | — | — | ? |
| *Chilabothrus schwartzi* | — | X | — | — | X |
| *Chilabothrus striatus* | — | X | X | X | X |
| *Chilabothrus strigilatus* | X | X | X | — | X |
| *Chilabothrus subflavus* | — | X | X | X | X |
| *Corallus cookii* | — | X | X | — | X |
| *Corallus grenadensis* | X | X | X | — | X |

large colonies of bat no longer do—particularly in the Bahamas and Turks and Caicos as well as in Puerto Rico, where cave tourism has led to a decline in bat occupancy and abundance (Gannon et al. 2005). Several caves in Cuba have been heavily transformed because of quarrying or for military uses, and the large bat populations that once inhabited those caves are gone. Although birds have been recorded in the diet of twelve species, only one boa species seems to be an avian specialist (*C. argentum*); the other species appear to be opportunistic regarding birds as prey. All eighteen boa species, however, include lizards in their diets (primarily *Anolis*, *Cyclura*, and *Iguana*). Of 972 prey records documented for West Indian macrostomatan snakes (boids, tropidophiids, colubrids, and dipsadids; ninety-seven species total), 716 (73.7%) were lizards (Henderson 2015). Birds constituted 1.1% of the sample and mammals 5.6%. That nearly all West Indian snakes include lizards in their diet at some time in their life is not surprising. About four hundred endemic lizard species occur in the Antilles, ranging in size from diminutive sphaerodactylids to massive iguanids, and some species can be found at spectacular population densities. In sharp contrast, the Late Quaternary West Indies harbored over one hundred endemic, nonvolant terrestrial mammalian species (or distinct island populations), of which only ten to thirteen species are extant. All these species are restricted to the Greater Antilles and the Bahamas Archipelago, where they occupy small fragmented ranges or single islands (Turvey et al. 2017). Fifteen of the eighteen species of boas include mammals in their diets (*Chilabothrus argentum*, *C. ampelophis*, and *C. gracilis* do not); we do not have any dietary data for *C. ampelophis*, but we suspect it is an anole specialist (Landestoy et al. 2021b). Of those fifteen, however, only three (*C. angulifer*, *C. striatus*, and *C. strigilatus*) currently have access to nonvolant mammalian species that are native to their islands, and only two (*C. angulifer* and *C. striatus*) have been documented exploiting that prey (Table 1.4). When the remaining twelve boa species prey on nonvolant mammals, the prey is always an introduced species such as *Rattus*.

The West Indies "witnessed the highest level of mammalian species extinctions anywhere in the world during the prehistoric Holocene and post-1500 AD historical era, and from a Late Quaternary fauna containing more than 100 endemic mammals" (Brace et al. 2015). According to Wing (2001) and Newsom and Wing (2004), rice rats were heavily exploited by prehistoric Amerindians and constituted a significant component of their dietary intake; evidence is suggestive of prehistoric overexploitation. Nevertheless, most or all of the Lesser Antillean rice rat populations likely survived until European arrival in the region (Turvey et al. 2010). That reality suggests that Lesser Antillean rice rat extinctions were driven by historical-era factors, such as the accidental introduction of *Rattus* spp. and, later, by the deliberate introduction of mongooses (Horst et al. 2001; Hedges and Conn 2012).

What we do not know is whether the extinction or extirpation of rodent species affected those species of

**Table 1.3.** Summary of current and hypothesized historical diets of Lesser Antillean boids (*Boa* and *Corallus*).

Prey taxa marked with one asterisk (*) are introduced and those marked with two asterisks (**) are extinct. Size represents the approximate maximum snout–vent length (SVL). See species accounts for additional information on diets. Real and potential prey bases are based on current knowledge of diets.

| Species | Distribution | Size (SVL) | Current prey base | Former prey base |
|---|---|---|---|---|
| *Boa nebulosa* | Dominica | ~3.0–4.0 m | *Iguana delicatissima*<br>*Gallus gallus**<br>Native birds<br>*Dasyprocta noblei**<br>*Mus musculus**<br>*Rattus* spp.*<br>*Brachyphylla cavernarum*<br>*Felis catus**<br>*Canis lupus familiaris** | *Iguana delicatissima*<br>Native birds<br>An as-yet undiscovered species of rodent (e.g., *Megalomys***, Oryzomyini sp.**)<br>Chiroptera spp. |
| *Boa orophias* | St. Lucia | ~3.0–4.0 m | *Iguana iguana*<br>*Gallus gallus**<br>Native birds<br>*Mus musculus**<br>*Rattus* spp.*<br>*Brachyphylla cavernarum*<br>*Felis catus**<br>*Canis l. familiaris** | *Iguana iguana*<br>Native birds<br>*Megalomys luciae***<br>Chiroptera spp. |
| *Corallus cookii* | St. Vincent | ~1.4 m | *Anolis* spp.<br>Native birds<br>*Mus musculus**<br>*Rattus* spp.* | *Anolis* spp.<br>Native birds<br>*Oligoryzomys victus*** |
| *Corallus grenadensis* | Grenada Bank | ~1.6 m | *Eleutherodactylus johnstonei**<br>*Iguana iguana*<br>*Anolis* spp.<br>Native birds<br>*Marmosa robinsoni*<br>*Mus musculus**<br>*Rattus* spp.* | *Iguana iguana*<br>*Anolis* spp.<br>Native birds<br>Oryzomyini sp.**<br>*Megalomys* sp.**<br>*Zygodontomys* sp.** |

**Table 1.4.** Inclusion of native and nonnative mammals in the present-day diets of West Indian boas.

| Species | Nonnative mammals in diet | Native mammals in diet |
|---|---|---|
| *Boa nebulosa* | Yes | No |
| *Boa orophias* | Yes | No |
| *Chilabothrus ampelophis* | No | No |
| *Chilabothrus angulifer* | Yes | Yes |
| *Chilabothrus argentum* | No | No |
| *Chilabothrus chrysogaster* | Yes | No |
| *Chilabothrus exsul* | Probably | No |
| *Chilabothrus fordii* | Yes | No |
| *Chilabothrus gracilis* | No | No |
| *Chilabothrus granti* | Yes | No |
| *Chilabothrus inornatus* | Yes | No |
| *Chilabothrus monensis* | Probably | No |
| *Chilabothrus schwartzi* | Probably | No |
| *Chilabothrus striatus* | Yes | Yes |
| *Chilabothrus strigilatus* | Yes | No |
| *Chilabothrus subflavus* | Yes | No |
| *Corallus cookii* | Yes | No |
| *Corallus grenadensis* | Yes | No |

**Table 1.5.** Last occurrence dates (common era, CE) for rice rats on Lesser Antillean islands known to support boa populations now or to have done so historically.

The information is based on historical records or calibrated radiometric dates from archaeological or paleontological horizons containing rice rat material. Modified from Turvey et al. (2010).

| Species | Island | Historical record | Calibrated radiometric date | Reference |
|---|---|---|---|---|
| *Antillomys rayi* | Marie-Galante | | 350–665 CE | Haviser 1997; Brace et al. 2015 |
| *Megalomys desmarestii* | Martinique | circa 1897 | — | Allen 1942 |
| *Megalomys luciae* | St. Lucia | Pre-1881 | — | Allen 1942 |
| *Oligoryzomys victus* | St. Vincent | 1892 | — | Allen 1942 |
| *Zygodontomys* sp. | Carriacou | — | 390–1280 CE | LeFebvre 2007; Fitzpatrick et al. 2009; Giovas et al. 2012 |
| Oryzomyini sp. | Grenada | — | ~200–750 CE | Mistretta 2019 |
| *Magalomys* sp. | Grenada | | ~200–750 CE | Mistretta 2019 |
| *Zygodontomys* sp. | Grenada | — | ~200–750 CE | Mistretta 2019 |

**Table 1.6.** Summary of current and historical diets of *Chilabothrus*.

Prey taxa marked with one asterisk (*) are introduced and those marked with two asterisks (**) are extinct. Size indicates the maximum recorded SVL in millimeters. See species accounts for additional information on diets. Real and potential prey bases are based on current knowledge of diets.

| Species | Distribution | Size (SVL) | Current prey base | Former prey base |
|---|---|---|---|---|
| *Chilabothrus ampelophis* | Hispaniola | 683 | Unknown | Unknown |
| *Chilabothrus angulifer* | Cuba | ~4900 | See table 5.3 | Table 5.3, plus numerous extinct native mammals |
| *Chilabothrus argentum* | Lucayan Archipelago | 1245 | Native birds; *Setophaga tigrina*; *Anolis sagrei* | Presumably native birds; *Anolis sagrei* |
| *Chilabothrus chrysogaster* | Lucayan Archipelago | 1500 | *Coereba flaveola*; Native birds; *Anolis scriptus*; *Spondylurus caicosensis*; *Hemidactylus mabouia**; *Mus musculus**; *Gallus gallus**; *Sphaerodactylus caicosensis*; *Sphaerodactylus underwoodi*; *Cyclura carinata*; *Aristelliger hechti*; *Leiocephalus psammodromus* | Native birds; *Anolis scriptus*; *Spondylurus caicosensis*; *Spondylurus turksae*; *Sphaerodactylus caicosensis*; *Sphaerodactylus underwoodi*; *Cyclura carinata* |
| *Chilabothrus exsul* | Lucayan Archipelago | 810 | *Osteopilus septentrionalis*; *Leiocephalus carinatus*?; *Anolis sagrei*?; *Amazona leucocephala bahamensis*?; *Mus musculus**? | *Osteopilus septentrionalis*; native lizards; native birds |
| *Chilabothrus fordii* | Hispaniola | 860 | *Mus musculus**; *Anolis cybotes*; *Anolis* spp.; *Aristelliger expectatus* | Native rodents; native lizards |
| *Chilabothrus gracilis* | Hispaniola | 905 | *Anolis cybotes*; *Anolis distichus*; native frogs? | *Anolis* lizards; native frogs? |
| *Chilabothrus granti* | Puerto Rico | 1112 | *Mus musculus**; *Anolis cristatellus*; *Sphaerodactylus* spp.; *Iguana iguana**; *Pholidoscelis exsul*; *Rattus rattus** | |
| *Chilabothrus inornatus* | Puerto Rico | ~2200 | *A. evermanni*; *A. gundlachi*; *A. cristatellus*; *A. cuvieri*; *Pholidoscelis* spp.; *Eleutherodactylus* spp.; *Gallus gallus**; *Columbina passerina*; *Bubulcus ibis*; *Monophyllus redmani*; *Erophylla sezekorni*; *Mormoops blainivillii*; *Pteronotus quadridens*; *Brachyphylla cavernarum*; *Chilabothrus inornatus*; *Cardisoma* spp.; Lampyridae; *Rattus rattus**; *Mus musculus** | Native birds; native lizards; bats; *Isolobodon portoricensis***; *Cyclura pinguis*; *Nesophontes edithae***; *Heteropsomys insulans***; *Puertoricomys corozalus***; *Elasmodontomys obliquus***; *Tainotherium valei*** |

*continues*

**Table 1.6 continued**

| Species | Distribution | Size (SVL) | Current prey base | Former prey base |
|---|---|---|---|---|
| *Chilabothrus monensis* | Isla Mona | 1255 | *Anolis monensis*; *Eleutherodactylus monensis*; *Rattus rattus**; native birds | *Anolis monensis*; *Eleutherodactylus monensis*; native birds |
| *Chilabothrus schwartzi* | Lucayan Archipelago | ~1200 | Unknown, likely native birds, lizards; *Rattus rattus**?; *Anolis brunneus*?; *Anolis sagrei*? | Unknown, likely native birds, lizards; *Geocapromys ingrahami*; *Cyclura rileyi* |
| *Chilabothrus striatus* | Hispaniola | 2489 | *Gallus gallus**; *Anolis* spp.; *Rattus rattus**; *Mus musculus**; *Felis catus**; *Solenodon paradoxus*?; native birds; bats; *Plagiodontia aedium* | *Solenodon paradoxus*?; *Plagiodontia aedium*; *Anolis* spp.; native birds; *Cyclura ricordi*; *Cyclura cornuta*; bats; other native mammals |
| *Chilabothrus strigilatus* | Lucayan Archipelago | 2330 | *Gallus gallus**; *Felis catus**; *Rattus rattus**; *Osteopilus septentrionalis*; *Anolis* spp.; native birds; bats; other native lizards; *Mus musculus**; *Cubophis vudii*? | *Geocapromys ingrahami*; *Osteopilus septentrionalis*; *Anolis* spp.; native birds; *Cyclura* spp.; other native lizards |
| *Chilabothrus subflavus* | Jamaica | 2050 | *Gallus gallus**; *Rattus rattus**; *Anolis valencienni*; *Cyclura collei*; *Eupsittula nana*; *Melanerpes radiolatus*; *Amazona collaria*; *A. agilis*; *Molossus molossus*; *Artibeus jamaicensis*; *Corvus jamaicensis*; *Rhinella marina** | *Oryzomys antillarum***; *Anolis* spp.; *Cyclura collei*; native birds; bats |

boas that included mammals in their diets. Was there a hiatus between the extinction or extirpation of native mammalian prey and the introduction and exploitation of alien mammalian species (e.g., *Mus* and *Rattus*) on islands that harbored boas? As noted previously, Turvey et al. (2010) considered it likely that most or all Lesser Antillean rice rat species survived until European contact, including several that survived at least until the late 1800s (and possibly into the early 1900s). Based on that scenario, *Boa orophias* on St. Lucia and *Corallus cookii* on St. Vincent both had now-extinct native rodents and extant introduced rodents as part of their prey base for several hundred years.

On the Grenada Bank, however, the fate of the three species of rice rats that occurred there is not as clear-cut (Table 1.5). All three species are known from the Early Ceramic and the "context where the specimens at Pearls were found supports the possibility that human translocation of rice rats occurred during the earlier occupation of Grenada" (Mistretta 2019). Today, adult *Corallus grenadensis* prey most frequently on introduced *Mus* and *Rattus* but have been known to take iguanas, birds, and opossums (*Marmosa robinsoni*) as well (Henderson 2015). It seems possible that the prey base for *Co. grenadensis* might have been devoid of mammals for a prolonged stretch. Several mammalian species were introduced to Carriacou and possibly Grenada by Ceramic Age peoples; juveniles of some of those species (e.g., agoutis and opossums) could have provided potential prey for *Co. grenadensis* prior to the arrival of European colonizers in the early 1600s.

Relative to boas in the Lesser Antilles, those that occur on the Greater Antilles have had access to a more diverse mammalian fauna, including sloths, insectivores, and primates in addition to rodents (Table 1.6). A wave of extinctions of large-bodied mammals (post 5000 years BP) implicates human-mediated factors; these extinctions might or might not have had any trophic impact on *Chilabothrus angulifer, C. inornatus, C. subflavus,* or *C. striatus*. The number of nonvolant mammals to become extinct in the Greater Antilles since European contact is extraordinary. The list includes insectivores (six species of *Nesophontes, Solenodon marcanoi*), rodents (including two species of *Boromys, Geocapromys columbianus,* two species of *Isolobodon, Oryzomys antillarum,* two species of *Plagiodontia*), and a primate (*Xenothrix mcgregori*) (Turvey et al. 2017).

*Chilabothrus angulifer* is the largest member of the genus and is the largest West Indian boa (~4.9 m; Tolson and Henderson 1993). Rodríguez-Cabrera et al. (2016a) provided interesting ecological scenarios for prehistoric *C. angulifer*. Pre-human and pre-Columbian *C. angulifer* had access to a wide variety of mammalian prey of various sizes, including small bats, nesophontid island-shrews, and echimyid rodents (<0.1 kg), medium-sized echimyid rodents (1 to <10 kg), and medium-to-large sloths (<10 kg to ≥200 kg), plus medium-to-large-sized birds with reduced flight capacity. Furthermore, "in part due to their greater conspicuousness and high rate of predation on domestic animals, large boas probably have been killed more frequently than their smaller relatives since the arrival of humans to the region, selecting against phenotypic traits advantageous under natural conditions in the past (large size)." Although Oviedo's 1535 (Fernández de Oviedo 1851a) observation of *C. angulifer* reaching 7.6–9.1 m total length (TL) was perhaps too suspect to consider, Gundlach's (1880) record of 6.4 m TL, that of Tolson and Henderson (1993) at 4.9 m, and finally 3.7 m in 2008 (Tolson in Rodríguez-Cabrera et al. 2016a), suggest that the maximum body size of *C. angulifer* appeared to have declined by 23–42% (Rodríguez-Cabrera et al. 2016a). Today *C. angulifer* is a trophic generalist, and humans continue to influence prey availability. (See the *C. angulifer* account for details.)

Whether the history of Hispaniolan *C. striatus* has followed a trajectory similar to that of *C. angulifer* is speculative. In 1515, Fernández de Oviedo (1851b; see also Rodríguez-Cabrera et al. 2016a) measured a snake that could only have been *C. striatus* at >6.1 m. Although that size seems unlikely based on the maximum total length currently known for this species (2.49 m), one wonders whether they might have reached modestly larger sizes historically with different prey availability. *Chilabothrus striatus* is known to prey on the endemic hutia, *Plagiodontia aedium*, as well as introduced rodents and native and introduced birds (Henderson et al. 1987).

On Jamaica, the rice rat *Oryzomys antillarum* was last observed in 1877 (Turvey and Helgen 2017). Widespread habitat destruction associated with sugarcane production and other cultivated crops, invasive rats (*Rattus* spp.), and introduced mongooses have likely contributed to its extinction, whether singly or cumulatively is unknown. Consequently, *C. subflavus* on Jamaica appears to have had access to now-extinct *O. antillarum* and introduced *Rattus* and *Mus* for several hundred years.

Puerto Rico is the only Greater Antillean island with zero surviving native nonvolant mammal species. The island previously had one species of island-shrew (*Nesophontes edithae*) and several endemic rodents (two spiny rats, *Heteropsomys insulans* and *Puertoricomys corozalus*, and two large-bodied hutia-like rodents, *Elasmodontomys obliquus* and *Tainotherium valei*, which likely reached about 13 kg in mass; Turvey et al. 2006, 2007). A Hispaniolan hutia (*Isolobodon portoricensis*) was introduced by Amerindians (Woods 1996). All of these species have since gone extinct. The LAD dates for these species are variable, and at least some species survived well past Amerindian colonization, which began 6000 years BP (Ayes 1995), possibly up to European colonization (Turvey et al. 2007). The large iguana *Cyclura pinguis* formerly occurred across Puerto Rico but was restricted to Anegada in historical times owing to habitat loss and predation by humans and introduced mammals (it has since been reintroduced elsewhere). All of these species could have been potential prey for the native boa *Chilabothrus inornatus*, while the boas *C. monensis* and *C. granti* feed almost exclusively on arboreal lizards, birds, and introduced murid rodents. Nearly the entirety of Puerto Rico (and also Vieques Island) was deforested around the beginning of the twentieth century, which likely dramatically altered both boa habitat and prey availability. Since that time the island has become reforested (54.8%; Marcano Vega 2019), with widespread populations of introduced mammals such as *Rattus* sp. and *Felis catus*. Mongooses (*Urva auropunctata*) were introduced to Puerto Rico and Vieques during sugarcane cultivation beginning around 1877 (Woods 1996) and have since flourished there; mongooses could have been partly responsible, along with near-complete deforestation, for the extirpation of boas on Vieques. But recent records of two *C. inornatus* (J.P. Zegarra, in litt. to RGR) and one *C. striatus* (Reynolds et al. 2014b) from that island suggest the possibility that boas could reestablish now that the forests have recovered.

The Lucayan Archipelago had a much more depauperate nonvolant mammalian fauna than the Greater Antillean islands to the south and east. Presently, only the Bahamas hutia (*Geocapromys ingrahami*) survives following its near extinction. This species is a medium-bodied (0.6 kg) rodent and was historically the most widespread terrestrial mammal in the region, occurring across most of the Bahamas banks, from the Little Bahama Bank (Stedman et al. 2007; but likely introduced by Lucayans, Oswald et al. 2020) to the Plana Cays east of the Crooked-Acklins Bank. This species undoubtedly served as a reliable and abundant food source for *Chilabothrus strigilatus*, a species now known to consume introduced mammals such as cats and rats (RGR, unpubl.), although it is unknown if other boas in the region would have used it as prey. *Chilabothrus exsul* is likely too small to take anything but newborn hutias, and *C. argentum* is largely arboreal and only known to consume birds (plus hutia likely did not occur on the Conception Bank; Oswald et al. 2020). Nothing is known of the food habits of *C. schwartzi*, and *C. chrysogaster* is not known to regularly consume mammals (Reynolds and Gerber 2012).

The Bahamas hutia nearly went extinct following European colonization of the Archipelago. The species was, indeed, thought to be extinct until what proved to be the last population was rediscovered on the remote East Plana Cay in the central Bahamas in 1966. Hutias have since been reintroduced to two cays in the Exuma Islands chain and, in the absence of predators, have reached high densities. The native predators of the hutia likely included the extant boa *C. strigilatus* and the extinct flightless 1-m-tall Chickcharnie Owl (*Tyto pollens*). *Chilabothrus strigilatus* is absent from the Plana Cays and presently uncommon in the Exuma Cays where hutias have been reintroduced, and the owl was driven extinct after European colonization (although tantalizing oral histories on Andros suggest that the species might have survived into the nineteenth century).

Presently the mammalian fauna of the Lucayan Archipelago is almost entirely composed of introduced species, including *Rattus*, *Mus*, *Procyon*, *Sus*, *Felis*, and *Canis*. All of these species, with the possible exception of all but newborn *Sus* and *Canis*, represent potential prey for some native boa species; however, the more likely outcome of interactions between these species and boas is that the boas become prey.

## *Chapter 2*

# COMPARISONS BETWEEN WEST INDIAN AND NEOTROPICAL MAINLAND BOAS

Thirty-seven species of boas are distributed on the Neotropical mainland (extending from south of the U.S.-Mexico border to the southern Chaco in Argentina), associated continental islands, or on one or more of at least ninety-two West Indian islands. Although all eighteen West Indian boas are endemic to specific islands or island banks, they nevertheless share a long evolutionary history with mainland boas, and two genera (*Boa* and *Corallus*) are shared between the mainland and the Antilles. More specifically, *Chilabothrus* diverged from their mainland relatives in the Miocene, while species of *Boa* and *Corallus* separated from their respective mainland relatives much more recently—likely in the Pleistocene. Acknowledging the close historical relationship between mainland and West Indian boas, we compare here (largely qualitatively) several biologically relevant aspects between the two boa faunas and especially focus on those that impact the natural history of the Antillean boa fauna today.

### GEOGRAPHIC RANGES

Accurately determining (or estimating) the area of the geographic range of a species is difficult to impossible; too many factors have to be taken into account. Here we estimate range sizes of West Indian boas and then compare them with those of some related mainland species. With few exceptions, for West Indian species we used the sizes of individual islands (in Powell and Henderson 2012) as our basis for range size (summing the areas if the boa has a multi-island distribution). This is, of course, a much closer approximation to recent historical range sizes, as many of these islands have suffered tremendous habitat loss or alteration that has reduced the effective ranges of many boa species. Determining range size for mainland taxa was necessarily more subjective. We used distribution maps and International Union for the Conservation of Nature (IUCN) Red List assessments whenever possible; however, Red List assessors were sometimes reluctant to provide data for estimated extent of occurrence (EOO). We believe that, in general, we have overestimated range sizes for West Indian species and underestimated those for mainland taxa and, while this is not ideal, it still provides an approach for comparing the ranges of these faunas. Historically, the largest geographic range of any West Indian boa (*Chilabothrus angulifer*) was likely smaller than the smallest range of any mainland species. Today that is not true. Largely owing to habitat destruction, at least two (*Corallus blombergii* and *Co. cropanii*), and possibly three (*Boa sigma*), mainland species likely have smaller geographic ranges than some Antillean species.

Cuba is the largest West Indian island (104,338 km$^2$), and *C. angulifer* is widely distributed on Cuba and Isla de la Juventud (2204 km$^2$), as well as over two dozen smaller satellite cays. Similarly, four species of *Chilabothrus* occur on Hispaniola (73,929 km$^2$) and associated satellite islands: the distribution of *C. striatus* is (at least historically) island-wide up to about 1220 m above sea level (asl), but the distributions of *C. fordii* and *C. gracilis* are much smaller and disjunct, owing to their more specific habitat preferences. We cannot yet fully assess the distribution of *C. ampelophis*, but it is probably very small and localized. On Puerto Rico, *C. inornatus* is widely distributed, but the total area of the island is only 8674 km$^2$, and the species is not (currently) established on satellite islands (but see the *C. inornatus* account). Elsewhere on the Puerto Rico Bank, *C. monensis* is restricted to Isla de Mona (55.8 km$^2$), and *C. granti* occurs on at least six islands (0.01–70.2 km$^2$; total area = 164 km$^2$; mean island area = 32.7 ± 29 km$^2$; but see the *C. granti* account) as well as in a single localized population on Puerto Rico. Although Jamaica has an area of 10,829 km$^2$, Newman et al. (2016) estimated that *C. subflavus* is restricted to less than 10% of the island's total area. Among the Lucayan species, the Bahamian *C. strigilatus* has by far the largest range with a total area of 8172 km$^2$ (occurring on at least twenty islands ranging in size from <1.0 km$^2$ to nearly 6000 km$^2$; mean = 408.6 ± 1319.6 km$^2$). The ranges of the other Lucayan species range in size from <7.0 km$^2$ for *C. argentum* to 3123 km$^2$ for *C. exsul* (it occurs on five islands, 4.2–1681 km$^2$; mean = 624.5 ± 831.3 km$^2$). Between those extremes are *C. schwartzi* (two islands, 252 km$^2$ and 497 km$^2$; total area = 749 km$^2$; mean = 374.5 ± 173.2 km$^2$) and *C. chrysogaster* (fifteen islands, 0.06–1552 km$^2$; total area = 2087.8 km$^2$; mean = 139.2 ± 394.3 km$^2$). No boa in the Lesser Antilles occurs on an island greater than 800 km$^2$; *Boa nebulosa*, *B. orophias*, and *Co. cookii* have single-island distributions, but *Co. grenadensis* occurs on eleven islands (Grenada plus at least ten of the Grenadines, with areas ranging from 0.35 km$^2$ to 32 km$^2$).

In sharp contrast to the Antillean species, some mainland boas have geographic distributions ranging from hundreds of thousands to millions of square kilometers. Based on range maps, we estimated the ranges of *Boa constrictor*, *Corallus hortulana*, and *Eunectes murinus* at 6.5–7.0 million km$^2$ and that of *Epicrates cenchria* at about 6.0 million km$^2$. The range of *Co. batesii* is likely also in the millions of square kilometers; that of *B. imperator* is estimated at 1,100,000 km$^2$ (Montgomery and da Cunha 2018). Smaller ranges (hundreds of thousands of square kilometers) include those of *Co. caninus*, *Co. ruschenbergerii*, *Epicrates alvarezi*, *Ep. assisi*, *Ep. crassus*, *Ep. maurus*, and *Eunectes notaeus*. We are reluctant to estimate the extent of the ranges of *B. sigma* and *Co. annulatus*, owing to more limited information about these species, but suspect both are about 100,000 km$^2$. Among the smaller ranges is that of *Eu. beniensis*, with an EOO of 47,285 km$^2$ (Muñoz et al. 2016). At the low end of areas of occurrence are two species of *Corallus*. Anthropogenic activities have reduced the rainforests west of the Andes in Colombia and Ecuador to <8.0% of their former distribution (Valencia et al. 2008), and *Co. blombergii* is now estimated (EOO) to occur over an area of about 4000 km$^2$ (Cisneros-Heredia 2016). Based largely on the map in Machado-Filho et al. (2011), we estimated the extent of occurrence for *Co. cropanii* in Brazil's Atlantic Forest at about 4500 km$^2$. This important domain has been fragmented and reduced to ~7.0% of its original expanse (Machado-Filho et al. 2011) and is in proximity to one of the world's largest human population centers (São Paulo).

## HABITATS

The landscapes of individual West Indian islands have been damaged by people for hundreds to thousands of years (Chapter 1). Natural habitat has been altered or obliterated in order to harvest timber, extract resources, plant crops, and make space for human habitations; more recently, habitats have been compromised for the tourist industry. Even islands that do not support a resident human population are exploited for their natural resources or altered by introduced species. Mainland habitats, likewise, have not been immune to ongoing habitat destruction (i.e., many millions of hectares per year) for timber, agriculture, and mining.

Like attempting to determine the sizes of geographic ranges accurately, assessing the areas of various habitat types is fraught with difficulties. To avoid those challenges as much as possible, we focused on forested areas in the West Indies and on the Neotropical mainland by using FAO (2015) forest resource assessments. We preface this by acknowledging that not all mainland boas are forest restricted (e.g., *Epicrates maurus*, *Ep. crassus*; *Eunectes* spp.). For those West Indian islands or island groups (e.g., the Bahamas) and those Central and South American countries that harbor boa populations, we used three criteria for our as-

**Table 2.1.** Mean values ± SD (range) of forest cover (in 2015) on West Indian islands (or island banks) that harbor boid populations and countries in Central America (+ Mexico) and South America that have populations of boids, as well as forest area designated for conservation of biodiversity in 2015.

All data are from FAO 2015.

| Comparison | West Indies | Central America | South America |
|---|---|---|---|
| Forest area (×1000 ha) | 615.2 ± 1031.8 (17-3200) | 10,786.3 ± 22,375.7 (265-66,040) | 74,766.5 ± 140,664.1 (8130-493,538) |
| Percent of land covered by forest | 45.9 ± 21.2 (3.5–80.0) | 40.3 ± 17.3 (12.8-62.1) | 59.1 ± 25.6 (9.9-98.6) |
| Primary forest area (×1000 ha) | 14.8 ± 29.0 (0–88) | 4812 ± 11.4 (0–33,056) | 39,478 ± 61,181.1 (1738-202,691) |
| Forest area designated for conservation of biodiversity (×1000 ha) | 303.7 ± 486.0 (2-1203) | 4527 ± 9532.7 (32-28,049.0) | 11,618.6 ± 14,044.0 (1764-46,969) |

sessment: (1) total forest area; (2) primary forest area; and (3) percent of land covered by forest. We calculated a mean ± SD for each of those three forest criteria in each of the three geographic regions. Looking only at percentages of land covered by forest, the West Indies (~46%) falls between Central America (~40%) and South America (i.e., only those countries in which boa species occur; ~59%). However, 46% of the total land area of the West Indies pales compared with 40% or 59% of areas on the mainland. When looking at total forest area (Table 2.1), and, especially, primary forest, the West Indies have ~15,000 ha compared with millions on the mainland.

## DIETS

Concurrent with their much larger geographic ranges, mainland boas also have access to prey bases that are taxonomically more diverse than those on West Indian islands, and West Indian boas are trophically much more reliant on native lizards and an introduced mammalian fauna. As noted in Chapter 1, most of the nonvolant mammalian prey base that West Indian boas likely exploited is now largely extinct.

By far the largest species of West Indian boa, *C. angulifer* is the closest ecological equivalent to species of *Boa* and almost certainly has the most taxonomically diverse diet of any West Indian boa. Rodríguez-Cabrera et al. (2020a) identified forty-nine prey species (351 prey items) based on unpublished data and the literature. Birds and mammals dominated; among nonvolant mammals, introduced species accounted for 73% of the prey diversity, but the native capromyid rodent *Capromys pilorides* was by far the most frequently exploited mammalian species. Bats represented important prey items for cave-associated boas <2.0 m snout–vent length (SVL). *Rattus rattus* was the introduced mammal most frequently consumed in natural situations. In anthropogenic habitats, boas >2.0 m readily took cats, dogs, and suckling ungulates such as pigs, goats, and sheep. Although a broad taxonomic array of birds was predated in both natural and altered habitats, the domestic fowl *Gallus gallus* was by far the most frequently consumed (60% of total avian prey) and was 69% of total prey in anthropogenic areas. Amphibians (three species; 6.1% of diversity and 2% of total prey records) and reptiles (six species; 12.3% of diversity and 2% of total prey records) were infrequently predated. Only five lizards (four species; 8% of diversity and 1.4% of total prey records) were documented.

We compiled a list of forty-nine prey species for the widespread mainland boa, *Boa constrictor*. Of those forty-nine taxa, only five (10.2 %) were introduced species. Of thirty-four prey species compiled for *B. imperator*, six (17.6%) were introduced species. *Boa imperator* frequently lives in closer proximity to humans than does *B. constrictor*, and opportunities to predate domesticated or human commensal prey are likely more commonplace. We believe, however, that both of these mainland species take many additional prey species that have so far eluded documentation or that we missed in our review of the literature. Quick et al. (2005) listed thirteen species identified from fifty-two items found in introduced *B. constrictor* in Aruba, Dutch Windward Islands. Birds represented the largest proportion (40.4%), followed by lizards (34.6%), and mammals (25%). All mammalian prey belonged to either one of three introduced species: *R. rattus* (17.3%), *Sylvilagus floridanus* (5.8%), and *Canis lupus familiaris* (1.9%). In sharp contrast, we have only eight prey records for the Dominican endemic *B. nebulosa*, and 75% are introduced species. On St. Lucia, *B. orophias* takes at least seven prey species of which 85.7% are introduced. Historically, both of the Lesser Antillean species of *Boa* had now-extinct rice rats as part of their prey bases (Table 1.3). We are confident that today they exploit more prey (e.g., native birds) than has so far been documented.

Other species of *Chilabothrus* vary in diet. Two species are trophically narrow: *C. gracilis* is an anole specialist, and *C. argentum* takes migratory birds and seabirds; we suspect *C. ampelophis* is also an anole specialist. *Chilabothrus chrysogaster* preys on a wide taxonomic array of lizards, including anoles, geckos, skinks, and iguanas; and adult *C. monensis* take small *Cyclura*, birds, and *R. rattus* (P.J. Tolson, in litt.). In a sample of twenty-eight prey items, small (= young) *C. striatus* took native lizards and birds, but 57% of the sample was introduced rodents taken by boas 600–1650 mm SVL (mean = 1166 ± 730 mm SVL; Henderson et al. 1987). These snakes also are known to take domestic fowl (*Gallus gallus*) and domestic cats (Schwartz and Henderson 1991). On Puerto Rico, the diet of *C. inornatus* is similar to that of *C. striatus* on Hispaniola. Of twenty-six digestive tracts containing prey, 76.9% contained introduced rodents (*Mus musculus*, but primarily *R. rattus*) and, of the thirty-four prey items recovered, twenty-two (64.7%) were those introduced rodents (Wiley 2003). Native anoles were also consumed, as well as domestic fowl, but introduced rodents are the most important prey in both free-ranging and cave-associated *C. inornatus* (Puente-Rolón et al. 2016). The diet of *C. subflavus* is less well known but includes native anoles, iguanas, and bats, as well as introduced domestic fowl and rodents (*R. rattus*; Henderson and Powell 2009). Other species, *C. fordii*, *C. granti*, and *C. strigilatus*, feed on native lizards and introduced rodents. We know little about the diets of *C. ampelophis*, *C. exsul*, and *C. schwartzi*.

Fewer prey records are available for species of *Epicrates* on the mainland. Based largely on information in Martins and Oliveira (1998) and Pizzatto et al. (2009), the widely distributed *Ep. cenchria* takes birds and mammals (primarily a wide taxonomic array of native rodents); the only introduced species were three *R. rattus* and a domestic fowl. Other prey species reported more recently in the diet of *Ep. cenchria* include a porcupine (Erethizontidae), a water rat (*Nectomys squamipes*), a squirrel (*Sciurus spadiceus*), a cricetid rodent (*Hylaeamys laticeps*), two antbird chicks (*Myrmotherula assimilis*), and two relatively large bird eggs (*Tinamus tao*) (Medeiros et al. 2009; Cassimiro et al. 2010; Leite and Dorado-Rodrigues 2017; Martins et al. 2018; Fiorillo and Batista 2019). Bat predation in caves has been repeatedly documented for *Ep. cenchria* and *Ep. maurus* (Lemke 1978; Ramos et al. 2012; Martin-Solano et al. 2016; Aya-Cuero et al. 2019). Although based on few prey records, lizards, birds, and mammals have been reported in the diet of *Ep. assisi* (Pizzatto et al. 2009). *Epicrates crassus*, like *Ep. cenchria*, fed primarily on native rodents (seventeen out of a sample of twenty prey items), but prey items also included one bird and one marsupial (Pizzatto et al. 2009). Ten prey items taken from *Ep. crassus* included the rodent *Galea spixii* and the lizard *Ameiva ameiva* (Vitt and Vangilder 1983).

*Corallus hortulana* is the most geographically and ecologically widespread species in the genus. Forty-seven prey species have been documented based on 139 prey records; many of those records are of birds, rodents, and bats not identified beyond class or order (Pizzatto et al. 2009; Henderson and Pauers 2012; Henderson 2015). We are assured by W. W. Lamar (in litt. to RWH, 9 March 2019) that *R. rattus* is predated by *C. hortulana* and likely is among the unidentified rodents, so we will increase the number of identified taxa to forty-eight. That said, *R. rattus* is the only introduced species (2.1%) in the sample. Only five prey items were lizards (9.6%). Although based on a much smaller sample size, a list of thirteen prey species was compiled for *Co. ruschenbergerii*; of those species, two were introduced (15.4%). In sharp contrast and based on eighty-four prey items, native lizards accounted for

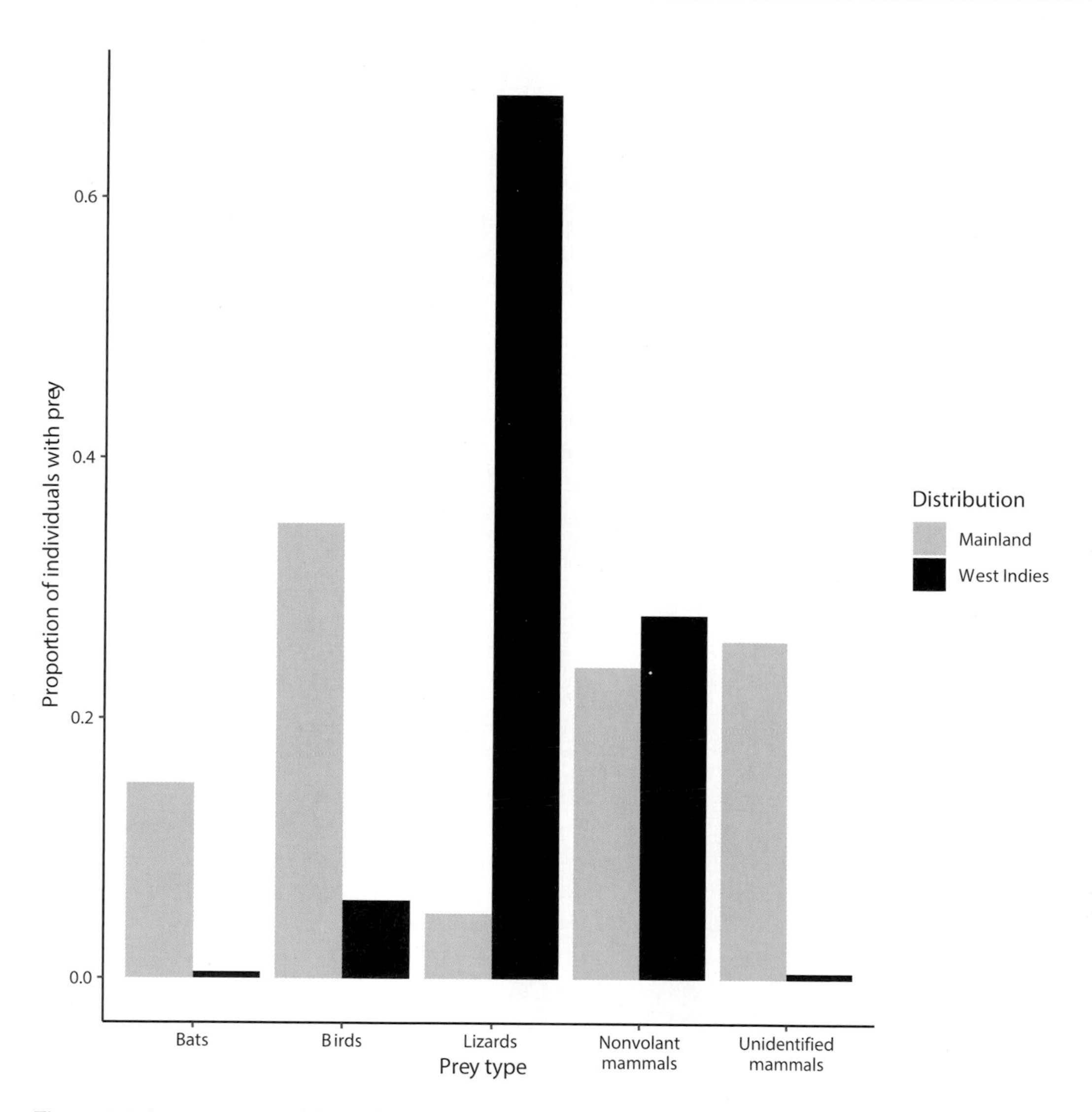


**Figure 2.1**. A comparison of diets of West Indian and mainland species of *Corallus*. The nonvolant mammals in the West Indies were composed of *Mus* and *Rattus*; on the mainland largely a wide variety of marsupials and rodents; (*n* = 271 prey records).

63.1% of the sample for the Grenada Bank endemic, *Co. grenadensis*; of twenty-two mammalian prey items, twenty-one (95.5%) were introduced rodents (Fig. 2.1).

Although no West Indian boa is morphologically or ecologically comparable to the Green Anaconda (*Eunectes murinus*), reviewing its prey composition is nevertheless instructive. Rivas (2020) identified twenty-six prey species in a sample of sixty-two prey items for *Eu. murinus* in the Venezuelan llanos. Seventeen of the prey species were birds, accounting for 45.2% of the prey items. Only four lizards (two species) were documented (6.5%). Ten mammals (16.1%; two species) and eleven turtles (17.7%; two species) were identified. None of the prey species were introduced. In a list of prey species compiled by Thomas and Allain (2021) for *Eu. murinus* across its range, only five (9.1%) were introduced. Bernarde and Abe (2010) have documented cattle (*Bos taurus*) and dogs in the diet of *Eu. murinus*. A sample of 220 prey items taken by the yellow anaconda (*Eu. notaeus*) in northern Argentina was composed largely of aquatic birds with lesser quantities of mammals and reptiles; none of the prey species were introduced (Miranda et al. 2017). Thomas and Allain (2021) compiled a list of forty-five species taken by *Eu. notaeus*; only one was introduced (2.2%).

In general, most mainland boas appear to be euryphagous. They have access to a wide taxonomic range of native vertebrate prey, including fishes, amphibians, lizards, snakes, crocodilians, hundreds of species of birds, and hundreds of species of mammals (including marsupials, edentates, rodents, bats, and primates, all of which are exploited by mainland boas). The most notable exceptions to euryphagy are *Co. batesii* and *Co. caninus*, both of which prey largely on marsupials and rodents (Henderson et al. 2013b; Henderson 2015). Although introduced species do fall prey to mainland boas, they account for a very small proportion of overall diets and appear to be largely limited to domestic fowl and species of *Rattus*.

## ABUNDANCE

With the exception of studies of insular populations of predominantly mainland species (e.g., *B. imperator*), virtually no data of which we are aware document the abundance (population size or density) of mainland boas. The exceptions to this lack of information for mainland taxa are the anacondas (for which there is no ecologically comparable species in the West Indies). To compensate for that lack of information, we compiled information on the abundance of boas relative to other snake species or to published encounter rates.

Dunn (1949) analyzed the relative abundance of snakes in a collection of 10,690 specimens more or less randomly collected in Panama during 1933–1945. Four boa species were included in the collection, and three of them were among the most frequently collected: *B. imperator* (388 specimens; 3.6% of entire collection), *Co. ruschenbergerii* (372; 3.5%), and *Epicrates maurus* (385; 3.6%); *Co. annulatus* was represented by only seven specimens (<1.0%). In terms of relative abundance, *B. imperator* and *Ep. maurus* were fairly constant regardless of where a collection originated; *Co. ruschenbergerii* was by far the most abundant boa among the collections from the Darién region (11.3% compared with 1.7% for *B. imperator* and 3.9% for *Ep. maurus*) but was uncommon elsewhere. Among the species collected more frequently than the boas were a vine snake (*Oxybelis aeneus*; 1440 specimens; 13.5%) and two pit vipers, *Porthidium lansbergii* (895; 8.4%) and *Bothrops asper* (824; 8.4%).

From January 1994 to July 1995, N. Vidal (in litt. to RWH, 22 November 2018) collected/rescued snakes as part of a large rescue program conducted to save animals threatened by the filling of a reservoir behind a newly constructed hydroelectric dam in French Guiana. Six boa species were collected, and, as in Dunn's total sample, several species were among the most frequently encountered snakes. Combining results from three sites (the dam, 30 km of road in primary forest, and everywhere else), Vidal collected/rescued seventy-three *B. constrictor*, thirty-eight *Co. caninus*, forty-two *Co. hortulana*, thirty-five *Ep. cenchria*, thirteen *Ep. maurus*, and twelve *Eu. murinus*. Interestingly, as in the Dunn collection, among the snakes collected more frequently than most of the boas were a vine snake (*Philodryas argentea*, 49) and a pit viper (*Bothrops atrox*, 107). Similarly, between June and September 1994, during the first stage of filling the Yacyreta Dam Reservoir on the Río Paraná along the Paraguayan–Argentinian border, approximately 1500 *Eu. notaeus* were rescued as they rested on branches of the flooded trees and bamboos (Waller et al. 2001). During the summer of 1962 in Sinaloa, Mexico, more than 150 *B. sigma* were encountered; most were from the southern part of the state and mostly on roads at night during the rainy season (Hardy and McDiarmid 1969).

*Corallus hortulana* can be common in some situations. In French Guiana, F. Starace (in litt. to RWH, 23 January 2018) reported encounter rates of seven or eight in about two hours, and, on one night, twelve to fourteen in three to four hours. In northeastern Peru over a 27-year period, W. W. Lamar (in litt. to RWH, 9 March 2019) found *Co. hortulana* to be as or slightly less common than Starace's observations in French Guiana. Henderson (2015) reported additional encounter rates for *Co. hortulana*; in general, one or two per hour on foot, or three to six per hour by boat. Working along a 1.5-km stretch of mangrove-lined creek on Batatas Island in Piauí, Brazil, P. da Costa Silva (in litt. to RWH, 3 September 2012) observed thirty-nine *Co. hortulana* in just under six hours, or about 6.5 per hour. Similarly, *Co. ruschenbergerii* can be common when surveyed by boat; in Venezuela's Orinoco River delta, Gilson Rivas (in Acosta Chaves et al. 2016) observed up to twenty on a single night, and twelve to fifteen along a 3-km stretch (in Henderson 2002). We cannot say whether the frequency of *Co. hortulana* encounters (or any species of *Corallus*) is due to its real abundance relative to other boas or to the brilliant reflection from its eyes when the beam of a headlamp hits them, thereby signaling its presence. Based on Dunn's Panama data, and on Vidal's data from French Guiana, we suspect it is a combination. Dunn's Panama sample of *Co. ruschenbergerii* probably was not collected at night, and Vidal collected many *B. constrictor* and species of *Epicrates* that do not generate a reflective eyeshine.

Table 2.2 presents information on encounter rates and population densities of Neotropical boas for both mainland and West Indian species. Schulte (1988) determined a population density of 2.70 *Co. batesii* per square kilometer (or 0.3/ha) in Peru, but he was skeptical of that value. Taylor et al. (2011) calculated a density of 11.02 *Co. ruschenbergerii* per square kilometer (or 0.1/ha) in Trinidad's Caroni Swamp. In the Venezuelan llanos, Rivas (2020) calculated a density of 0.16 *Eu. murinus* per hectare, and Micucci and Waller (2007) determined a density of 30–60/km$^2$ (0.3–0.6/ha) for *Eu. notaeus* in northern Argentina. In sharp contrast, densities for insular populations of *B. imperator* off Belize and Honduras range from 0.5 to 16.4/ha.

In the West Indies, population densities for boas ranged from 2.08/ha (for *B. orophias*), ~12/ha for *C. chrysogaster*, to >100/ha for *C. granti*. Encounter rates ranged from 0 to 11.3 per person-hour (p-h) in *Co. grenadensis*, 0.1 to 7.0/p-h in *C. chrysogaster*, and 2.06 ± 1.74/h in *C. angulifer*. On a single night in the Grenada Grenadines, Henderson et al. (2013a) encountered 50 *Co. grenadensis* in just under 3 hours, and the mean time between captures/observations was 3.4 ± 0.6 minutes. Over a span of 5 years, RWH and his collaborators marked 254 *Co. grenadensis* along a 1-km stretch of road in the Grenada Grenadines; population density estimates for boas at their site ranged between 21.4 and 31.5 boas per hectare. Fourteen *C. angulifer* were encountered in one night at a cave (Rodríguez-Cabrera et al. 2015); Linares Rodríguez et al. (2011) counted up to fifteen and seventeen individuals in separate caves; and Berovides Álvarez and Carbonell Paneque (1998) repeatedly found over twenty individuals in a single cave, up to a maximum of twenty-eight boas observed in June. On Big Ambergris Cay on the Caicos Bank, >1200 *C. chrysogaster* have been marked, and on three consecutive nights in August 2019, thirty-seven boas were collected in ninety minutes, forty-seven in 100 minutes, and fifty-six in 110 minutes (RGR, unpubl.).

## PROXIMITY TO HUMAN ACTIVITY

Islands, by their very nature, are highly susceptible to perturbations. With increasing human population density comes increasing pressure on the environment. Habitats can be dramatically altered or altogether eliminated. Higher human population densities can mean more grazing livestock and more predatory cats and dogs. More roads mean more vehicular traffic and more fatal encounters with wildlife, including boas. And more people almost certainly translate to an increase in the number of people who are intolerant of snakes and who mistakenly believe the only good snake is a dead snake.

Avoiding human activity or evidence of human presence is difficult on most West Indian islands (Fig. 2.2). Using human population density (number/km$^2$) as our measurement (data from World Bank Group 2018), a comparison between densities on West Indian islands and countries on the Neotropical mainland that harbor boa populations is instructive. The mean population density for islands harboring boas is 227 ± 121.33 (range = 37–398). The Bahamas and Turks and Caicos are somewhat outliers owing to their lower human population densities; if they are eliminated, the mean population density increases to 261.73 ± 95.89 (range = 99–398). In comparison, on the South American mainland the mean density among countries that harbor boas is 27.63 ± 21.23 (range = 4–67) and that for Central America is 109.71 ± 98.03 (range = 16–308). In the Greater Antilles, Haiti has the highest human population density and Cuba the lowest; in the Lesser Antilles, Grenada has the highest and Dominica the lowest densities for humans (Table 2.3). In South America, Guyana and Suriname have the lowest densities (4) and Colombia the highest (44). Relative to South America, human population densities are much higher in

**Table 2.2.** Encounter rates (snakes per person-hour [p-h]) and densities (number ± SD/ha) of Neotropical boas. Unless otherwise stated, encounter rates were determined during searches made on foot and density measures are identified using an asterisk.

| Species | Locality | Encounter rate or density* ($n$ = sampling periods) | Source |
|---|---|---|---|
| *Boa imperator* | Cozumel Island, Mexico | 0.11 ± 0.03/10 km of road transect* | Romero-Nájera et al. 2006 |
| | Cays off Belize (4.52–24 ha) | 0.5–16.4/ha | Boback 2005 |
| | Cayo Cochino Pequeño, Honduras (0.64 $km^2$) | 632 ± 143 (total population size)<br>9.9/ha* | Reed et al. 2007 |
| *Boa orophias* | St. Lucia | 2.08/ha* | Daltry 2009 |
| *Corallus batesii* | Amazonian Peru | 2.7/$km^2$* | Schulte 1988 |
| *Corallus cookii* | St. Vincent: Various | 0.0–2.83 ± 0.12/p-h ($n$ = 23) | Powell et al. 2007 |
| *Corallus grenadensis* | Grenada: Pearls | 2.7 ± 2.31/p-h (range 0–11.3; $n$ = 26) | Powell et al. 2007 |
| | Grenada: Grand Etang | 0.23 ± 0.61/p-h (range 0–2; $n$ = 22) | Powell et al. 2007 |
| | Grenadines: Union Island | 1.32 ± 0.86/p-h | Quinn et al. 2011 |
| *Corallus hortulana* | Amazonian Peru | 3–6/p-h (by boat) | W. W. Lamar in Henderson 2002 |
| | Brazil: Piauí: Batatas Island | 6.6/p-h ($n$ = 1; by boat) | P. da Costa Silva (in Henderson 2002) |
| *Corallus ruschenbergerii* | Trinidad: Caroni Swamp | 11.022/$km^2$* | Taylor et al. 2011 |
| | Venezuela: Delta Amacuro: Orinoco Delta | 12–15 along 3 km* (by boat) | G. Rivas in Henderson 2002 |
| *Chilabothrus angulifer* | Cuba (cave associated) | 2.06 ± 1.74/h<br>9–28/cave<br>2–17/cave<br>1–14/cave | Alfonso Álvarez et al. 1998<br>Berovides Álvarez and Carbonell Paneque 1998<br>Linares Rodríguez et al. 2011<br>Rodríguez-Cabrera et al. 2015) |
| *Chilabothrus chrysogaster* | Turks Islands: Big Ambergris Cay (400 ha) | 0.92/p-h at night<br>0.44/p-h by day<br>5.04/ha<br>~12/ha* | Reynolds and Gerber 2012<br>This volume |

*continues*

**Table 2.2 continued**

| Species | Locality | Encounter rate or density* (*n* = sampling periods) | Source |
|---|---|---|---|
| *Chilabothrus granti* | Puerto Rico Bank: Cayo Diablo | >100/ha* | Tolson 1988 |
| | | 105–125/ha* | Tolson 1991c, 2004 |
| | Puerto Rico Bank: Culebra | 0.72/p-h | Garcia 1992 |
| | | 0.01/h | Puente-Rolón 2001 |
| *Chilabothrus inornatus* | Puerto Rico | 5.23/ha*, 44/ha* | Tolson 1997 |
| | | 5.6/ha* | Rio-López and Aide 2007 |
| | | 1.24/ha* | Mulero Oliveras 2019 |
| *Chilabothrus monensis* | Isla Mona | 120/ha* | Tolson 2000 |
| *Chilabothrus subflavus* | Jamaica | 0.004/p-h (day and night) | Koenig 2019 |
| *Eunectes murinus* | Venezuela (llanos) | 0.16/ha* | Rivas 2020 |
| *Eunectes notaeus* | Argentina | 30–60/km$^{2}$* | Micucci and Waller 2007 |
| | | 1.34/h (by canoe; 2–3 km/h) | T. Waller (pers. comm., 4 April 2019) |

**Figure 2.2.** A collection of skins of the Cuban boa, *Chilabothrus angulifer*, from the North Carolina Museum of Natural Sciences. The tag in the middle reads: "One of the seven sneaks [*sic*] collected in a tree hole 'La Caoba Farm' Gibara Oriente Province (= Santiago de Cuba Province). Cuba. July 3–1952. Col: M. Díaz-Piferrer." Photo by R. Graham Reynolds.

**Table 2.3.** Data for countries/islands in the West Indies that harbor extant boid species.

Land area, forest area 2015, percent change in forest area (1990–2015), primary forest area 2015, forest area designated for conservation of biodiversity, and land devoted to agriculture are from FAO 2015. Population densities (1961 and 2017) are from the World Bank Group 2018. A dashed line (—) indicates that no data were available.

| Country | Land area (×1000 ha) | Population density 1961 (per km²) | Population density 2017 (per km²) | Forest area 2015 (×1000 ha) | Percent change 1990–2015 | Primary Forest Area 2015 (×1000 ha) | Area designated for conservation of biodiversity (×1000 ha) | Agricultural area 2015 (×1000 ha) |
|---|---|---|---|---|---|---|---|---|
| Bahamas | 1001 | 11 | 39 | 515 | 0 | 0 | — | 14 |
| Turks and Caicos | 43 | 6 | 37 | 34 | 0 | — | — | — |
| Cuba | 10,654 | 68 | 110 | 3200 | 45.7 | 0 | 531 | 6279 |
| Dominican Republic | 4832 | 71 | 223 | 1983 | 35.1 | — | 1203 | 2352 |
| Haiti | 2756 | 143 | 398 | 97 | −0.8 | 0 | 4 | 1840 |
| Jamaica | 1083 | 152 | 267 | 335 | −0.4 | 88 | 72 | 444 |
| Puerto Rico | 856 | 271 | 376 | 496 | 8.4 | 0.051 | 72.7[a] | 236.7[b] |
| Dominica | 75 | 81 | 99 | 43 | −0.3 | 26 | — | 25 |
| St. Lucia | 61 | 149 | 293 | 20 | −0.1 | 17 | 10 | 10.6 |
| St. Vincent and The Grenadines | 39 | 211 | 282 | 27 | 0.1 | 0 | — | 10 |
| Grenada | 34 | 268 | 317 | 17 | 0 | 2 | 2 | 8 |

[a] Castro-Prieto et al. (2019)
[b] Puerto Rico Planning Board Program of Economic and Social Planning, Subprogram of Economic Analysis

Central America; Belize has the lowest (16) and El Salvador the highest (308). The density for Mexico is 66. The human population density for the world is 58.

In summary, the West Indies provide vastly different challenges for its boa fauna relative to those on the mainland (Table 2.4). Whereas the distributions of mainland boas may cover hundreds of thousands to millions of square kilometers (albeit with restrictions imposed by habitat alterations and destruction), West Indian boas occur on islands that range in size from <1 km$^2$ to just over 100,000 km$^2$ and with the same restrictions as the mainland. Geographic ranges of species of Antillean boas and of individual boa populations are dictated by island size and available habitat. Habitat is constrained by island area but also by the history of land use, current rates of deforestation (unavailable for most West Indian islands), and the amount of land devoted to the conservation of biodiversity (Table 2.1). Because we used forest area as our primary criterion of suitable habitat, mainland species obviously have the potential of utilizing much larger areas of forested habitat.

In general, West Indian boas prey on native lizards and introduced mammals. Only the Cuban *C. angulifer* preys frequently on nonvolant, native mammals (as well as introduced domestic fowl). On the mainland, although introduced *Rattus* and *Gallus* are occasionally predated, boa diets are composed largely of a broad taxonomic array of native birds and mammals. Because of constraints imposed largely by island size, West Indian boas must live in close proximity to humans; consequently, the fact that domestic animals, human commensals (*Mus, Rattus*), or both are important dietary components of many West Indian boas is not surprising.

Despite the constraints of island size, severely altered habitats, a mammalian prey base comprised largely of introduced species, and the constant proximity of human activity, many boas in the West Indies occur at population densities unheard of on the mainland (boas are not the only island-dwelling Neotropical snakes that occur at relatively high densities; e.g., Table 4 in Hileman et al. 2017). We attribute this to a suite of factors: (1) relative absence of competitors, (2) relatively few natural predators, (3) a prey base of lizards that can occur at phenomenal densities, (4) a prey base of introduced rodents that occurs in both natural and anthropogenic situations, and (5) the apparent ability (adaptability) of most West Indian boas to live in close proximity to humans.

West Indian boas have obviously adapted to altered habitats and introduced prey species and predators. They go about their lives on their respective islands in the same way boas go about theirs on the mainland.

**Table 2.4.** Summary of comparisons between West Indian and mainland boas

| Comparison | West Indies | Mainland |
|---|---|---|
| Geographic Range | Small: from <10 km$^2$ to just over 100,000 km$^2$ | Large: usually hundreds of thousands to millions of square kilometers |
| Habitat | Virtually all altered to some degree by human activity | Severely altered to vast expanses relatively untouched by human activity |
| Prey base | Largely introduced species (especially mammals) | With few exceptions (e.g., *Rattus rattus*), native species |
| Abundance | Frequently high population densities | Apparently low population densities (except for insular populations) |
| Human proximity | Usually live in proximity to humans | Although some species are sometimes encountered in human proximity, their large geographic ranges allow them to often live away from the vicinity of humans. |

**Table 2.5.** Longevity of some West Indian boa species compared with other New World boas. Records are from public collections (zoos) or verified breeder records.

| Species | Longevity (years, months) | Sex | Location of record | Citation |
|---|---|---|---|---|
| **West Indian Boas** | | | | |
| *Chilabothrus angulifer* | 27, 0 | F | Czech Republic | Frynta et al. 2016 |
| *Chilabothrus chrysogaster* | 14, 2 | M | Houston Zoological Gardens | Snider and Bowler 1992 |
| *Chilabothrus exsul* | 21, 8 | M | Houston Zoological Gardens | Snider and Bowler 1992 |
| *Chilabothrus fordii* | 24, 5 | F | Zoo Atlanta | Snider and Bowler 1992 |
| *Chilabothrus granti* | 39, 0 | M | Knoxville Zoological Society | P. Colclough, pers. comm. |
| *Chilabothrus inornatus* | 23, 9 | M | Baltimore Zoo | Slavens and Slavens 2003 |
| *Chilabothrus monensis* | 10, 2 | F | Toledo Zoological Society | Slavens and Slavens 2003 |
| *Chilabothrus striatus* | 22, 1 | M | Philadelphia Zoological Gardens | Snider and Bowler 1992 |
| *Chilabothrus striatus* | 18, 5 | F | Gladys Porter Zoo | Slavens and Slavens 2003 |
| *Chilabothrus strigilatus* | 10, 11 | M | Fort Worth Zoological Park | Snider and Bowler 1992 |
| *Chilabothrus subflavus* | 24, 3 | M | Dallas Zoo | Slavens and Slavens 2003 |
| *Corallus cookii* | 14, 3 | F | San Diego Zoo | Snider and Bowler 1992 |
| **Other New World Boas** | | | | |
| *Boa constrictor* | 40, 4 | F | Private breeder (Shorrock) | Snider and Bowler 1992 |
| *Boa imperator* | 29, 11 | F | Philadelphia Zoological Gardens | Snider and Bowler 1992 |
| *Charina bottae* | 20, 5 | F | Private breeder (Hoyer) | Snider and Bowler 1992 |
| *Corallus annulatus* | 14, 9 | M | Fort Worth Zoological Park | Snider and Bowler 1992 |
| *Corallus caninus* | 16, 10 | M | Chicago Zoological Park | Slavens and Slavens 2003 |
| *Corallus hortulana* | 15 | F | Zoo Atlanta | Snider and Bowler 1992 |
| *Epicrates cenchria* | 31, 0 | — | Philadelphia Zoological Gardens | Snider and Bowler 1992 |
| *Epicrates crassus* | 16, 11 | F | Staten Island Zoological Society | Snider and Bowler 1992 |
| *Epicrates maurus* | 31 | M | Los Angeles Zoo | Snider and Bowler 1992 |
| *Eunectes deschauenseei* | 23, 2 | F | Philadelphia Zoological Gardens | Slavens and Slavens 2003 |
| *Eunectes murinus* | 31, 9 | — | Basel Zoo | Biegler 1966 |
| *Eunectes notaeus* | 23, 7 | M | Lincoln Park Zoo | Slavens and Slavens 2003 |
| *Lichanura trivirgata* | 18, 8 | — | Private breeder (Hensley) | Snider and Bowler 1992 |

Relative to mainland boas, perhaps the single most remarkable "trait" of West Indian species is their ability (of necessity) to coexist with humans. This in no way implies that humans are more tolerant of snakes in the West Indies, although they may be; only two of hundreds of islands harbor dangerously venomous snakes. Moreover, although the only reason many people need to kill a snake is the fact that it is a snake, informed people may give them a pass more often than on the mainland, where potentially harmful species are abundant and widely distributed.

Despite apparently higher population densities in the West Indies, because of their island-confined geographic ranges, and because of their relatively long life spans and slower reproductive rates (Table 2.5), boa populations are likely more susceptible to environmental perturbations. The negative impact of clear-cutting 1000 ha of forest on an island that may have only 20,000–30,000 forested hectares is likely much greater than on the mainland, where forested areas may be hundreds of thousands to millions of hectares. Finally, it is important to remember that a higher population *density* for a given West Indian species does not equate to more *individuals* of that species than for a congener on the mainland, where geographic ranges are vastly greater.

## *Chapter 3*

# BOAS AS INVADERS IN THE WEST INDIES AND BEYOND

As naturalists, we sometimes think of boas as primeval representatives of intact forests and wild scrubland, where they might often represent a large (or the largest) predator—akin to jaguars or harpy eagles. We also are keenly aware of the conservation concerns faced by many boa species and the dire prospects faced by those that are endangered or critically endangered. Nevertheless, there are several boa species that are expanding their ranges thanks to human transport and their popularity in the pet trade. Boa species are becoming established in various locations in the Caribbean as well as on the mainland boundaries of the Caribbean. Most nonnative boa sightings are what are termed waifs, or one-to-several individuals that have escaped from captivity and managed to survive for a time in the wild but do not represent a reproductively established population. Yet we are witnessing successful novel island colonizations by nonnative boas, sometimes with devastating consequences for local wildlife.

By far the most widely introduced boa species are *Boa imperator* and *B. constrictor* (for taxonomic comments on these species see Reynolds and Henderson 2018). These two similar species, the former from Central America and the latter from South America, are immensely popular in the pet trade because of their general ease of care and breeding, as well as the many color morphs that selective breeding has produced. Further, many local populations of these species exhibit regional color and size variations and thus have been widely collected (Green 2011; Montgomery et al. 2015) and reproduced in captivity as "locality specific" morphs or lineages. Such specialty animals can fetch thousands of U.S. dollars per animal and represent a widespread and potentially lucrative trade. Thus, captive boa constrictors have found their way to most Caribbean islands with large human populations. That is when these animals might be released accidentally or intentionally and can rapidly establish reproductive populations in the wild. We will examine six case studies of *Boa* introductions below.

## PUERTO RICO

A large tropical island with lush forests, Puerto Rico represents an area that seems particularly susceptible to the introduction of nonnative tropical reptiles. Indeed, a major invasion of *B. constrictor* is presently underway on the island (Reynolds et al. 2013b). This invasion was initiated by the escape of at least one (and possibly several) litters of offspring from female boa constrictors housed in outdoor chicken-wire cages at the Mayagüez Zoo, located in western Puerto Rico. Boas began turning up around the zoo and near the campus of the University of Puerto Rico–Mayagüez in the mid-2000s. Subsequent study by ARPR and others revealed that boas were becoming common in the patches of forest surrounding the city of Mayagüez and, further, that the boas were spreading up into the mountains and along Highway 2, a major corridor connecting Mayagüez to cities along the northern coast of Puerto Rico. Genetic analysis of several dozen boas captured from these areas revealed that the animals were very closely related and, further, that they were genetically very similar to boa constrictors in U.S. zoos (Reynolds et al. 2013b). Thus, the genetic data and capture histories corroborate the anecdotal report of the accidental introduction of related boas at the Mayagüez Zoo. *Boa constrictor* is fairly easy for experienced searchers to locate with survey effort in the wild in western Puerto Rico, and they appear to be expanding their range to the north and east (Berensten et al. 2015). Dietary analysis indicates that the boas are mostly eating introduced rats, which is good news for the native bird populations. Nevertheless, rats are the primary diet of the native Puerto Rican boa (*Chilabothrus inornatus*), hence dietary overlap and resource competition might negatively impact the native species (ARPR, unpubl. data).

## COZUMEL, MEXICO

*Boa imperator* was first reported from the island of Cozumel in the literature in 1999 (Martínez-Morales and Cuarón 1999), where the species had not been previously recorded despite its proximity to mainland Mexico (Vázquez-Domínguez et al. 2012). Most researchers have concluded that the species was first introduced to the island via the intentional release of between six and thirty boas used during filming of a movie in 1971. After filming wrapped, the boas were released into the wild and subsequently proliferated. This population has since been the focus of several studies examining their ecological impact on the island (Cuarón et al. 2004, 2009; González-Baca 2006; Romero-Nájera et al. 2007) as well as their genetic diversity and origins (Vázquez-Domínguez et al. 2012; Suárez-Atilano et al. 2019). Boas are now common on the island although cryptic—Martínez-Morales and Cuarón (1999) reported encounter rates of 0.18 boas per 10 km of forest surveyed, while Romero-Nájera et al. (2007) found a similar encounter rate of 0.11 boas per 10 km of forest surveyed. Boas use all available habitat types on the island, although they are most common in the interior forests away from human habitation (Romero-Nájera et al. 2007). There is no difference in boa abundance relative to vegetation type on Cozumel nor are there seasonal changes in abundance or detectability (González-Baca 2006; Romero-Nájera et al. 2007).

Genetic data suggest that the released animals were likely obtained from several localities in mainland Mexico, thus bringing substantial genetic diversity to the newly established population relative to other *Boa* introductions (Suárez-Atilano et al. 2019; see comparisons in Bushar et al. 2015). The Cozumel population of *B. imperator* has both a higher effective population size (by an order of magnitude) and a higher allelic diversity (Vázquez-Domínguez et al. 2012) than either the Aruba or Puerto Rico populations of *B. constrictor* (Reynolds et al. 2013b; Bushar et al. 2015).

Negative impacts to the endemic fauna of Cozumel have been well documented (Cuarón et al. 2004, 2009), and boas appear to select larger prey items preferentially rather than smaller abundant prey (González-Baca 2006). The boa population is threatening both endemic rodents (*Peromyscus leucopus cozumelae*) and birds (*Toxostoma guttatum*). Interestingly, Mexican law prohibits the destruction of this species, which is protected in Mexico, hence no eradication can be undertaken on Cozumel, and the species is likely going to remain on the island for the foreseeable future (Suárez-Atilano et al. 2019).

## ARUBA

Aruba, Curaçao, and Bonaire are continental islands located between 32 and 80 km from mainland South America (Venezuela). *Boa constrictor* occurs on some nearby continental islands, including Isla Margarita (Fuenmayor et al. 2005; Ugueto and Rivas 2010) and Trinidad and Tobago (Murphy 1997) to the east, but was not recorded elsewhere on major islands off the coast of Venezuela historically. *Boa constrictor* was not considered a part of the native fauna of Aruba (although it might have been present and gone extinct) and was not documented from the island until 1999 (Quick et al. 2005). Since that initial documentation, the species has exploded in population size, and in one decade rangers from the Parke Nacional Arikok were capturing as many as 741 boas per year from the wild (fig. 2 in Bushar et al. 2015). Boas have now been found across most of the island (Quick et al. 2005; Bushar et al. 2015), and a concentrated control effort has resulted in the capture and humane euthanasia of thousands of boas. This is important given the likely impact these boas are having on the native bird populations

(Quick et al. 2005; also see Romero-Nájera et al. 2007). Boas on Aruba were found to take both endothermic and ectothermic prey—a sample of forty-seven boas had fifty-two prey items in the stomachs; 40.4% were birds, 34.6% were lizards, and 25% were mammals (Quick et al. 2005). Genetic analysis of the boa nuclear and mitochondrial DNA suggested that the population was founded from at least three unrelated individuals from northern South American populations, and that Aruba boas are not related to *B. constrictor* individuals established on Puerto Rico nor were there as many founders as in the Cozumel *B. imperator* population (Bushar et al. 2015).

## FLORIDA, USA

*Boa constrictor* was first introduced to south Florida via the pet trade in the 1960s, with at least one population likely becoming reproductively established by the 1970s (King and Krakauer 1966; Krysko et al. 2016). A reproductive population located at the Charles Deering Estate (a county park) in Miami-Dade County was reported in 1992 (Snow et al. 2007). In 1996, sixty-nine boas were removed from this park, and subsequent years (up to 2005) yielded an additional twenty-seven individuals that were also removed (Snow et al. 2007). These individuals were mostly young-of-year juveniles, although individuals representing all size classes were found. Boas at Deering were found to use a variety of habitat types, from maritime hammocks to landscaped vegetation (Snow et al. 2007).

*Boa constrictor* has also been found in numerous other south Florida locations, including Everglades National Park and Big Cypress National Park. These are considered waifs, and the only likely breeding populations are presently restricted to Miami-Dade County and the Florida Keys (Hanslowe et al. 2018). Boas were first verifiably reported in the Keys in 2003, and since that time as many as twenty-four boas of all age classes and both sexes have been reported from eleven islands in the archipelago from Key Largo to Key West (Hanslowe et al. 2018). Interestingly, only one verifiable report (on iNaturalist.org) has been recorded since 2014. An Argentine boa (*B. c. occidentalis*) was also found in south Florida in 2009, but this subspecies is not established (Krysko et al. 2011, 2016).

## ST. CROIX, U.S. VIRGIN ISLANDS

The most recent establishment of a *Boa* on Caribbean islands was the report of a "red-tail boa" from St. Croix in 2010 (Coles, in Golden 2017). Since that time boas have expanded rapidly in many parts of the island, and all size classes are now found regularly (Angeli et al. 2019). Genetic analysis determined that these boas belonged to the species *B. imperator* (Golden 2017; Angeli et al. 2019) and not *B. constrictor* (the species established on Puerto Rico). The establishment of *B. imperator* on the island is somewhat remarkable given the high density of introduced small Indian mongoose (*Urva auropunctata*) there (Horst et al. 2001). These predators are likely responsible for causing the extirpation of native Crucian snakes and would perhaps have been expected to slow the growth of a population of *Boa*. Nevertheless, at least eighty-eight boas have been removed from the island since 2012, and reports exist from many locations in the western half of the island, with the highest density found along the far western portion of the island (Angeli et al. 2019). The species is expected to be capable of spreading island-wide in the near future, although it is currently unknown what possible impacts this new invasion might have on the Crucian fauna.

## ANTIGUA AND BARBUDA

It is likely that a species of *Boa* occurred historically on the island of Antigua (see the Antigua boa species account). This species (or subspecies) is known from subfossil material from two archeological sites on the island and is accepted as having gone extinct, probably in the last millennium. Nevertheless, individual *Boa* sp. have shown up in recent years on the island, and while some are likely escaped pets (M. Friedman, in litt. to RGR), it is yet unknown whether the species might reestablish a new population on the island. At least one large ("5 foot long" [1.5 m]) individual *Boa*, of unknown species, was captured in 2019 at Horsford Hill, and another was spotted near Turtle Bay a few kilometers southeast (Joseph 2019).

### OTHER BOID SPECIES

While *Boa* is the most widely introduced boa taxon in the region, other boa species have also started to become established outside their native ranges. A particularly worrisome situation is presently unfolding in Puerto Rico, where numerous exotic species are turning up in the wild. Puerto Rico has a flourishing illegal exotic pet trade, and this has led to a rapidly growing number of nonnative reptile species being introduced to the island. The Puerto Rico Department of Natural Resources maintains a captive animal facility in Cambalache State Forest where at any given time the staff might be maintaining representatives of one-to-two-dozen reptile species that have been captured from the wild—from pythons to caiman to small exotic snakes and lizards. In addition to *B. constrictor*, the boids *Epicrates maurus* and *Ep. cenchria* have also been captured on Puerto Rico after presumably having escaped captivity. These medium-sized boas are native to Central and South America and are exceedingly common in the U.S. pet trade. Several anacondas (*Eunectes* sp.) have also been reported from the wild, although it is yet unclear whether they are green (*Eu. murinus*) or yellow (*Eu. notaeus*) anacondas (ARPR, pers. obs.).

West Indian boas, while presently much more infrequently introduced outside their native ranges, are nevertheless turning up in unusual places. In 2009 a Hispaniolan boa (*Chilabothrus striatus*) was found dead on a road on the island of Vieques, Puerto Rico (Reynolds et al. 2014b). This animal was mistakenly identified as the native Puerto Rican boa (*C. inornatus*), which had not been seen on Vieques in many decades. Subsequent examination (Reynolds et al. 2014b) determined that the animal was, in fact, an adult female Hispaniolan boa and a likely escaped captive animal. Vieques was almost completely deforested during the early nineteenth century, and hence any native boas that occurred there were almost certainly extirpated. Now, the island has a lush secondary forest that would seemingly provide good boa habitat, although the healthy mongoose population might prevent establishment of boas on the island (but see the St. Croix account above).

In 2014, a boa was captured by the Miami-Dade Fire Rescue near a car rental facility in Miami, Florida, and identification was determined through examination and genetic analysis (by RGR) to be a Bahamas boa (*Chilabothrus strigilatus*). This is a well-represented species in the pet trade owing to its variable color patterns and it being widely poached for the pet trade in the 1970s and 1980s. As of now, it is unclear whether the animal came to Miami accidentally in cargo or whether it represents an escaped pet from a Miami residence. It is thought to be the latter, as the car rental facility does not ship vehicles abroad (Krysko et al. 2016). Two Puerto Rican species (*C. inornatus* and *C. granti*) were brought to Florida in cargo shipments and were discovered prior to escape (Krysko et al. 2016). In 2014 an adult *C. granti* was found in a shipment of electronic waste arriving to Tampa, Florida, from Puerto Rico. As there is only a single known population of this species on the main island of Puerto Rico, it is unclear whether the specimen came from another island (such as Culebra or St. Thomas) or whether the electronic supplies were transported from near the only Puerto Rican population. A third, but less likely, possibility is that it came from an unknown population of Virgin Islands boas. Also in 2014, a Puerto Rican boa (*C. inornatus*) was collected from another shipment of electronic waste arriving to the same Tampa, Florida, port (Krysko et al. 2016). Such a curious circumstance could be the result of storage of these discarded electronics near boa habitat, or it could be a small-scale attempt to smuggle animals from Puerto Rico to the U.S. mainland. Other booid species have also been found in Florida, including species of *Epicrates*, *Eryx*, *Eunectes*, *Acrantophis*, and *Calabaria* (summarized in Krysko et al. 2016), all of which are represented by escaped pets and are not yet reproductively established in the wild.

Native species are often moved around to areas where they have not previously been seen. This has been particularly apparent in Puerto Rico, where two native boa species (*C. inornatus* and *C. granti*) are occasionally reported outside their known present-day ranges. In 2014 an adult female *C. inornatus* was found on the island of Culebra, Puerto Rico, by Puerto Rico Department of Natural Resources rangers. While *C. granti* is present on Culebra, no *C. inornatus* had been reported there in recent decades. Genetic analysis showed that the individual was closely related to individuals from the karst region of northeastern Puerto Rico, where the species is common (RGR and ARPR, unpubl. data). Similarly, in 2010 an adult female *C. in-*

*ornatus* and an individual of unknown sex were found on the island of Vieques (J. P. Zegarra, pers. comm. to RGR, 20 May 2019), where no boas (besides the single *C. striatus*) have been seen since the near complete deforestation of that island in the early 1900s. While this likely represents a recent reintroduction, analysis of genetic data to confirm this are currently occurring (RGR, unpubl. data). Because no other boas have been reported from Vieques and Culebra, it is not likely that there are reproductive populations of *C. inornatus* on those islands.

The endangered *C. granti* is also subject to poaching and, interestingly, intentional misrepresentation of its range on social media. A series of Facebook posts over the last few years (2016–2018) have purported to show adult and juvenile *C. granti* from locations such as Ciales, Puerto Rico—a mountainous region where the species has never been recorded. Indeed, this species is known from only a single tiny locality (nowhere near Ciales) on the main island of Puerto Rico despite years of intensive searches elsewhere. It is unclear why someone would want to mislead others into thinking that boas occur where they do not—although it might be an attempt to cover up poaching activity.

Boas continue to be popular as pets and, undoubtedly, we will continue to witness new introductions of boa species to areas outside their native ranges. While the ability of some boas to colonize new areas quickly in the face of all the usual threats from people and predators is promising from the standpoint of potential reintroduction campaigns, it is our hope that these introductions do not come at the cost of damage to threatened native boa species or other wildlife. We encourage readers to report sightings of nonnative boas to the relevant authorities so that new introductions can be tracked and controlled early on.

*Corallus grenadensis* approaching *Anolis aeneus* at Westerhall (St. David Parish), Grenada. The anole had been sleeping and apparently unaware of the boa; it was awakened by the photographer. Photo by R. Sajdak.

## *Chapter 4*

# CONSERVATION OF THE WEST INDIAN BOA FAUNA

Since the arrival in the Antilles of Amerindians, herpetofauna have been both beneficiaries and victims of human activities. As noted by Newsom and Wing (2004), Fitzpatrick and Keegan (2007), and Rick et al. (2013), the impact of humans on their environments in the pre-Columbian West Indies should not be underestimated. With the arrival of Europeans, habitat transformations accelerated, and many alien plants and animals were introduced. The eighteenth and nineteenth centuries saw the clear-cutting of forests to make way for agriculture (especially sugarcane plantations), reducing habitats available for forest-adapted species. Today, burgeoning human population growth coupled with rampant development (often tourism-driven) on many islands does not bode well for many segments of the herpetofauna. Henderson and Powell (2001) estimated that human-introduced resources benefited at best 5–10% of the herpetofauna (almost entirely reptiles), whereas at least 50% (including most amphibians) have been negatively affected. Habitat alterations, however, have had an impact to some degree on 100% of the native West Indian herpetofauna (Table 4.1).

## THREATS TO THE WEST INDIAN BOA FAUNA

### *Introduced Invasive Predators and Competitors*

Over the past 500+ years (and likely long before that), alien species have entered West Indian islands with human assistance, intentional or inadvertent, and, as a group, have had a devastating impact on the herpetofauna. Introduced plants alter the very nature of habitats; nonnative carnivores consume frogs, lizards, and snakes; large mammalian herbivores severely alter habitats; and invasive snakes (e.g., *Boa* spp.; see Chapter 3) may compete with and/or consume native species.

The small Indian mongoose (*Urva auropunctata*) was introduced initially to Jamaica in 1872 (Espeut 1882) and subsequently became established on more than twenty additional islands, some as late as the early twentieth century (Hoagland et al. 1989). Exclusive of the Lucayan Archipelago, Dominica, and the Grenadines, mongooses occur on most West Indian islands that also harbor boas. They have been implicated in the extirpations of populations and even extinctions of several species of squamate reptiles, including snakes (especially species of the dipsadid genera *Alsophis* and *Erythrolamprus*). However, whereas West Indian boas are largely crepuscular and nocturnal, mongooses are diurnal. This is not to say they do not interact on those islands where boas and mongooses do occur, but encounters may be rare. Barbour (1914) declared *Chilabothrus subflavus* "almost exterminated by the ravages of the mongoose." We must, however, be cautious when it comes to Barbour's assessments of abundance given the absence of systematic surveys. Grant (1940) also believed *C. subflavus* numbers had been reduced by mongooses and feral cats. According to Lazell (1964), in reference to *Boa orophias* on St. Lucia, "[A]nyone who has ever seen both a mongoose and a [boa]—even a young one—the idea that mongooses could exterminate the latter must seem a shade ludicrous." On Cuba, remains of anoles and snake scales (possibly a dipsa-

**Table 4.1.** Summary of IUCN Red List assessments for West Indian boas.

NE = Not Evaluated; LC = Least Concern; NT = Near Threatened; VU = Vulnerable; CR = Critically Endangered; DD = Data Deficient. Some of these assessments are from "in review" or "in press" Red List assessments at the time of writing.

| Species and summary | NE | LC | NT | VU | EN | CR | DD |
|---|---|---|---|---|---|---|---|
| *Boa nebulosa* | — | X | — | — | — | — | — |
| *Boa orophias* | — | — | — | — | X | — | — |
| *Chilabothrus ampelophis* | X[a] | — | — | — | — | — | — |
| *Chilabothrus angulifer* | — | X | — | — | — | — | — |
| *Chilabothrus argentum* | — | — | — | — | — | X | — |
| *Chilabothrus chrysogaster* | — | — | X | — | — | — | — |
| *Chilabothrus exsul* | — | — | — | X | — | — | — |
| *Chilabothrus fordii* | — | — | X | — | — | — | — |
| *Chilabothrus gracilis* | — | — | X | — | — | — | — |
| *Chilabothrus granti* | — | — | — | — | X | — | — |
| *Chilabothrus inornatus* | — | X | — | — | — | — | — |
| *Chilabothrus monensis* | — | — | X | — | — | — | — |
| *Chilabothrus schwartzi* | — | — | — | — | X | — | — |
| *Chilabothrus striatus* | — | X | — | — | — | — | — |
| *Chilabothrus strigilatus* | — | X | — | — | — | — | — |
| *Chilabothrus subflavus* | — | — | — | X | — | — | — |
| *Corallus cookii* | — | — | X | — | — | — | — |
| *Corallus grenadensis* | — | X | — | — | — | — | — |
| Totals | 0 | 6 | 5 | 2 | 3 | 1 | 0 |

[a] *Chilabothrus ampelophis* is likely CR according to Landestoy T. et al. (2021b).

did) have been found in the stomachs of mongooses (Borroto-Páez 2011). Projections based on ecological niche modeling suggest that climate change will facilitate the spread of *U. auropunctata* (Louppe et al. 2020). Since the species is already widespread in the area, this prediction may not be as devastating for the West Indies as in more northern locales.

Introduced common opossums (*Didelphis marsupialis*; Fig. 4.1) are abundant on the St. Vincent and Grenada Banks (as well as other islands) and occupy a variety of arboreal and terrestrial habitats. As generalist carnivores, they are known to prey on the arboreal boa *Corallus grenadensis* (Henderson 2015) and presumably eat other snakes, lizards, and frogs. As many as eighteen opossums have been observed in one night along a 1.0-km stretch of road on Carriacou.

Domestic cats (*Felis catus*) are among the most effective predators of native vertebrates. García et al. (2001) found that 33% of the stomachs of feral cats on Isla de Mona contained the remains of reptiles. On Cuba they are well-known predators of anoles, both in natural and urban areas (Borroto-Páez 2009, 2011). They also prey upon bats at cave entrances during the nocturnal exodus (Borroto-Páez 2011; Mancina 2011), competing with *Chilabothrus angulifer* for a prey resource. A similar situation has been reported with feral cats at Culebrones Cave in Puerto Rico, where *C. inornatus* also forage for bats (Rodríguez-Durán et al. 2010). Nearly 80 years ago Grant (1940) blamed feral cats for the decline of *C. subflavus* on Jamaica, and on St. Thomas, U.S. Virgin Islands, three *C. granti* were rescued from house cats (Tolson and Henderson 1993). Cats have been observed preying on *C. strigilatus* on Eleuthera, Bahamas (RGR, unpubl. data). But occasionally the tide reverses. At least one predation event has been documented of a cave-associated *C. angulifer* on one of the feral cats that regularly used caves to forage for bats (R. Martínez in Rodríguez-Cabrera et al. 2020b), and *C. strigilatus* has been observed feeding on entire litters of kittens on Long Island, Bahamas (RGR, pers. obs.). Further, cats might control rat populations where the two species occur, which has been suggested to reduce the chance of *C. granti* being eaten by either species (Tolson 1991c).

Dogs (*Canis lupus familiaris*), in collaboration with cats, are responsible for the near extirpation of the large iguanid lizard *Cyclura carinata* from Pine Cay in the Caicos Islands (Iverson 1978). From an estimated adult population of nearly 5500 iguanas (15,000 total population in June–July 1974), by mid 1976, no iguanas were flushed during five transect surveys, and spoor evidence of only five iguanas was found in June 1976. We would be surprised if dogs did not occasionally also prey upon young boas. On Cuba at least, feral dogs may compete with *C. angulifer* for food resources such as hutia. On several cays north of Cuba, the extirpation of hutias (*Capromys pilorides*), iguanas (*Cyclura nubila*), and other reptiles has been associated with dog introductions (Borroto-Páez 2009, 2011). On many of those cays and in mountain zones, scats of feral dogs containing hutia hairs (*Ca. pilorides*, *Mesocapromys melanurus*, and *Mysateles prehensilis*) and even eviscerated hutia carcasses (*Ca. pilorides*) are found relatively often (Borroto-Páez 2009, 2011; TMRC, unpubl. data). Hutias in particular represent the most important prey of medium-sized to large *C. angulifer* in natural habitats (see species account). On some cays north of central Cuba, the current scarcity of the formerly common *C. angulifer* has been associated with feral dog packs in the area, with the subsequent extirpation of hutias and iguanas (see species account). Hutias and solenodon are also negatively impacted by human-induced land use changes in the Dominican Republic, resulting in their ranges being restricted to remaining higher-quality protected areas (Kennerley et al. 2019).

**Figure 4.1.** *Top*: Introduced common opossums (*Didelphis marsupialis*) occur on many West Indian islands (the one pictured was on Carriacou, Grenada Grenadines), and they are known to eat small *Corallus grenadensis* as well as other small reptiles. Photo by R. Sajdak. *Bottom*: Goats (*Capra hircus*) and abundant invasive plants are tremendously damaging to natural ecosystems in the Caribbean. This image is from Cockpit Country, Jamaica, a critically important area for *Chilabothrus subflavus*. Photo by Brent Newman.

Likely introduced inadvertently during early European settlement, rats (*Rattus norvegicus* and *R. rattus*) are predators on eggs and young of reptiles and may modify habitats by exploiting seeds and seedlings. Tolson and Henderson (2006, 2011) suggested that small species of snakes were most vulnerable, and Tolson (1988) found that *Chilabothrus granti* was absent at all sites on the Puerto Rico Bank that did not have a rat predator to control rodent populations, thus rendering those sites unsuitable for boas. When *C. granti* was reintroduced to islands where rats had been eradicated, boa populations flourished. At least one island (Cayo Ratones), however, was recolonized by rats and the boa population, once five hundred strong, disappeared in less than 30 years (USFWS 2018; also see species account). It is not presently known whether rats directly preyed upon the boas or outcompeted them (or both).

Grazers such as goats (*Capra hircus*; Fig. 4.1), cattle (*Bos taurus*), sheep (*Ovis aries*), and donkeys (*Equus asinus*) can quickly eliminate palatable vegetation from an area, degrading habitats that once provided food and cover for native frogs and reptiles. On Carriacou and Petite Martinique (Grenada Grenadines), we have observed the efficient manner in which goats have eliminated ground vegetation and the low understory in patches of woodland, as well as in the beautiful stand of forest above Chatham Bay on Union Island (St. Vincent Grenadines). Repeated boa surveys in goat-browsed depression forest on Isla de Mona failed to find evidence of *Chilabothrus monensis* (Tolson et al. 2007). Additionally, pigs (*Sus scrofa*) destroy native vegetation by rooting and may consume snakes (Tolson and Henderson 2006, 2011). Netting and Goin (1944) believed that the introduction of pigs to Stranger's Cay on the Little Bahama Bank was responsible for the extirpation of *C. exsul*, although it is more likely that the species that was lost was actually the Bahamas racer (*Cubophis vudii*) as boas do not occur on any other islands north of Little Abaco. These grazers have significantly altered habitats on some islands in the Turks and Caicos, rendering them dusty and devoid of the typical vegetation, undoubtedly impacting boa populations. Whereas Big Ambergris Cay has a larger population of *C. chrysogaster*, South Caicos, a mere 21 km distant, has none (RGR pers. obs.).

Although neither a predator nor a competitor, the introduced cane toad, *Rhinella marina*, has been implicated in the deaths of at least one species of West Indian boa. Based on either direct or circumstantial evidence, Wilson et al. (2011) provided documentation of the deaths of several *Chilabothrus subflavus* in Jamaica. Cane toads secrete a bufogenin toxin that can be fatal if ingested by species that have not coevolved with *R. marina* or related species. The individuals of *C. subflavus* that fell victim to the toxin either ingested or attempted to ingest *R. marina.*

Many habitats in the West Indies are dominated by nonnative plants, and invasive species of plants can negatively impact reptiles. Wilson et al. (2006) noted that invasive plants effectively alter habitats and prey bases in much the same way as introduced herbivores.

*Habitat Loss and Fragmentation*

Habitat destruction is the major cause of species extinction (e.g., Tilman et al. 2002). The West Indies have a long history of habitat alterations, beginning with the first human inhabitants ~7000–4000 years BP (see Chapter 1). Forests were cleared for timber harvest, agriculture, and residential development. Some small islands had virtually all forest eliminated to facilitate the planting of sugarcane and cotton; larger islands fared somewhat better, but over hundreds of years, all lost the vast majority of their primary forests (Chapter 2). Today, tourism is critical to the economic stability of many islands, and habitats are severely altered for the construction of resorts, restaurants, golf courses, and all the associated infrastructure of development.

The West Indies are considered a biodiversity hotspot, where only 11.3% of the original vegetation remains in the region (Myers et al. 2000); a more realistic estimate of intact vegetation might be 10% (Hedges 2006). Despite the exploitation by native frogs, lizards, and snakes of introduced coconut palms, banana, cacao, and coffee trees, as well as human edifices and associated debris, relatively few species have benefited from human-influenced resources (Henderson and Powell 2001), although edge-loving treeboas (*Corallus*) might be more abundant now than when West Indian forests were intact (e.g., Henderson 2015). Introduced orchard and forest trees (e.g., mahogany) or "fence-row" species (e.g., oleander) may simulate natural vegetation patterns, but complex habitats are used more frequently by more species, and habitats associated with clear-cuts are rarely used by the herpetofauna (Henderson and Powell 2001). Although *Chilabothrus angulifer* was primarily considered to be a forest species (Sheplan and Schwartz 1974), it now frequently occurs in highly disturbed areas with adequate cover (Tolson and Henderson 2011; Rodríguez-Cabrera et al. 2016a; 2020b). High concentrations of

**Figure 4.2.** *Left*: A large adult female *Chilabothrus strigilatus ailurus* from southern Cat Island, Bahamas. Notice the large black scar on the neck, which was caused by a wildfire. Photo by R. Graham Reynolds, courtesy of Mark Keasler. *Right*: Roads can be a major source of mortality for boa populations. An adult *Chilabothrus c. chrysogaster* killed on a road in March 2011 near Mugjin Harbour, Middle Caicos, Turks and Caicos Islands. Photo by R. Graham Reynolds.

*C. fordii* and *C. striatus* have been found in the vicinity of Haitian villages where little original vegetation remained, and *C. chrysogaster, C. exsul, C., granti, C. striatus, Corallus cookii,* and *Co. grenadensis* have been remarkably abundant in human-altered edge habitats rather than forest interiors (Tolson and Henderson 2011). In the Dominican Republic, the major threat to the four endemic species of *Chilabothrus* are habitat fragmentation or loss (S. J. Incháustegui, in litt. to RWH, 29 May 2020). Nevertheless, *C. gracilis* and *C. striatus* were found to be abundant in an overgrown failed apartment village development in eastern Dominican Republic near Miches (RGR, unpubl. data). *Chilabothrus subflavus* appears to prefer more heavily forested situations, and the widespread deforestation and habitat fragmentation (by commercial bauxite mining and selective logging) have been implicated in its decline (Newman et al. 2016). Alternatively, reforestation on Puerto Rico has benefitted *C. inornatus,* and the species is now widespread on the island, including in proximity to human dwellings and in small forest patches surrounded by urbanization (Reynolds and Henderson 2018; Mulero Oliveras 2019). According to Tolson and Henderson (2011), "The most serious consequences seem to accrue when environmental insults follow in close succession, such as the deforestation of the Virgin Islands that was followed by large-scale annual burning that accompanied the cultivation of sugar cane. Such activities may explain why [*C. granti*] is limited to the east end of St. Thomas (Nellis et al. 1983)." Without a doubt, boas suffer simultaneously from multiple impacts that include changes to vegetation, human impacts, and introduced species, among others (Fig. 4.2).

*Accidental Killing*

For terrestrial wildlife, roads have become a major source of accidental mortality (Fig. 4.2). Reptiles, especially larger species, are frequently killed while crossing, or perhaps thermoregulating, on roads. On Dominica, for example, it is not unusual to encounter road-killed *Boa nebulosa*. One hundred and thirty-two surveys along a 30-km stretch of road on Puerto Rico documented forty-seven dead introduced *B. constrictor* over a 10-month period (J. P. González-Crespo and ARPR, unpublished). Conversely, RWH has logged approximately 8000 km on Grenada's roads but personally has documented only one dead *Corallus grenadensis* and fewer than five live treeboas; road-killed *Co. cookii* are equally rare on St. Vincent. That both species of *Corallus* are strongly arboreal likely contributes to the apparent low numbers of dead individuals. In the Bahamas, Reynolds et al. (2016b) suggested that the significant road mortality they observed should be considered crucial information when preparing a conservation assessment and action plan for the boa

*Chilabothrus exsul*. On Cuba, *C. angulifer* is also frequently found dead on the roads, especially on those roads close to forested areas (TMRC, unpubl. data). It is possible that some boas use roads to thermoregulate since the asphalt remains warm at night, where they become easy victims of vehicles. As resident and transient human populations increase, so too will paved roads, automobile traffic, and wildlife mortality. Sadly, of course, some boa deaths on roads are not accidental; rather, the driver has made a conscious decision to kill the snake. In Colombia, estimates suggest that no fewer than 50,000 snakes die annually because of vehicular traffic (Lynch 2013). On Cuba, sugarcane-associated *C. angulifer* frequently fell victim to combine harvesters, and several carcasses could be found in those sugarcane fields that were burnt prior to harvesting. Fortunately, the sugarcane industry in Cuba was dismantled in the early 2000s, with the subsequent reduction in the impact it had on *C. angulifer* populations. In the Dominican Republic, *C. striatus* is regularly found dead on roads, often in narrow strips of riparian vegetation growing along drainage ditches bordered by large agricultural fields. This suggests that boas use these small vegetation corridors but suffer mortality when needing to cross roads between the strips (RGR, unpubl. data).

*Intentional Killing*

Boas are intentionally killed every day on every West Indian island where they co-occur with humans, for no other reason than for being a snake. Exploitation of West Indian frogs and reptiles has a long history. Several species of snakes are exploited for so-called medicinal purposes. On Dominica, *Boa nebulosa* is killed so that its fat may be processed for "snake oil" (Malhotra et al. 2011). Puerto Rican *Chilabothrus inornatus* is likewise sacrificed for "snake oil" (Reagan 1984) as well as for meat for human consumption (Bird-Picó 1994). On Cuba *C. angulifer* is hunted for its meat, fat (for "curing" hemorrhoids and respiratory problems), and its hide (for belts, wallets, and shoes; Rivalta González et al. 2003; Linares Rodríguez et al. 2011). In the Dominican Republic, *C. striatus* is likewise killed in order to extract oil from its fat; martial arts practitioners and ballet dancers can be under the impression that it is good for easing joint discomfort (S. J. Incháustegui, in litt. to RWH, 19 May 2020). The species is also killed because it will prey on domestic fowl, and, occasionally, it is killed for human consumption (S. J. Incháustegui, in litt. to RWH, 19 May 2020). Treeboas (*Corallus grenadensis*) on Grenada are possibly being used in traditional medicine and cuisine by some members of the resident Asian diaspora community. Certainly, residents of West Indian islands (with the notable exceptions of Martinique and St. Lucia) must be aware that the snakes that occur on their islands are not a threat, especially since no one on those islands has ever died from snakebite. Nevertheless, countless conversations with St. Vincentians and Grenadians suggest that, of those who do encounter snakes, the snake has a better than 50/50 chance of being killed. When 511 people were surveyed and asked what came immediately to mind about snakes on Grenada, the most frequent responses mentioned killing, danger, and fear. When informed that no venomous (i.e., dangerous) snakes occurred on Grenada, 67.2% of 125 people interviewed responded that that did not change how they felt about snakes (Rusk in Henderson and Powell 2018). We have no data regarding the numbers of snakes killed over a span of time on any West Indian island that harbors boas, but absolute numbers of snake deaths at the hands of humans will not approach those for Colombia, where it is estimated that more than 100 million snakes are killed each year (Lynch 2013).

Although not outright killing, *Co. grenadensis* has traditionally been used by participants in Jab-Jab (or Devil Mas) during Grenada's annual carnival. With their bodies covered in used motor oil, grease, stale molasses, or mud, and wearing hats adorned with cattle horns, some Jab-Jab participants carried live or dead boas (Martin 2007) with which to frighten carnival-goers with the hope of extorting money in exchange for a quick departure. Live boas routinely had their mouths sewn shut or their teeth removed; either practice is a death sentence for the boa. Supposedly, the practice of using live boas has been outlawed, but the law is rarely enforced (Campbell 2018; see the *Co. grenadensis* account). In lieu of treeboas, other wildlife (e.g., opossums) is sometimes used to frighten, repulse, or intimidate carnival-goers (Henderson and Powell 2018).

The paleros are the practitioners of any of the branches of the Regla de Palo, a religion that originated in the Bantu ethnic groups of sub-Saharan Africa. They were the descendants of the kingdom of Manikongo (or Mwene Kongo) in the region that today extends between Nigeria and Angola; they were represented among the first slaves that arrived in Cuba. In this religion the *maja* (*C. angulifer*) is used as the Cuban exponent of the Mboma Ñoca (the serpent), either as a religious garment or as a purifier, or is considered a guardian and messenger; the boa's fat, skin, and teeth are also used in magic spells (Bolívar et al. 2013).

The largest boa species usually get involved in human-predator conflicts because they have the potential to affect the local economy of farmers and the countryside residents in general. The boas take advantage of food concentrations found in these situations, such as concentrations of domestic fowl. *Chilabothrus angulifer* has demonstrated a high incidence of predation on domestic animals, including poultry, cats, and dogs (see species account).

In the Dominican Republic, *C. striatus* has been reported preying upon several bird species kept in aviaries (Ottenwalder 1980; Henderson et al. 1987). Sheplan and Schwartz (1974) note that *C. strigilatus* is known as "fowl snake" throughout the Bahamas and commented on the possible relationship of its abundance on certain islands to the extensive poultry-raising industry that once existed, for example, on Eleuthera (see species account). Domestic fowl in Puerto Rico are consumed as well by *C. inornatus* (Rivero 1978; Wiley 2003) and increasingly by the expanding introduced *B. constrictor* population. The above seems reason enough to make many people, particularly from the countryside, kill boas in every encounter. That some boas exhibit defensive strategies that make them frightening or repulsive to people might also contribute to their being killed (Table 4.2)

**Table 4.2.** Defensive strategies of West Indian boas.

We assume that crypsis and constricting behavior apply to all these species.

| | Throat inflation | Retreat/ hiding | Balling posture | Voiding cloaca | Body rotation | Striking/ biting | Hissing | Tail stiffening |
|---|---|---|---|---|---|---|---|---|
| *Boa nebulosa* | — | X | — | X | — | X | X | — |
| *Boa orophias* | — | X | — | — | — | X | X | — |
| *Chilabothrus ampelophis* | — | — | X | X | — | X | — | X |
| *Chilabothrus angulifer* | — | X | X | X | — | X | X | — |
| *Chilabothrus argentum* | — | — | X | X | — | — | — | — |
| *Chilabothrus chrysogaster* | — | X | X | X | — | X | — | — |
| *Chilabothrus exsul* | — | X | X | X | — | — | — | — |
| *Chilabothrus fordii* | — | X | — | — | — | — | — | — |
| *Chilabothrus gracilis* | — | X | X | X | — | — | — | — |
| *Chilabothrus granti* | — | — | X | X | — | — | — | — |
| *Chilabothrus inornatus* | X | X | X | X | — | X | X | — |
| *Chilabothrus monensis* | — | X | X | X | — | X | — | — |
| *Chilabothrus schwartzi* | — | — | X | X | — | — | — | — |
| *Chilabothrus striatus* | — | X | X | X | — | X | — | — |
| *Chilabothrus strigilatus* | — | X | X | X | — | X | — | — |
| *Chilabothrus subflavus* | X | — | — | X | — | X | X | — |
| *Corallus cookii* | — | X | X | X | X | X | — | — |
| *Corallus grenadensis* | — | X | X | X | X | X | — | — |

*Restricted Population Sizes and Ranges*

All West Indian boas are endemics, and many are geographically limited to either a single island or a small area on a larger island; as a consequence, they are vulnerable to stochastic events, including inadvertent or intentional alterations of already limited habitats. For some species, this is of primary concern. *Chilabothrus argentum* occurs in a dwindling habitat of less than 0.1 $km^2$ on an island <7.0 $km^2$, the only place the species occurs. Similarly, *C. ampelophis* is presently known from an area of <10 $km^2$; the only known specimens were all found within 1 km airline distance of each other.

*Climate Change*

The earth is undergoing human-mediated climate changes that are adversely affecting biodiversity on a global scale (e.g., Vinnikov et al. 1999; Gibbons et al. 2000). Subtle but insidious responses to a warming global climate by frogs and reptiles include species with temperature-related sex determination that could develop skewed sex ratios and lowland species moving to higher elevations, where they might displace already stressed high-elevation endemics (e.g., Huey et al. 2009, 2012). Species of boas on low-lying islands, such as in the Lucayan Archipelago and many of the Grenadines, are likely especially vulnerable to even subtle changes in, for example, a climate warming scenario (e.g., Penman et al. 2010; Gunderson and Leal 2012) or changes in sea level (Murray-Wallace and Woodroffe 2014). Rising seas following the last glacial cycle inundated vast areas of what was once almost certainly boa habitat. For example, the Great Bahamas Bank, when emergent, would have had an area of 103,000 $km^2$; now the land on the bank amounts to <10,000 $km^2$. Although these changes occurred much more gradually than the current rate of human-induced climate change, boas lost vast amounts of terrestrial habitat. Although currently there are no larger studies implicating human-induced climate change adversely affecting West Indian boas, it would be naïve to believe they will be immune to imminent future changes. Some impacts are likely already being seen. The island of Little Ambergris Cay on the Caicos Bank has a population of *C. chrysogaster* but is so low lying that it is rapidly fragmenting into a group of smaller islands separated by tidal creeks (RGR, unpubl. data). The only known location of *C. argentum* lost about a hectare of land during Hurricane Joaquin (RGR, unpubl. data).

*Natural Disasters*

The threat of hurricanes is an annual event in the Caribbean. These storms can have substantial effects on vegetation, forest structure, and biotic communities; and climate change might increase their frequency and intensity. Because the islands have been subjected to hurricanes for thousands of years, however, they probably have caused few extinctions or even extirpations. On the other hand, because of the intermittent nature and unpredictability of hurricanes, native species might not have evolved behaviors adaptive to surviving hurricanes (but see Donihue et al. 2020; Rabe et al. 2020).

On Grenada, Henderson and Berg (2005) documented short-term effects of Hurricane Ivan (7 September 2004, Category 3, winds of 200 km/h). Treeboas (*Corallus grenadensis*) at Mt. William were observed foraging in the crowns of trees lying on the ground, while others foraged in leafless trees and shrubs (Henderson 2015). Data collected during later visits strongly indicated that boas had moved away from the severely damaged site, presumably to others in which vegetation structure was more similar to that before the hurricane struck. What we do not know is how many animals, especially arboreal species, perish during a hurricane; the numbers could be substantial—and effects on populations already small for various other reasons could be catastrophic.

In Puerto Rico's Luquillo Experimental Forest, post–Hurricane Georges (21 September 1998; sustained winds of 184 km/h), the mean monthly movement of *Chilabothrus inornatus* increased (pre-hurricane 10.7 ± 1.1 m/d/mo; hurricane 14.9 ± 1.3 m/d/mo; post-hurricane 16.8 ± 2.2 m/d/mo; Wunderle et al. 2004). The home range size in males changed little between pre- and post-Hurricane periods but increased for two females, although this increase might be related to post-parturition behavior (Wunderle et al. 2004).

Volcanic eruptions on West Indian islands, although less frequent than hurricanes, can be more devastating. Each eruption has the potential of killing virtually everything in its path and is an important and natural source of catastrophic disturbance to ecological communities. The 1979 eruption of La Soufrière on St. Vincent was considered less extensive than that of 1902 (Fiske and Sigurdsson 1982). We must wait to learn the short- and long-term effects of the April 2021 eruption of Soufrière. Although *Co. cookii* would not have occurred near the summit of La Soufrière (1234 m asl), lava flow from these eruptions would have devastated habitat and wildlife at lower elevations, and ashfall might also impact these areas.

*Alien Pathogens and Parasites*

Many parasites infecting native amphibians and reptiles (e.g., ticks, presumably *Aponomma* spp.) also are native and probably evolved in conjunction with their hosts. Encountering boas with tick infestations, from one or two to dozens, is not unusual (e.g., Grant 1940; Henderson 2015). Surveys on *Chilabothrus inornatus* documented a snake that was unhealthy and had forty-five ticks (*Ornithodoros puertoricensis*) removed from its body (Abreu-Rodríguez and Moyá 1995). At Cueva de Culebrones in Puerto Rico during the early 1990s it was very common to see boas with ticks as the result of the presence of cattle in open areas (ARPR, pers. obs.). At least two endoparasites have been reported in *C. angulifer*: a sporozoan *Haemogregarina* sp. and the cestode *Ophiotaenia nattereri* (see species account). Other parasites undoubtedly have been introduced with the many alien species that have become established. Currently, snake fungal disease, caused by the fungus *Ophidiomyces ophidiicola*, is ravaging populations of snakes in the southeastern United States (Lorch et al. 2016). The disease has impacted a broad taxonomic array of snakes, and *C. inornatus* is the first West Indian boa to test positive for the fungus (E. Mulero Oliveras, pers. comm. to ARPR, 6 May 2019). As of this writing, the impacts of parasites and fungus on native boas are unknown but will hopefully become the focus of future study (see Epilogue).

*Environmental Pollution*

Although we have no documented evidence demonstrating the impact of pesticides on any West Indian herpetofauna, anecdotal accounts suggest that they have had a deleterious and sometimes substantial effect on some species of Antillean frogs and reptiles. Sugarcane dominated island economies until the late-nineteenth century. Today, banana cultivation is the principal wide-scale cash crop on larger islands, and plantations cover many thousands of hectares in the Greater and Lesser Antilles, including formerly forested hillsides. This industry, responsible for as much as 17% of the GDP of some countries and 50% of the jobs (FAO 2019), has introduced the subsidized use of chemical fertilizers and extensive applications of pesticides and herbicides. Martin (2007) suggested that pesticides have had a detrimental effect on Grenada's snake populations. Powell and Henderson (2007) and Tolson and Henderson (2011) noted that these chemicals may not have directly affected the herpetofauna but surely had a negative impact on the arthropodan prey of many frogs and lizards. Short of finding animals dead in the field, we may never be able to assess the full impact of pesticide pollution (Henderson and Powell 2009).

*Depleted or Shifting Prey Bases*

An estimated 77–78% of the native rodent fauna of the West Indies is now extinct; more specifically, 93% are extinct on Hispaniola and 100% in the Lesser Antilles (Morgan and Woods 1986). Prior to the arrival of Europeans and the introductions of *Mus* and *Rattus*, Antillean populations of *Corallus* likely fed on now-extinct populations of murid rodents (*Megalomys*, Oryzomyini, and *Zygodontomys*; Henderson 2015; Mistretta 2019). However, *Oligoryzomys victus* and *Rattus* spp. likely coexisted on St. Vincent for as many as 200–300 years (Turvey et al. 2010; Henderson 2015). Rock iguanas (*Cyclura*) are important prey for some species of *Chilabothrus*, but *Cyclura* populations have declined on many islands (Alberts 2000). According to P. J. Tolson (in Tolson and Henderson 2011), "[T]he loss of *C. pinguis* from the Puerto Rico Bank may be mitigated somewhat by the introduction of *Iguana iguana*, which is heavily used as prey by [*Chilabothrus*]." Overhunting of hutias (mainly *Ca. pilorides*) in many natural areas of Cuba might be an important factor pushing medium-sized to large *C. angulifer* toward anthropogenic environments in search for alternative prey (Rodríguez-Cabrera et al. 2020b). Once these boas reach farms and small villages, domestic animals become a major portion of their diet (see species account).

*Human Overpopulation*

The West Indies already includes several of the most densely populated countries in the world (e.g., Barbados, St. Martin/St. Maarten), and the human population on all major islands that harbor boa populations exhibited substantial increases in human population between 1961 and 2017 (World Bank 2018). For example, the population density in the Turks and Caicos went from 6 people/km$^2$ to 37/km$^2$; the Dominican Republic from 71 to 223/km$^2$, and St. Lucia from 149 to 293/km$^2$ (see Table 2.3). The growth of the human population on Antillean islands (as elsewhere) aggravates many of the factors already negatively affecting the herpetofauna. More people need more housing, more housing requires more development, and more development means that more habitats will be transformed to accommodate those additional people. Development also may require that building materials be imported from other islands or, more likely, from the United States. Those materials may harbor pathogens, frogs, lizards, or snakes that are alien to the West Indies, thereby introducing species that may eat, compete with, hybridize with, or infect with equally alien parasites and pathogens species native to that particular island. Such habitat alterations render species already stressed by habitat contraction even more vulnerable to extirpation or extinction (Henderson and Powell 2009). Destruction of natural habitats to make room for urban development pushes the most adaptable boa species toward urbanized areas, increasing the human–wildlife conflict and hence decreasing the survival possibilities of the snakes.

*Commercial Exploitation*

Schlaepfer et al. (2005) noted that "species with restricted ranges [and] high levels of endemism (e.g., small island species) . . . could be detrimentally affected by even a small number of individuals being removed from the wild." Legal commercial exploitation of West Indian frogs and reptiles appears to be of minimal concern (Powell et al. 2011), and most hobbyists and breeders are motivated by a genuine love of the animals rather than exploitation. Most island governments do not allow commercial collection and exportation of native wildlife. We are, however, aware of live individuals of the arboreal boa *Corallus grenadensis* being illegally taken from Grenada and the St. Vincent Grenadines. On the German website for Terraristik, Union Island *Co. grenadensis* were advertised for sale in February 2018, and individuals of the St. Vincent endemic *Co. cookii* were advertised for sale on the internet as recently as November 2018. Additionally, five *Co. grenadensis* (no locality given) were offered for sale in September 2016 at €600 (= US$680) each. According to G. Gaymes of the St. Vincent and the Grenadines Department of Forestry (in litt.to RWH, 15 January 2019), they have never issued permits (including under the Convention on International Trade in Endangered Species of Wild Fauna and Flora [CITES]) for the exportation of species of *Corallus*. Both species of *Corallus* are protected in their respective countries and both are listed on CITES Appendix II and require special permits for exportation.

Based on data from CITES and reported in TRAFFIC and FFI (2019), 106 *Co. grenadensis* and 204 *C. angulifer* were exported between 2015 and 2019, and they accounted for the majority in the trade in endemic Caribbean reptiles. Nearly all of the *Co. grenadensis* were exported either by the Netherlands or Dutch Sint Maarten and were described as captive bred (CITES data, courtesy of W. Outhwaite, in litt., 14 April 2020). The boas were likely smuggled out of their country of origin and then shipped, with documentation, from Sint Maarten.

## APPLIED CONSERVATION OF WEST INDIAN BOAS

Happily, there are many individuals and organizations in the Caribbean and further abroad who care deeply about mitigating threats to West Indian boas. What follows is an incomplete overview of the hundreds of people involved in this work across numerous countries, but we want to highlight some of this work and some of these initiatives, if for no other reason than to end this chapter on a high note and with a breath of optimism.

*Bahamas*

The Bahamas National Trust (BNT) and the Bahamas Ministry of Environment and Housing are working on national conservation priorities, and the five species of boas found in the country are included among these

**Figure 4.3.** *Left*: Scott Johnson has been a major force in boa conservation and education in New Providence, Bahamas. Photo by R. Graham Reynolds. *Top right*: University students conduct radio tracking studies of *Chilabothrus chrysogaster* in the Turks and Caicos Islands. These studies have provided crucial data on home range size and movement patterns of the species, as well as provided training for upcoming generations of conservation scientists. Photo by R. Graham Reynolds. *Bottom right*: Veterinarians from the North Carolina Zoo conduct important health assessments on *Chilabothrus granti* in the U.S. Virgin Islands. Photo by Dustin Smith.

priorities. Recent work has included the establishment of a robust education campaign and boa relocation task force on the island of New Providence, staffed by volunteers from the BNT, Ardastra Gardens, and interested locals (Fig. 4.3). In 2019 the BNT hosted the IUCN Conservation Planning Specialist Group and scientists and conservationists from around the Bahamas, the University of North Carolina Asheville, and the North Carolina Zoo to create a 5-year Species Conservation Action Plan for the endemic species *Chilabothrus argentum*, the world's most endangered boa species.

*Turks and Caicos*

For 14 years, RGR and G. P. Gerber of the San Diego Zoo Wildlife Alliance have led focused study and applied conservation of *Chilabothrus chrysogaster* in the Turks and Caicos in partnership with scientists from the Department of Environment and Coastal Resources and the Turks and Caicos National Trust. Having obtained data from >1200 individuals, these researchers have demonstrated the ecological importance of the species to the ecosystems in the TCI as well as identified critical habitat and thus far prevented the introduction of exotic predators to the last remaining boa stronghold in the archipelago.

*Cuba*

For several decades P. J. Tolson has led conservation and research efforts for the Cuban boa (*Chilabothrus angulifer*). Centered at Guantanamo Bay, this work has yielded great insight into movement and reproductive biology of this species. A new cadre of scientists based in Cuba, including TMRC, LMD, and others, are extending this work across the island and generating a tremendous amount of information on the biology and conservation status of Cuban boas.

**Figure 4.4.** Urbanization and forest clearing encroaching upon the last remaining habitat for *Chilabothrus granti* on the main island of Puerto Rico. Photo by R. Graham Reynolds.

*Jamaica*

For a number of years, the Windsor Research Centre has been the heart of boa conservation in Jamaica. Susan Koenig has hosted numerous students and researchers over the years studying Jamaican boa biology, and these studies have led to invaluable natural history and conservation information. At least one Fulbright Fellowship, one doctoral dissertation, and one master's thesis have been focused entirely or in part on *Chilabothrus subflavus*. With the recent news that the Goat Islands will be preserved instead of developed, the National Environment and Planning Agency (2007) and partners are planning to use Great Goat Island as a wildlife sanctuary and reintroduction site, possibly for *C. subflavus*.

*Virgin Islands*

Virgin Islands boas (*Chilabothrus granti*) have been successfully reintroduced in the U.S. Virgin Islands, demonstrating the potential for applied conservation activities. A consortium of conservationists is working to conserve *C. granti* in Puerto Rico and the U.S. and British Virgin Islands. Building off decades of foundational work from 1985 to 2004, this new cadre of scientists, students, zoo biologists and veterinarians, government officials, and conservationists are working to protect the remaining populations of the species, increase local outreach, and reestablish lost populations across the region (Fig. 4.4).

*Puerto Rico*

All three boa species in Puerto Rico: the Puerto Rican boa (*Chilabothrus inornatus*), the Mona boa (*C. monensis*), and the Virgin Islands boa (*C. granti*) are of some conservation concern. Despite 50 years passing since *C. inornatus* became protected by law, the species is still considered endangered by some. However, there is a consensus that the species is more abundant than previously thought. Early protection combined with the approval of different local laws and regulations focused on land protection have benefited *C. inornatus* populations in some parts of Puerto Rico. Nevertheless, boas in Puerto Rico, Isla de Mona, and the Virgin Islands are still facing the same threats that drove them onto the endangered species list, plus a host of new threats, such as snake fungal disease and the introduction of exotic invasive snake species. As a professor at University of Puerto Rico–Mayagüez and an active conservationist, ARPR and collaborators such as J. P. Zegarra are working on developing and overseeing research projects focused on *C. inornatus* and *C. granti* to continue recovery efforts.

**Figure 4.5.** *Top left and bottom left*: Rescued boas (*Corallus grenadensis*) being set free by Serpent and Snake Relocation Program (SSRP) team members. *Top right*: Veterinarian Dr. Marie Rush taking a blood sample from a *Co. grenadensis* with Isidore Monah providing illumination. *Bottom right*: Dr. Rush removing an unappreciated *Co. grenadensis.* from a tree. The boa's eye reflection is clearly visible. Photos by Marie Rush and SSRP.

*Lesser Antilles*

In 2010, while teaching at St. George's University, veterinarian Dr. Marie Rush (e.g., Rush et al. 2020) started the Grenada Serpent and Snake Relocation Program. She recruited her veterinary students and Grenadians who were interested in conservation and protecting *Corallus grenadensis* to respond to complaints of boas on private property. Working out of St. George's, Marie and her team would capture the boa and relocate it to appropriate habitat away from its former location (Fig. 4.5). Just as importantly, Marie or a team member would explain to the property owner that boas were harmless and were providing a great service by controlling rat populations that could negatively impact their livestock and produce. Over time, the number of landowners who would routinely kill any boa they encountered declined from about 80% to 30%. Elsewhere in Grenada, Isadore Monah performed the same service from the more centrally located Pearls area. Grenada Bank treeboas are fortunate to have Marie, Isadore, and their younger recruits as ambassadors.

*Caribbean and Latin American Boa Specialist Group*

Founded by RGR and ARPR in 2013, this regional group of the IUCN Snake Specialist Group (formerly the Boa and Python Specialist Group) holds occasional scientific meetings focused on sharing data and encouraging collaboration among researchers studying boas in the Western Hemisphere. Look for announcements about upcoming meetings in publications such as *Herpetological Review.*

**Figure 5.1.** A Bimini boa (*Chilabothrus strigilatus fosteri*) from South Bimini, Bahamas. Photo by Katie Grudecki.

# *Chapter 5*

# SPECIES ACCOUNTS

## THE LUCAYAN ARCHIPELAGO

This chain of islands, stretching some 1360 km from the Abacos to the Turks Islands, includes twenty-one island banks supporting thousands of large and small islands administered by the Commonwealth of the Bahamas and the Turks and Caicos Government. Most are quite flat, with the highest point in the archipelago reaching only 63 m asl at Mount Alvernia, Cat Island, Bahamas. Composed of oolitic islands formed atop carbonite platforms (banks), it is likely that islands in this archipelago capable of sustaining terrestrial vertebrates only became emergent during the early Pleistocene (based on historical biogeographic reconstructions). Several islands are presently inundated, such as Hogsty Reef, as well as the Silver, Navidad, and Mouchoir banks (four total banks, under disputed jurisdiction by the Turks and Caicos and the Dominican Republic), which would have brought the island chain to within 60 km of Hispaniola during periods of lower sea level. These latter banks are significant, as they likely facilitated the dispersal of reptiles (including boas) to the archipelago from Hispaniola. The Lucayan Archipelago, despite being of relatively recent origin, supports five species of boas—all members of the genus *Chilabothrus* (Fig. 5.1). Three of these species are recently recognized. One species was recently discovered (*C. argentum*), while two subspecies were found to be distinct and recognized as separate species (*C. strigilatus* and *C. schwartzi*).

## CONCEPTION BANK BOA, SILVER BOA

*Chilabothrus argentum* Reynolds, Puente-Rolón, Aviles-Rodriguez, Geneva, and Herrmann 2016

### *Taxonomy*

This species was first described in 2016 following its formal discovery in 2015 (Reynolds et al 2016c). Like other Lucayan boas, the species is derived from Hispaniolan ancestors, although it is not yet clear if it evolved in situ on the Bahamas islands, but that is a likely scenario. The species is closely related to the recently elevated *C. schwartzi* on the Crooked-Acklins Bank and appears to be part of the eastern Bahamas lineage of boas descended from Hispaniolan colonists in the Pleistocene.

### *Etymology*

The species' epithet is Latin for silver, representing both the coloration of adult animals and that the first individual was found in a silver palm (*Coccothrinax inaguensis*). The species' known range is an island bank presently uninhabited by humans, hence there is no known colloquial name. Other Lucayan boas in this genus are called "fowl snakes."

### *Description*

This species is known from a total of only forty-eight individual boas (several of which have been recaptured at some point) that have been examined in the wild (Fig. 5.2). It is a small member of the genus *Chilabothrus*, and males and females might exhibit very slight sexual size dimorphism, with females being heavier ($P = 0.03$)

**Table 5.1.** Scale count and morphological comparisons among the five *Chilabothrus* species from the Lucayan Archipelago.

Data are from Sheplan and Schwartz 1974, Buden 1975, and Reynolds et al. 2016c. Squamation is reported as ranges. SSD = sexual size dimorphism. Meristics for *C. chrysogaster* include both subspecies.

| Character | *C. argentum* | *C. chrysogaster* | *C. exsul* | *C. schwartzi* | *C. strigilatus* |
|---|---|---|---|---|---|
| Ventrals | 275–282 | 242–277 | 236–251 | 277 | 231–261 |
| Subcaudals | 82–91 | 74–95 | 69–75 | 95 | 69–85 |
| Loreals | 1 | 1–3 | 1 | 1 | 1 |
| Circumorbitals | 9–10 | 9–13 | 10 | 9–10 | 10–11 |
| Supralabials | 11–12 | 12–16 | 13 | 12–13 | 11–15 |
| Suboculars | 0–1 | 2–3 | 3 | — | 2–3 |
| Postoculars | 4 | 3–5 | 6 | — | 4 |
| Preoculars | 1–2 | 2 | 1 | — | 2 |
| Midbody dorsal | — | 39-47 | — | 36–38 | 31–39 |
| Dorsal blotches | None to very few, light gray | Light gray to dark brown | Light gray to dark brown | — | Light gray to black |
| Spot/stripe dimorphism | No | Yes | No | No | No |
| Ventral color | Pure cream | Cream with gray stippling | Pure cream | — | Cream to gray to brown, mottled or with blotches |
| Maximum female SVL (mm) | 1245 | 1500 | 810 | — | 2180 |
| Maximum male SVL (mm) | 1035 | 865 | 635 | — | 2330 |
| SSD | No | Yes | Yes? | — | No |

and longer ($P = 0.04$) on average than males (Fig. 5.3). The maximum size for a female is 1245 mm SVL and 610 g in mass, while the average size is 939 mm SVL with a mass of 231 g. Maximum male size is 1035 mm SVL with a mass of 310 g, and the average male size is 854 mm SVL with a mass of 174 g. A single neonate has been observed, identified by the presence of an umbilical scar, and was 387 mm SVL and 16.5 g in mass. Two juveniles, diagnosed by the presence of juvenile coloration (see Figs. 5.3 and 5.4) and no obvious umbilical scar, were 394 and 483 mm SVL and 15 and 24 g in mass, respectively. There is 1 loreal scale, ventrals are 275–282, subcaudals are 82–91, supralabials are 11–12, and there are 10 circumorbitals (Table 5.1).

The overall color is silvery gray to light tan (Fig. 5.2), and dorsal and lateral scales are infused with mottled gray-to-brown pigment and taper to a somewhat narrow point, especially on the flanks but also on the dorsum. Like some other *Chilabothrus* species, juveniles undergo slight ontogenetic color change, from a light brownish-orange coloration with lots of patterning (Fig. 5.3) to a silver-gray adult coloration with reduced patterning. Juveniles appear to have "eyespot" patterning on the tops of their heads (Fig. 5.4), composed of a pair of dark circles formed by 4–6 dark-gray pigmented scales on each side of the upper portion of the head. Conception Bank boas over 540 mm SVL had already transitioned to gray dorsal coloration, suggesting that the ontogenetic color change occurs around 500 mm SVL. Adult Conception Bank boas are light

**Figure 5.2.** Adult female *Chilabothrus argentum* resting in a *Bursera simaruba* tree. Photo by R. Graham Reynolds.

**Figure 5.3.** Juvenile *Chilabothrus argentum*; note the distinct pattern that is usually lost as the animals age. Photo by Anthony Geneva.

**Figure 5.4.** *Left*: Head detail of a juvenile *Chilabothrus argentum*, showing the darker eyespots toward the back of the head. *Right*: Juvenile *C. argentum* showing tail coiling ability. Photos by Alberto R. Puente-Rolón.

gray to light khaki brown or tan dorsally, with an unpatterned and cream-colored venter. Some irregular markings are present on the dorsum, consisting of patches of darkly or lightly pigmented scales singly or in groups. This patterning is usually subtle, although is much more apparent in the juveniles and occasionally in some adults. Most adults at arm's length appear largely unpatterned but upon close examination show hints of faint dorsal markings. The tail is darkly pigmented, tapers to a very fine point, and is prehensile and capable of being tightly coiled (Fig. 5.4). The pupil is black and the iris is mottled gray in both adults and juveniles, and like most species in the genus the eyes are noticeably large relative to the head size in juveniles and appear to "bulge" out of the head (Fig. 5.3). The tongue is dark gray to black with white tips.

**Figure 5.5.** *Chilabothrus argentum* is highly arboreal. *Left*: Adult male *Chilabothrus argentum* exhibiting climbing behavior. Photo by R. Graham Reynolds. *Right*: Typical resting posture for *C. argentum* in a *Coccoloba uvifera* tree. Photo by Alberto R. Puente-Rolón.

*Distribution*

This species is endemic to the small (<7 km$^2$) Conception Island Bank, although biogeographic study suggests that it likely occurred elsewhere in the Bahamas, such as Rum Cay 22 km to the southeast. Rum Cay, although currently supporting a small settlement of humans and covered with lush coppice forest and mangroves, was previously one of the most densely populated islands in the Bahamas (Craton and Saunders 1992). During the early nineteenth century Rum Cay supported a major salt-raking enterprise and was likely severely environmentally impacted during this period. Hence, it is plausible that if boas did occur there, they might have been extirpated during this time. On Conception Bank, the species occurs in only one small area (<0.1 km$^2$) and does not occur bank-wide. Conception Island has at different times in history been referred to as San Salvador, Little San Salvador, and Little Island and has had some sustained human presence and visitation in the last two centuries, although it has been uninhabited for many decades now.

*Habitat*

*Chilabothrus argentum* is found in closed-canopy tropical scrub forest, which consists of a mixture of "coppiced," or short-canopy, trees such as *Bursera simaruba* and *Metopium toxiferum* and understory *Coccothrinax inaguensis* palms. Boas are regularly found adjacent to the beach, or even on vegetation overhanging the sea. One individual was observed to traverse the beach at night (Reynolds et al. 2016c). This species appears to be restricted to dense scrub and has not been found in open coastal vegetation with limited arboreal habitat. Diurnal refuges likely include tree holes, although one boa was found underneath palm litter during the day. Boas are most frequently spotted at night while moving through forest undergrowth and canopy (Fig. 5.5).

*Abundance*

At the only location where they occur, this species exists at a relatively high density, but the limited range on the Conception Island Bank severely limits population size. An ongoing capture–mark–recapture study suggests that as few as 128 (± 37) adult boas exist. In total, forty-eight boas have been marked in this population

**Figure 5.6.** Young adult male *Chilabothrus argentum* just after attempting to capture a migratory songbird (*Setophaga tigrina*). The bird had been sleeping on the branch. Photo by R. Graham Reynolds.

as of 2017, and seven individuals have been recaptured, with one individual having been captured four times over 2 years in a 50-m$^2$ area.

*Activity and Trophic Ecology*

This species is active nocturnally and actively seeks prey by moving slowly in branches or assuming ambush positions. It is thought that adults largely feed on sleeping birds that they hunt arboreally, although no actual prey items have been recovered from boas. Conception Bank boas are occasionally observed traversing the ground between trees but have not been observed to hunt on the ground. One individual crawled several meters across open beach from the forest and onto RGR, who was asleep at the water's edge (Reynolds et al. 2016c). Potential prey items include nesting seabirds and migratory songbirds, and it is likely that food availability fluctuates widely during the year.

Juveniles likely feed on the abundant lizard, *Anolis sagrei*, although this has not been directly observed. Only one prey stalk has been observed—a young male (635 mm SVL) was observed attempting to capture a sleeping Cape May Warbler (*Setophaga tigrina*), a migratory species. The boa slowly approached the bird from underneath, tongue flicking, until the boa was directly touching the bird with its tongue. It lunged upward, grabbed the bird in its jaws, and attempted to loop coils over the bird to subdue it. Apparently, its jaw grip was not sufficient to hold the bird, and the warbler escaped leaving behind some feathers in the snake's mouth (Fig. 5.6).

*Predators and Defensive Behavior*

No natural predators are presently known, and external and internal parasites are unknown. It is likely that raptors such as Merlins and Ospreys, which are common on the Conception Island Bank (RGR, pers. obs.), would consume boas if spotted. Nocturnal predators such as owls are unknown on the island.

Because no predation attempts have been observed and no deceased boas have been found, there is limited information on defensive behaviors. Conception Bank boas do not move when approached and are loath to strike when handled, although, like most boas, they excrete the contents of the cloaca when handled. Juveniles will ball up loosely when handled, although they will readily uncoil and do not maintain the ball posture for long.

*Reproduction*

Little is known of the reproductive biology of this species, although it is live-bearing and likely resembles the reproductive biology of other members of the genus (see *C. chrysogaster* account). One neonate was observed 13 July 2017 and had an obvious umbilical scar, suggesting it had been born in late June or early July. Whether this is typical is unknown, but would suggest possibly earlier parturition than other *Chilabothrus*, which typically give birth in the late summer.

*Conservation Status*

This species is protected from export by the Bahamas Wild Animals Protection Act of 1968 and is listed as Critically Endangered on the IUCN Red List (Reynolds 2017). It is the most endangered species of boa globally. The single known population is in serious risk of extinction in the next few decades and likely consists of only a few hundred adults. The biggest threat is the small area of suitable habitat available on the Conception Island Bank. It is shrinking with each large hurricane that strikes the island and causes loss of island area and saltwater incursion into the forest. Small populations are also vulnerable to extinction owing to inbreeding depression or stochastic population decline

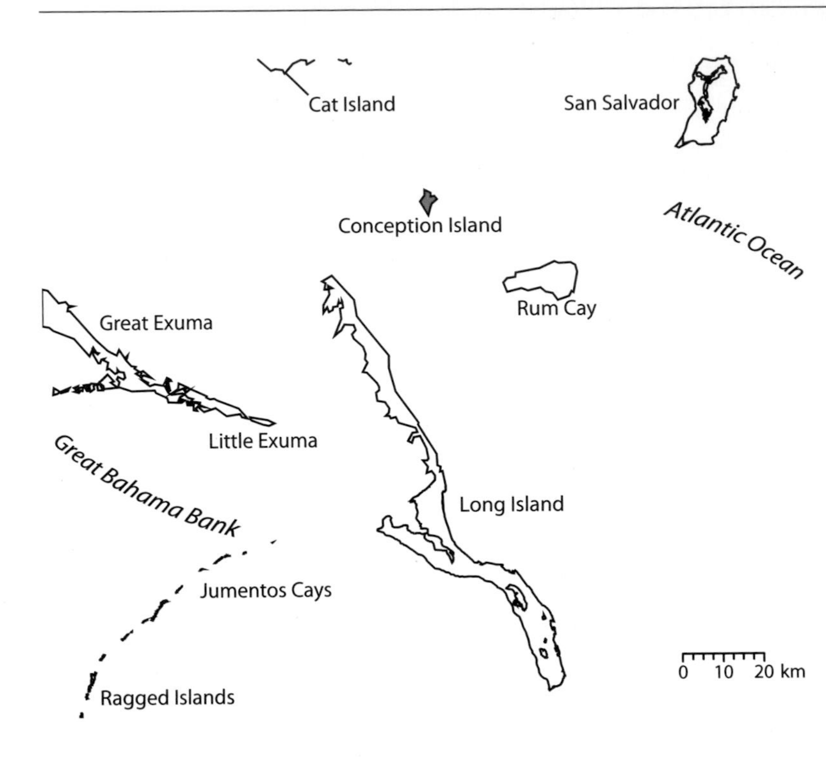


**Figure 5.7.** Range of *Chilabothrus argentum*, located entirely on the Conception Island Bank in the central Bahamas.

(Madsen et al. 1996). There are invasive mice on the island and possibly other introduced species such as rats and cats. Poaching is also a concern, although it has not yet been documented. The population is sufficiently difficult to find and access so that poaching is currently unlikely. Poaching laws are strongly enforced in the Bahamas and can lead to very large fines, summary confiscation of boats and equipment, and jail time.

Historically, humans have intermittently occupied Conception Island—Lucayan sherds are present as well as the ruins of a small settlement from around 1900. Otherwise, the island has likely served as a stopover for fishers, although a double shipwreck left survivors stranded there for several weeks in 1812. Collectively all these people left indelible marks on Conception Island, now visible as a lack of large trees, building ruins, invasive plants, and old sisal patches.

Finally, the Conception Bank is used as a narcotrafficking drop point, and this type of illegal activity could lead to damage to the population. For example, traffickers use flares to illuminate drop sites, and these could cause devastating forest fires on the island that would further endanger the species. Figure 5.7 shows the distribution of *Chilabothrus argentum*.

## TURKS AND CAICOS BOA

*Chilabothrus chrysogaster* (Cope 1871)
- Turks Island boa (*Chilabothrus chrysogaster chrysogaster*) (Cope 1871)
- Inagua boa (*Chilabothrus chrysogaster relicquus*) (Barbour and Shreve 1935)

### *Taxonomy*

This species is the oldest cladogenically separate lineage of the extant Lucayan boas, having diverged approximately 5 million years ago (MY) from Hispaniolan ancestors (Reynolds et al. 2013a). It is not clear when the species reached the Bahamas and Turks and Caicos Banks, although the Silver, Mouchoir, and Navidad Banks (now submerged) might have facilitated colonization from Hispaniola. The species appears to have colonized the region separately from other Lucayan boas (Reynolds et al. 2013a).

Turks and Caicos boas are represented by two subspecies: the Inagua boa (*C. chrysogaster relicquus*) and the Turks Island boa (*C. c. chrysogaster*) (Reynolds 2012). Very little is known of the former subspecies, while the latter is well studied on the Caicos Bank (but

**Table 5.2.** Museum specimens of *Chilabothrus chrysogaster relicquus.*

The SVL is as measured by RGR or Sheplan and Schwartz (1974). ASFS = Albert Schwartz Field Series; KUH = University of Kansas Biodiversity Institute Herpetology Collection; MCZ = Museum of Comparative Zoology, Harvard University; UMMZ = University of Michigan Museum of Zoology.

| Museum # | Field # | Date | Type | Locality | Collector | SVL | Sex |
|---|---|---|---|---|---|---|---|
| KUH 260080 | ASFS X4815 | 1949 | — | Matthew Town | S.W. Erickson | 1310 | F |
| KUH 260081 | ASFS V20458 | 1970 | — | ~16 km N Matthew Town | Anonymous local | 810 | M |
| UMMZ 170392 | — | 1973 | — | — | — | — | — |
| UMMZ 170393 | — | 1973 | — | — | — | — | — |
| MCZ 37891 | — | 1934 | Holotype | Sheep Cay, Inagua | T. Barbour | 864 | M |

not the Turks Bank, where a single small population is presently known). Hence most of the information in this account comes from ongoing studies of populations of *C. c. chrysogaster* on the Caicos Bank.

*Etymology*

The species epithet *chrysogaster* is latinized Greek meaning "yellow/orange stomach" and references the light-colored ventral surface, which is in reality generally cream colored, although variation in mottling exists. This species is colloquially referred to as "fowl snake" in the Bahamas and "rainbow boa" in the Caicos Islands. It is also known as the Southern Bahamas boa. The subspecific epithet *relicquus* is a misspelling of the Greek *relicqus*, meaning relic. This name was applied to the first individual of that subspecies that was found on Sheep Cay, a small cay adjacent to Great Inagua, and was (at the time in the 1930s) thought to contain the only remaining members of the subspecies (Sheplan and Schwartz 1974).

*Description*

This species is generally characterized by relatively small size (<1.5 m SVL) and pronounced sexual size dimorphism, with females reaching much greater length and mass than males (Reynolds and Gerber 2012). Dorsal coloration consists of a gray to tan background with a series of darker blotches dorsally. The venter is cream colored with a suffusion of gray speckling. Neonates are born bright orange with plain venters, transitioning to adult coloration in the first year.

*Chilabothrus chrysogaster relicquus*: This subspecies is known from only five museum specimens plus anecdotal observations in the field (Sheplan and Schwartz 1974; Buden 1975; this volume). It is not clear which specimens Sheplan and Schwartz (1974) used to describe the subspecies, as different specimens are listed in their account and in their list of specimens examined. Additionally, not all museum specimens are readily located in database searches, hence we list them in Table 5.2 (see also Fig. 5.8). The largest specimen (University of Kansas Biodiversity Institute Herpetology Collection [KUH] 260080) is a 1310 mm SVL female, and the largest male is 864 mm SVL (MCZ 37891). This subspecies is not known to be as variable in coloration as *C. c. chrysogaster.* Individuals have distinct angular dark blotches dorsally and laterally that are distinct against the lighter background colors (Fig. 5.9), and a faint postocular stripe is sometimes present. Sheplan and Schwartz (1974) note that *C. c. relicquus* is distinct from *C. c. chrysogaster* in having more apparent dorsal blotches, but this distinction does not hold following extensive characterization of color pattern variation in the latter taxon (Reynolds and Gerber 2012; Reynolds et al. 2020a). The iris is light gray streaked with dark lines. There are 1–2 loreal scales, dorsal scales at midbody are 46–47, and ventrals are 269–275 (Sheplan and Schwartz 1974). There are modally 17 infralabials and 2 supralabials entering the eye (Table 5.1).

*Chilabothrus chrysogaster chrysogaster*: This taxon is a small-sized member of the genus *Chilabothrus*

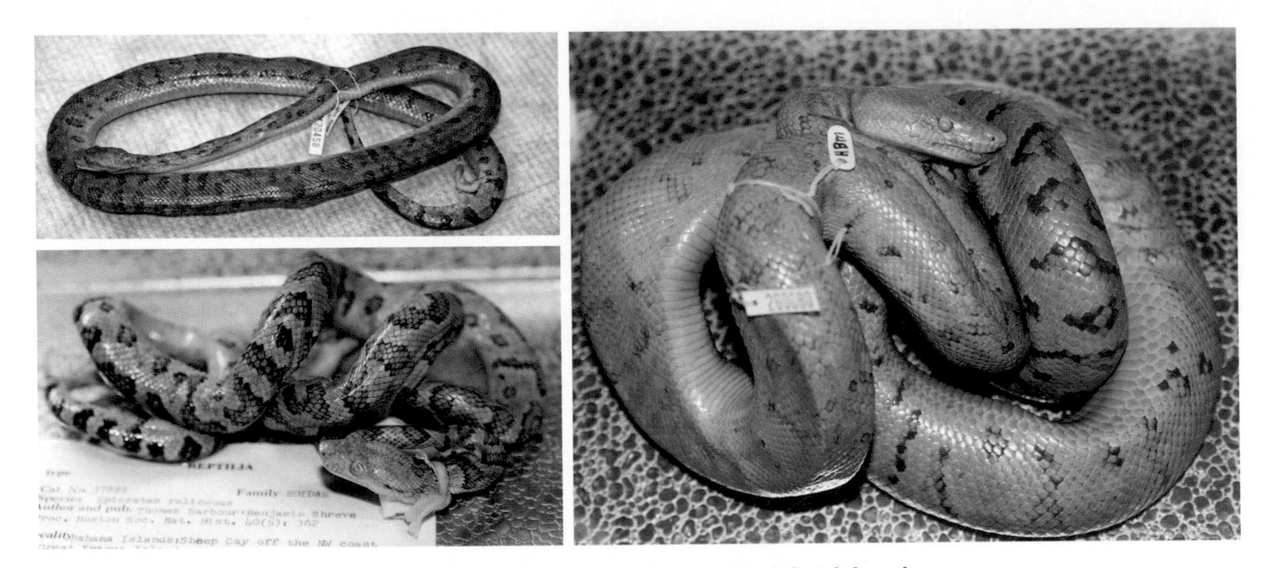

**Figure 5.8.** Museum specimens of *Chilabothrus chrysogaster relicquus*. *Top left*: Adult male KUH 260081 collected in 1970. *Bottom left*: Adult male MCZ 37891, the holotype specimen collected in 1934 on Sheep Cay. *Right*: Adult female KUH 260080 collected by S. W. Erickson in 1949. Photos by R. Graham Reynolds.

**Figure 5.9.** *Left*: Inagua boa (*C. c. relicquus*) from near Matthew Town, Great Inagua, Bahamas. *Right*: Head detail of an Inagua boa (*C. c. relicquus*) from near Matthew Town, Great Inagua, Bahamas. Note the two loreal scales. Photos by Joseph Burgess.

but exhibits some of the most significant sexual size dimorphism in the genus (Figs. 5.10–5.12). Adult females can reach just over 1500 mm SVL (Reynolds et al. 2011c) and 1675 g in mass when gravid (RGR, unpubl. data), although maximum female sizes around 900–1100 mm SVL and 200–350 g in mass are far more characteristic of what constitutes a "large" individual of this species. Males are generally 575–750 mm SVL with a mass of 75–125 g, with a record male SVL of 865 mm and mass of 197 g (Reynolds and Gerber 2012). An ongoing study has examined over 1200 individuals in the Big Ambergris Cay population (as of 2021; RGR, unpubl. data) and found a mean female size of 721 mm SVL, tail length of 130 mm, and mass of 123 g. Males are considerably smaller than females (mass: $n = 1166$, $P < 0.001$; SVL: $n = 1216$, $P < 0.001$), with mean SVL of 652 mm, tail length of 125 mm, and mass of 70 g (Fig. 5.10). Females have significantly longer tails than males ($n$ = 1206, $P < 0.001$), although the difference is slight (130 mm versus 125 mm, respectively). Neonates are born at approximately 280–300 mm SVL and 6–8 g in mass (Reynolds and Deal 2010). On the Caicos Bank, young born in the fall reach a size of 350 mm SVL and 9–16 g in mass by March, when they also begin to lose juvenile coloration (see below). There are 1–3 loreal scales. Additional scale counts are based on six specimens examined by Sheplan and Schwartz (1974) and several dozen examined by Buden (1975). Dorsal scales at midbody are 39–43, ventrals are 245–277, and subcaudals are 74–95; there are modally 15 infralabials, 12–16 supralabials, 11 circumorbitals, and there are 3 supralabials entering the eye (Table 5.1).

The tail tapers to a fine tip and is somewhat prehensile and capable of coiling (see Fig. 5.4 in *C. argentum* account). Many adult snakes have lost the tip of the tail, sometimes as much as 30–40 mm or more. Tail tips seem to heal well, and missing tips don't appear to impede the animals. Males frequently have relatively large spurs, but variation in this trait does not allow consistent diagnosis of sex without probing. Similarly, male tails are often longer relative to SVL, but this is not a significantly different trait between sexes and thus is also an unreliable way to sex animals. Turks and Caicos boas have some distinct squamation characteristics from other Lucayan and Hispaniolan boas (Table 5.1), including a higher number of loreals and circumorbitals. The dorsal scales are elongate, with a more rounded distal tip than those of *C. schwartzi*, which are distinctly sharp and pointed, especially on the flanks.

Coloration in this subspecies is extremely variable (Reynolds et al. 2020a). Generally, Turks and Caicos boas have a light-colored venter with faint gray to brown freckling, a light gray dorsal background color, and a series of darker blotches or stripes dorsally (Fig. 5.13). The presence of blotches or stripes forms a distinct polymorphism, referred to as "striped" and "spotted," although until recently these color patterns were not well defined. Stripes consist of two or four dark lines four scales wide running longitudinally from the base of the head along the dorsum and lateral sides to the tip of the tail. Spots (blotches) are highly variable in size, shape, and distribution, with various sizes and shapes usually not extending to the lateral aspect but confined to the dorsum. Striped morphs are restricted to the Caicos Bank and occur at about a 15% frequency relative to spotted morphs in the Big Ambergris Cay population (Reynolds et al. 2020a). Both striped and spotted patterns generally comprise darker scales forming the border of the pattern, while slightly lighter scales fill the space between pattern borders. A dark postocular stripe one-scale wide occurs from the back of the orbit to the jaw, where the stripe tapers to

**Figure 5.10.** Dramatic sexual size dimorphism in *Chilabothrus c. chrysogaster.* Typical adult male on the left, typical adult female on the right, both from Big Ambergris Cay, Turks and Caicos Islands. Photo by R. Graham Reynolds.

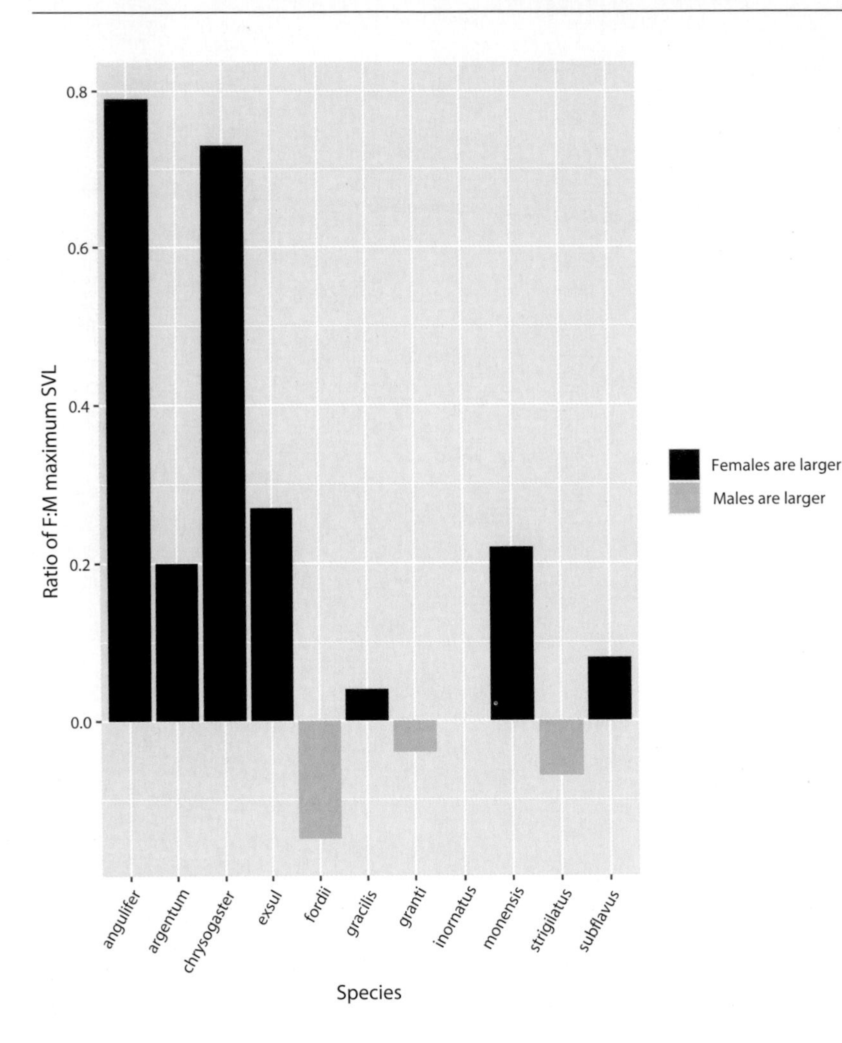


**Figure 5.11.** Some members of the genus *Chilabothrus* exhibit sexual size dimorphism, with females (black) or males (gray) reaching larger maximum sizes (SVL) than the other sex. These bars represent the ratio of female to male maximum size difference normalized to 1.0. A number above 1.0 means that females grow larger, below 1.0 means that males grow larger. *Chilabothrus angulifer* and *Chilabothrus c. chrysogaster* have the most dramatic sexual size dimorphism, with females reaching much larger sizes than males. Some species do not have enough data on body size for accurate comparison.

a fine point over the quadrate bone. The pupil is black, and the iris is a mottled light gray. The tongue ranges from light pink to dark grayish red among individuals, although nearly always with white tips. Neonates are born with an orange coloration on both the ventral and dorsal sides (Fig. 5.14), which transitions to gray on the dorsum and cream on the venter between about 300–400 mm SVL. Despite the seemingly bright neonate colors, this orange coloration is surprisingly cryptic on vegetation stems such as seagrape (*Coccoloba uvifera*).

Since the species' original description, coloration was thought to be minimally variable except for the striped/spotted polymorphism and blotch variation. Recently, extensive study of boas on the Caicos Bank has shown that these populations exhibit a huge range of coloration (Reynolds and Gerber 2012; Reynolds et al. 2020a). These authors documented three dorsal color components: background color, pattern color, and pattern type. Boas on the Caicos Bank exhibit background colors ranging from light gray to light brown to copper to red, with pattern colors from dark gray to dark brown to black. Dorsal patterns range from nearly completely patternless to irregular spotting, saddles, single spot row, paired spot rows, dorsal stripes, both dorsal and lateral stripes. Even this is a generalization of colors. Caicos Bank boas can exhibit colors on a range across these color and pattern types, with the surprising ability occasionally to retain juvenile orange and red colors—known as incomplete ontogenetic color change that results in red or copper adults (Figs. 5.13, 5.15). Currently the Caicos Bank boas are the only populations known to exhibit such a range of colors—Turks and Caicos boas from the Inagua Bank and Turks Bank only exhibit the classic light

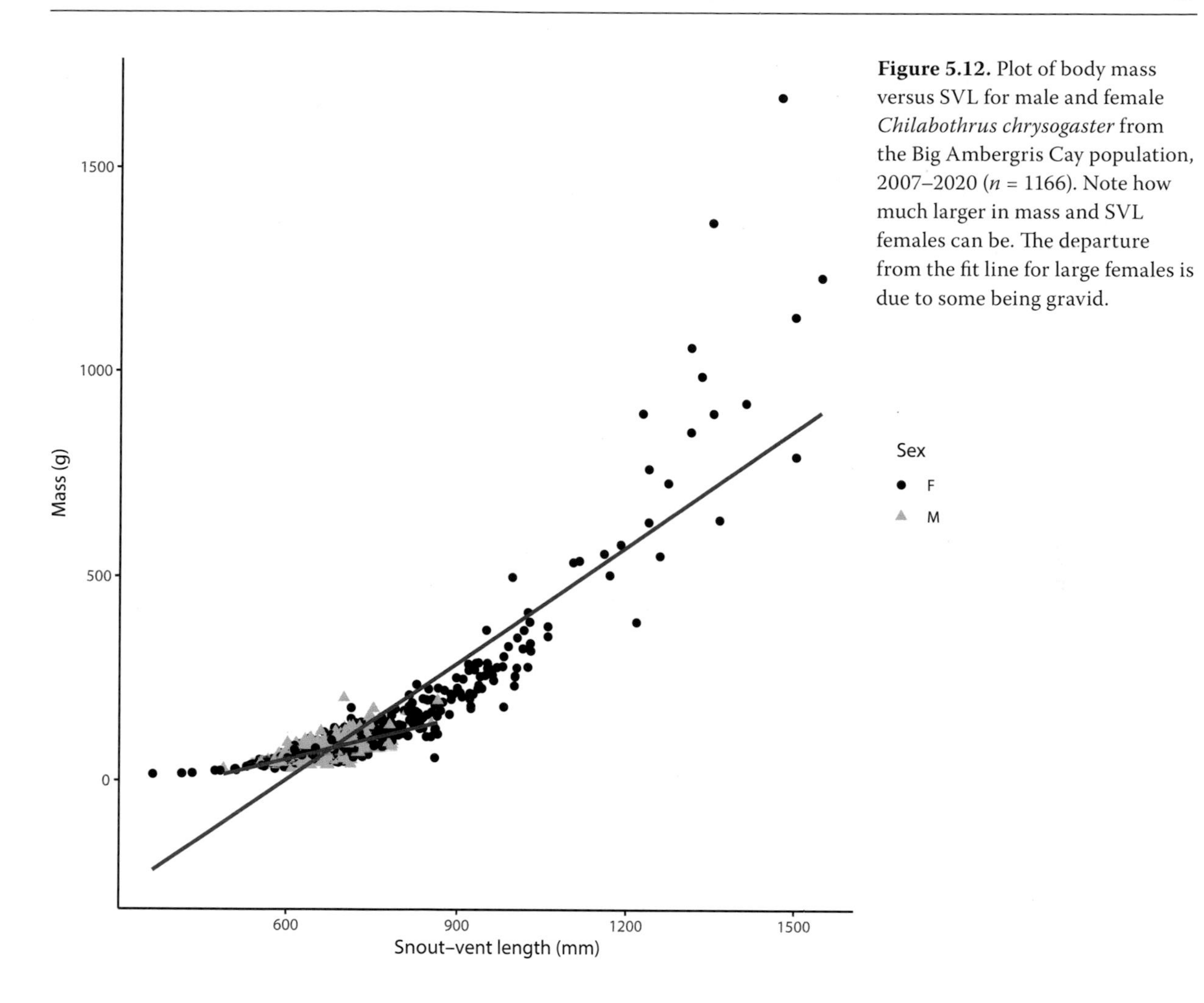


**Figure 5.12.** Plot of body mass versus SVL for male and female *Chilabothrus chrysogaster* from the Big Ambergris Cay population, 2007–2020 ($n$ = 1166). Note how much larger in mass and SVL females can be. The departure from the fit line for large females is due to some being gravid.

gray background, dark gray spotting pattern, with no striped individuals reported.

*Distribution*

*Chilabothrus c. relicquus*: The Inagua boa is known from areas adjacent to Matthew Town, Great Inagua, and from Sheep Cay, a low-lying island nearly connected to Great Inagua at low tide (Sheplan and Schwartz 1974). Its distribution elsewhere on Great Inagua is not well characterized, and no extensive formal surveys have been conducted. Boas do not occur on Little Inagua (Buckner et al. 2012), which is on a separate bank.

*Chilabothrus c. chrysogaster*: The Turks Island boa occurs on the Turks Bank and Caicos Bank (Reynolds 2011c, 2011d; Reynolds et al. 2011b; Reynolds and Gerber 2012; Reynolds et al. 2020a). On the former, it is thought to have been extirpated from most islands, including Grand Turk from which there are historical records. However, a single population was discovered in 2009 on Gibbs Cay, a very small island east of Grand Turk, that probably supports fewer than fifty individuals (Reynolds and Niemiller 2009, 2010; Reynolds 2011a, 2011b). This is the only know remaining population on the Turks Bank, and it is genetically similar to Caicos Bank populations (Reynolds 2011c, 2011d; Reynolds et al. 2011b). Other populations might persist on some of the Turks Cays, although significant vegetation damage caused by donkeys, goats, and rats over the last two to three centuries has greatly degraded habitats on islands such as Salt Cay and Cotton Cay, and boas most likely do not occur on these islands (RGR, pers. obs.). Boas were not found on surveys of Long Cay in 2009 (note that there are islands

**Figure 5.13.** *Top left*: Spotted morph adult female *Chilabothrus chrysogaster* from North Caicos. Photo by Matthew Niemiller. *Top right*: Spotted morph adult male *C. chrysogaster* from Big Ambergris Cay. Photo by R. Graham Reynolds. *Bottom left*: Striped morph adult male *C. chrysogaster* from Big Ambergris Cay. Photo by R. Graham Reynolds. *Bottom right*: Copper striped morph adult female *C. chrysogaster* from North Caicos. Photo by B. Naqqi Manco.

**Figure 5.14.** Spotted and striped morph juvenile *Chilabothrus chrysogaster* from North Caicos and Big Ambergris Cay, respectively, showing typical neonate coloration. Photos by R. Graham Reynolds.

**Figure 5.15.** Rare copper-red color morph of *Chilabothrus chrysogaster* from North Caicos, Turks and Caicos Islands. This morph color has been observed on North Caicos and Middle Caicos. Photo by Jeff Lemm.

called "Long Cay" on both the Turks Bank and the Caicos Bank).

On the Caicos Bank, the species is known from nine islands, including the large islands of Providenciales as well as North, Middle, and East Caicos (Reynolds and Gerber 2012). West Caicos has suitable habitat but has lots of feral cats. It has not been thoroughly surveyed, and we are not aware of reports of boas there. The species is also found on some smaller islands, such as Dellis Cay and Parrot Cay, both located in the Caicos Cays between Providenciales and North Caicos, as well as Joe Grant's Cay near East Caicos. Boas have not been found under extensive survey effort on other Caicos Cays, such as Water Cay, Pine Cay, and Little Water Cay. Similarly, no boas have been found recently on South Caicos, despite historical records, and they are thought to have been extirpated there. Long Cay (near South Caicos) has good habitat and a reintroduced *Cyclura carinata* population following cat eradication (Mitchell et al. 2002), but boas were not detected during nocturnal surveys in 2018. Nevertheless, there are historical records from this island (University of Michigan Museum of Zoology [UMMZ] 117027), and the population might persist. On the southeastern margin of the Caicos Bank, boas occur on Little Ambergris Cay (actually three smaller cays joined by mangrove creeks) and Big Ambergris Cay. The densest population of this species occurs on Big Ambergris Cay, a 400-ha island with an excellent variety of habitats and large populations of other squamates like *Cyclura*, *Anolis*, and *Leiocephalus*. Other small cays around the margin of the Caicos Bank are likely too small to support boa populations.

*Habitat*

Turks and Caicos boas occupy habitats ranging from dense closed-canopy tropical dry forest on North Caicos to scrub vegetation on Big Ambergris Cay and Great Inagua. They can also occasionally be found foraging in dunes or in rocky ironshore areas very close to the sea (Fig. 5.16). On larger islands they are frequently found near old wells or other freshwater sources in coppice forest or scrub; indeed, this is the type of situation in which individuals have recently been located on Great Inagua. They prefer rocky areas with dense vegetation, rocks laying on the surface to provide cover, or subterranean crevices and interstices in which to seek diurnal refuge (Fig. 5.17). Old rock walls overgrown with vegetation are also a favored habitat, and large flat rocks lying on the surface are commonly used as diurnal refugia providing a cooler temperature than the ambient air temperature (Reynolds and Gerber 2012). Large females often rest during the day at the base of thick bushes, partially under fallen leaves, such that they are nearly invisible to a passerby. They can occasionally be found underneath palm fronds, and will occupy sandy *Coccothrinax*-dominated forest, although at much lower density than in coppice forest. Radio-tracked boas have been observed foraging in bushes and shrubs, on the ground in open rocky areas, in coppice forest, in red mangroves (*Rhizophora mangle*), and in coastal wrack (Fig. 5.16). Juveniles are always found in arboreal situations, although they are most often seen when climbing in vegetation less than a meter or two high.

**Figure 5.16.** *Top images*: Typical diurnal resting areas beneath large shrubs and palms on Big Ambergris Cay. Radio-tracked adult female (>1 m SVL) *Chilabothrus chrysogaster* are hidden underneath the vegetation in these images. *Bottom left*: *Chilabothrus chrysogaster* will forage along the wrack line on Big Ambergris Cay. *Bottom right*: Radio tracking demonstrated that *C. chrysogaster* will forage in mangroves at night but will not spend the day there. Note the depth of the water (1 m) in the image; a juvenile blacktip shark was observed swimming just underneath the boa. Photos by R. Graham Reynolds.

**Figure 5.17.** *Top*: *Chilabothrus chrysogaster* capturing and consuming an adult male *Leiocephalus psammodromus* in the morning along a roadside on Big Ambergris Cay. *Bottom*: Prime habitat for *C. chrysogaster* on Big Ambergris Cay. Note the large rocks on the surface and native scrub vegetation. This habitat supports a high density of lizards. Photos by R. Graham Reynolds.

*Abundance*

Because the subspecies *C. chrysogaster relicquus* occurring on Inagua has not been systematically surveyed, we do not know how much of Great Inagua is occupied, nor do we have any information regarding abundance. Individuals can occasionally be found under survey effort by experienced herpetologists, although focused surveys have not been conducted, and the subspecies is considered quite rare (Sheplan and Schwartz 1974).

On larger islands on the Caicos Bank, Turks Island boas are generally uncommon, although some local populations of higher density are known from areas of closed-canopy tropical dry forest. The privately owned island of Big Ambergris Cay contains the best-studied population of the species and one of the longest sustained studies (annually for over 14 years as of 2021) of any species of *Chilabothrus* (Reynolds et al. 2020a). Over this time researchers have obtained data on over 1200 boas on Big Ambergris Cay, finding a slightly female-biased sex ratio ($\chi^2$ = 3.0376, $P$ = 0.08). This 400-ha island is estimated to have a population size of at least 4765 boas, or a density of about 12 boas per hectare—an almost unique population density of boas anywhere in the Caribbean. Dozens of boas can be found in a short nocturnal search on this island, with a maximum of 52 boas found in 3 hours of searching on one night in August (RGR, unpubl. data).

*Activity and Trophic Ecology*

Turks Island boas are well studied on the Caicos Bank. They are largely nocturnal, with occasional feeding activity observed in the early to midmorning or early evening. One individual, likely a male, ~600 mm SVL, was observed capturing and swallowing a male *Leiocephalus psammodromus* on the side of a road (Fig. 5.17). Boas can be active year-round, although more individuals are active on any given night during warmer periods. Large females are infrequently seen in cooler months and are much more readily found in late summer. Smaller individuals have been found to be active in nearly all weather conditions, including during tropical storms. A 20-month radio tracking study of *C. c. chrysogaster* on Big Ambergris Cay ($n$ = 19) showed that home ranges of adult females are between 0.7 and 1.6 ha on average (RGR, unpubl. data).

Foraging boas have a higher body temperature than ambient conditions, although males and females do not differ in body temperature when foraging (Reynolds and Gerber 2012). They are mostly terrestrial active foragers and search rock crevices slowly for sleeping lizards. Individuals are frequently observed stalking active lizards, such as *Aristelliger*, which involves slow movements toward the prey followed by long periods of motionless waiting and limited tongue flicking. A similar behavior is employed when searching for sleeping *Leiocephalus* and *Spondylurus*, although the boas move more frequently and poke their rostra in and out of crevices near the edge of the rock underneath which the prey is sleeping. Once prey is detected, the snake will occasionally thrash to push itself under the rock or cover object where the lizard is hiding. Touching a stalking boa on the anterior third of the body elicits a rapid lateral strike, while touching the boa near the tail results in fleeing behavior. The anterior third of the body is also the region where most scars occur on the boas, suggesting that their stalking behavior positions prey items to encounter this area on the boa in order to facilitate a rapid lateral strike followed by throwing the coils over the prey (Fig. 5.18). More rarely, boas have been observed lying motionless with their heads resting on rocks or rock walls, using limited to no tongue flicking (Fig. 5.18). This behavior is likely an attempt to capture the nocturnally active gecko *Aristelliger hechti*, which frequents rock walls. Although largely a terrestrial forager, on islands with larger trees the species is quite capable of climbing into the canopy (Reynolds and Manco 2013) and is known to feed on roosting birds such as Bananaquits (*Coereba flaveola*) or domestic fowl. Juveniles frequently stalk bushes and trees for sleeping anoles. Interestingly, this species is known to be attracted to lizard carrion, but only in the Big Ambergris Cay population. Road-killed iguanas sometimes draw boas in, and we have on consecutive nights seen two individual boas attempting to consume the same juvenile road-killed *Cyclura carinata*. Other boas have been observed consuming flattened and dried carcasses of both *Cyclura carinata* and *Leiocephalus psammodromus* on at least six occasions.

There are almost certainly age and sex effects related to trophic ecology in this species. Neonates are very small (<10 g) and arboreal and likely feed

**Figure 5.18.** *Chilabothrus chrysogaster* are highly saurophagous. *Top left*: Adult female consuming an invasive *Hemidactylus mabouia* on Big Ambergris Cay. Photo by Joseph Burgess. *Top right*: Adult male consuming an *Anolis scriptus*. Photo by Anonymous. *Bottom*: Ambush positioning of an adult male *C. chrysogaster*, awaiting an *Aristelliger hechti* to pass by on Big Ambergris Cay. Photo by R. Graham Reynolds.

on *Sphaerodactylus* geckos or on young *Anolis* lizards. Yearlings frequently occupy large rock piles where they likely feed on the abundant *Sphaerodactylus, Aristelliger*, and introduced *Hemidactylus* geckos (Reynolds et al. 2017, 2020b). Males are of medium size and are almost exclusively saurophagous, feeding on *Anolis, Aristelliger, Spondylurus*, and *Leiocephalus* lizards (Figs. 5.17, 5.18). Females are capable of growing large enough to feed on small *Cyclura* iguanas (up to 1 kg) and also readily consume birds, a diet that is facilitated by positive allometric growth of head length and width. This suggests that the species is engaged in ontogenetic niche partitioning related directly to body size and indirectly related to sex. Female boas have routinely been observed consuming hatchling and juvenile *Cyclura carinata*, and large females can capture and consume adult female and small male *Cyclura*. Adult male *Cyclura*, which can reach over 3 kg, are too large to serve as prey.

Nothing is known regarding the feeding ecology of Inagua boas, although presumably they resemble Turks Island boas.

*Predators and Defensive Behavior*

Predators of juvenile boas likely include night herons, which frequently hunt in the same areas where the boas forage. Other natural predators are unknown. External parasites are rare, and internal parasites are unknown. We have observed encysted cactus spines 4 mm in length inside the flank of an individual, although this type of injury is apparently infrequent. Adults frequently have scars likely imposed by bird beaks, although most are on the anterior half of the animal and thus suggest injuries sustained during prey capture. Many adults are missing the tip of the tail, although the reasons for this are unclear and could be related to either predation attempts or injury by shifting rocks.

Defensive behaviors include occasionally attempting to flee when approached, particularly when near a hole or crevice. Individuals in the open remain motionless instead of fleeing. Adults of this species are largely docile and rarely defensively strike when handled, although when they do, they usually only strike once and then resume attempts to flee. Most strikes are elicited when attempting to pick up a foraging boa by the anterior half of the body, and such strikes are probably predatory strikes and not defensive strikes (see above). Neonates can be irascible and will strike repeatedly. When handled the species will void the contents of the cloaca, which can constitute a copious amount of fluid in larger females. Juveniles and smaller individuals will coil into a ball when handled, although they will readily uncoil once placed back on the ground.

*Reproduction*

Like most *Chilabothrus* species, female Turks and Caicos boas are thought to reproduce every other year, likely owing to the necessity to accumulate sufficient energetic reserves for reproduction. Breeding likely takes place in the spring, and boas are particularly active from March through August in some populations. Courtship is unknown from the wild, but in captivity consists of "tail-searching" and physical contact between potential mates. The hemipenis of *C. c. chrysogaster* (KUH 260040) is 20 mm in length, with a bilobed sulcus and six scalloped flounces (Buden 1975). In *C. c. relicquus*, on a partially exposed hemipenis (KUH 260081), there are three flounces on the undivided asulcate surface with a basal papilla near the first flounce (Sheplan and Schwartz 1974).

Gestation likely requires between 225 and 250 days, and neonates are most frequently observed in the late fall and early winter between October and January (Tolson 1980; RGR, unpubl. data). Large females are not very active in colder weather from December to February. Litters consist of nine to twenty-eight young (Buden 1975; Tolson and Henderson 1993), born with a mean SVL of 290 mm and mean mass of 7.0–8.4 g. Parturition in one wild-caught female lasted for ~7.5 hours (Buden 1975). Neonates and young-of-year are very rarely observed in the wild—they are quite cryptic. Among over 1200 boas recorded during 14 years of fieldwork on Big Ambergris Cay, RGR and collaborators have found fewer than twelve neonates. A single litter can contain both striped and spotted individuals, although frequencies of the coloration in litters and the relationship to maternal pattern morph are at present unknown. This species will reproduce in captivity (J. Murray, in litt. to RGR), although raising young to adulthood is challenging. Captive lifespan of at least 14 years has been recorded (Snider and Bowler 1992).

*Conservation Status*

As one of the better-studied West Indian boas, a fair amount is known regarding the conservation concerns for the species. This species is listed on CITES Appendix II and is protected by the Bahamas Wild Animals Protection Act of 1968. It is listed as Near-Threatened on the IUCN Red List (Reynolds and Buckner 2021). The single largest threat to the species appears to be feral cats, which will readily consume them. They are also routinely killed by humans, and most human encounters result in the death of the snake. A host of other perturbations might also influence the survival of the species, including development and road mortality. Snakes are routinely found dead on roads, as they cross roads slowly at night and their narrow bodies are hard for motorists to discern. The road crossing North and Middle Caicos is particularly known for snake mortality (RGR, pers. obs.). During the peak of development on Big Ambergris Cay from 2006 to 2008, boas were routinely found dead on sand roads.

This species can tolerate some development as long as it retains native vegetation, loose rocks, and a relatively low density of feral cats. For example, boas are infrequently found in and around the extensive Grace Bay development as well as around the heavily developed Leeward neighborhood and golf course on Providenciales, Turks and Caicos. They can also be readily found in gardens on North and Middle Caicos. Nevertheless, the long-term survival of those populations is questionable, as gardeners and groundskeepers rou-

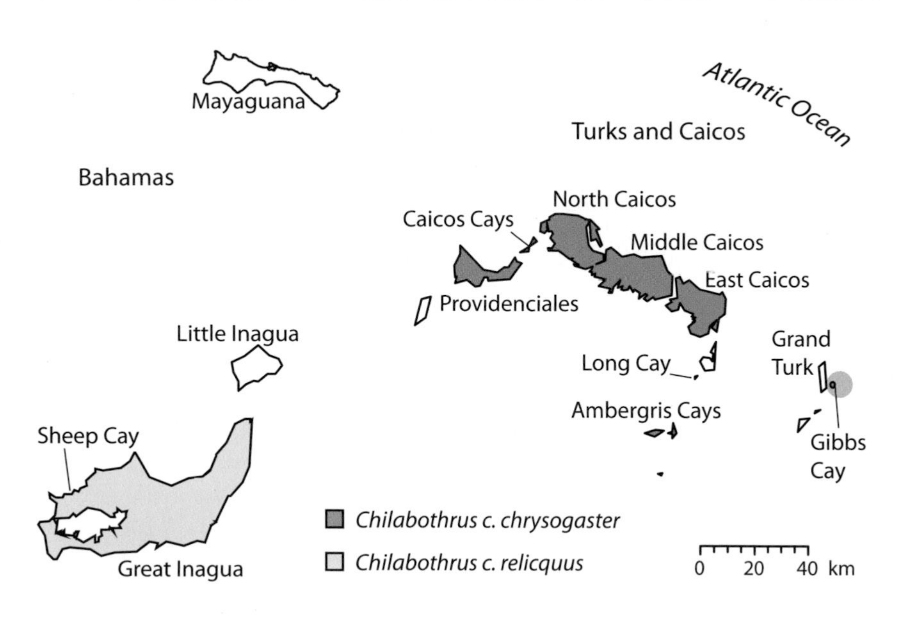


**Figure 5.19.** Range of *Chilabothrus chrysogaster* in the Bahamas and Turks and Caicos islands. *Chilabothrus c. relicquus* (yellow) is found in the Bahamas on Great Inagua Island and Sheep Cay. *Chilabothrus c. chrysogaster* (green) is found on nine islands on the Caicos Bank and one island (Gibbs Cay) on the Turks Bank.

tinely kill snakes, and road and cat mortality likely combine to drive populations toward extirpation.

The densest population of this species, and among the densest of any West Indian boa species, occurs on Big Ambergris Cay, on the southeastern end of the Caicos Bank. This island is undergoing development into vacation properties and has experienced a dramatic amount of construction and laying of roads and other infrastructure over the last two decades. Thus far, the boa population seems to have withstood this disturbance, but the development plans for the island include eventual construction of hundreds of homes. The 2008 Global Financial Crisis caused a ~10-year pause in development, which likely greatly benefitted the boas. Indeed, we have found a much higher frequency of large females over the last 10 years (2010–2020) than the decade preceding (2004–2010), when development was occurring rapidly and hundreds of people were on the island each day. If the planned development scenario unfolds, the boa population will likely dramatically suffer from increased human activity and disturbance. No invasive predators were established on Big Ambergris Cay as of March 2018, although despite some biosecurity measures rats and mice apparently arrived in the summer of 2018. A single feral cat was humanely captured and removed from the island in 2008.

On the Caicos Bank, the species has been extirpated from South Caicos, and possibly from Long Cay, Pine Cay, and the Water Cays (Reynolds 2011c, 2011d). Ongoing conservation work by the authors and others is examining the feasibility of reintroducing populations to some islands following invasive predator removal. However, there is an alarming uptick in the number of introduced species of reptiles and amphibians on the Caicos Bank (Reynolds 2011a, 2011b; Reynolds and Niemiller 2011; Reynolds and Riggs 2011a, 2011b; Reynolds et al. 2011a, 2011d, 2011e). On the Turks Bank, the species is extirpated from Grand Turk but has never been reported (or surveyed) on islands other than Grand Turk and Gibbs Cay.

Next to nothing is known of *C. c. reliquus* on Great Inagua, although this subspecies is likely impacted by similar disturbances such as cats and development. On Inagua, industrial development for salt production has driven most of the ecological disturbance around Matthew Town, although feral grazing mammals have damaged the native vegetation in the interior of the island. A large portion of the island is contained in the Inagua National Park, hence the boa population is likely spared from extensive development, although vegetation changes and feral cats (and grazers) could still cause significant impact on this subspecies. Figure 5.19 shows the distribution of *Chilabothrus chrysogaster.*

## ABACO ISLANDS BOA, NORTHERN BAHAMAS BOA

*Chilabothrus exsul* (Netting and Goin 1944)

### *Taxonomy*

This species was described in 1944 following the collection of a specimen in 1942 near Blackwood (= Blackrock) Settlement north of Treasure Cay, Abaco Island, Bahamas (Netting and Goin 1944). Like other Lucayan boas, the species is derived from Hispaniolan ancestors, although it is not yet clear if it evolved in situ on the Bahamas islands. The species is closely related to *C. schwartzi* on the Crooked-Acklins Bank and *C. argentum* on the Conception Island Bank. There are no recognized subspecies.

### *Etymology*

The species' epithet is Latin for *exiled*, although the original description did not provide a reason for choosing the name. Presumably, it is because the species was thought to be restricted to the Little Bahamas Bank, and thus away from the much larger Great Bahamas Bank. The species is often referred to locally as "fowl snake" and is commonly (incorrectly) thought to be the same species that occurs elsewhere in the Bahamas.

### *Description*

This species is one of the smallest members of the genus and has the second shortest maximum SVL of any West Indian boa, behind *C. ampelophis*. The maximum size for a female is 810 mm SVL, and maximum male size is 566 mm SVL, although a male boa captured 8 August 2015 by RGR near Cherokee, Abaco Island, measured 635 mm SVL and represents a new size record. Mass in life is largely unknown—a gravid female weighed 234 g and gave birth to nine offspring and had two unfertilized ova and a stillborn offspring, all weighing a combined 62 g. Sexual size dimorphism is thought to exist, with females obtaining a slightly larger body size (length and mass) than males. Neonates are known to vary between 231 and 280 mm SVL and 4.5 and 7.3 g at parturition. There is one loreal scale, ventrals are 236–251, subcaudals are 69–75, supralabials are 13, and circumorbitals are 10 (Table 5.1).

The dorsal and lateral scales are infused with mottled gray-to-dark brown pigment and taper to a somewhat narrow point, with sharper points on lateral scales relative to dorsal scales. Some scales consist of several colors, with darker pigment deposited either distally, proximally, or evenly. Most adult *C. exsul* are gray or brownish dorsally with an unpatterned (aside from some light gray "freckling") and cream-colored venter

**Figure 5.20.** Adult male *Chilabothrus exsul* from near Cherokee, Great Abaco, exhibiting a typical adult coloration. Photo by R. Graham Reynolds.

(Fig. 5.20). Some individuals have a browner color middorsally, which fades to gray on the flanks. In the direct sunlight this gives a sort of burnt copper sheen (Fig. 5.21). Regular and well-defined ovoid or saddle-shaped markings are present on the middorsum, consisting of dozens of patches of darkly pigmented scales that contrast with the dorsal background color. Smaller dark circular or diamond patches occur irregularly along the flanks, and occasional single dark scales are found across the dorsal surface. Rarely, orangish red scales are found scattered across the body, often on the flanks near the tail. This coloration and color pattern continues uninterrupted to the tip of the tail, which tapers to a very fine point. The tail is not prehensile and does not coil tightly. The pupil is black, and the iris is dark orange in juveniles and mottled grayish-brown to reddish-brown in adults.

Some variation in coloration has been documented recently, including an unpatterned male (525 mm SVL; RGR, pers. obs.) found dead on Queen's Highway south of Marsh Harbour. The persistence of reddish pigment on some scales suggests that individuals with red dorsal coloration might exist. If so, this would be an example of incomplete ontogenetic color change, which is documented in other species (see *C. chrysogaster* account). A single albino or leucistic individual was found in the wild on one of the Abaco Cays in 2006—apparently a very large adult female based on blurry photographs taken in situ and examined by RGR and S. T. Giery. Like some other *Chilabothrus* species, juveniles undergo dramatic ontogenetic color change, from a light orange coloration with lots of patterning (Fig. 5.22) to a silver-gray adult coloration with similar

**Figure 5.21.** *Top*: Adult female *Chilabothrus exsul* from near Cherokee, Great Abaco, exhibiting a browner coloration than is typical in the species. Photo by R. Graham Reynolds. *Bottom*: Typical adult female coloration from Little Harbour, Abaco. Photo by Sean T. Giery.

**Figure 5.22.** Neonate *Chilabothrus exsul* found crossing a road near Little Harbour, Great Abaco. Photo by R. Graham Reynolds.

dorsal patterning. Abaco boas over 411 mm SVL had already transitioned to gray or brownish dorsal coloration. Like most species in the genus the eyes are noticeably large relative to the head size in juveniles and appear to "bulge" out of the head when viewed from above. The tongue is black with white tips.

*Distribution*

This species is endemic to the eastern end of the Little Bahama Bank, where most of the range is on the Abaco Islands (= Great Abaco plus Little Abaco islands). *Chilabothrus exsul* is also found on several satellite islands to the east of Abaco, including Green Turtle Cay, Man-o-War Cay, Sandy Cay, Elbow Cay, and Tilloo Cay (Schwartz and Henderson 1991; Buckner et al. 2012; Krysko et al. 2013). The species has been suggested to occur on the eastern end of Grand Bahama Island (Sheplan and Schwartz 1974; Tolson and Henderson 1993), and a captive specimen at the Rand Nature Center in Freeport was anecdotally captured on Grand Bahama, although we have been unable to generate additional information on that specimen. To our knowledge no other records or photos exist for boas from Grand Bahama, although formal surveys for this species have never been conducted there.

*Habitat*

Abaco Islands boas, while popularly associated with Caribbean Pine (*Pinus caribaea*) forest, actually prefer coppice forest and reach much higher abundance in this type of habitat. Coppice forest on Abaco is largely restricted to the coastal zone along the eastern edge of the island and is replaced by pine forest and then mangrove marl moving to the west. In coppice forest, boas prefer rocky limestone hillsides, where they presumably retreat into crevices and underground interstices during the day. Despite this preference for coppice, the species can be found in pine forest but is frequently localized to "islands" of exposed limestone that dot the pineland landscape, and boas are much easier to find following a forest burn of the understory scrub (RGR, pers. obs.). The species is not known to frequent mangrove forest, although they occasionally occur within a few meters away from mangroves on the cays or on the eastern end of Abaco, particularly on limestone hillsides that run down to the shore. The species is apparently able to persist in pockets of coppice surrounded by urbanization, as it remained common (prior to Hurricane Dorian) in and around the settlements of Marsh Harbour and Treasure Cay. In addition to natural cover objects such as limestone slabs, these boas can be found in piles of wood, stacks of PVC pipes, under debris such as swept-up palm fronds, and in old cisterns (Tolson and Henderson 1993; RGR, pers. obs.). Old rock walls are also a favored habitat, particularly in coppice forest, and the species will also use crevices under concrete slabs and sidewalks. Netting and Goin (1944) encountered *C. exsul* emerging from cavities in coral to warm in the sun at the margin of a clearing on Great Abaco.

*Abundance*

No estimates of abundance have been published, and the species is generally considered to be rare and patchily distributed. Sheplan and Schwartz (1974) attempted to find individuals via road surveys, finding only a single dead individual. Reynolds et al. (2016b) found a total of eleven boas over the course of 1420 km of road transect surveys across Abaco Island, an encounter rate of 0.008 boas/km. Dedicated surveys on foot are far less productive, although Netting and Goin (1944) suggested that three boas were readily found basking on coppice outcrops in clearings in virgin Caribbean pineland. No other abundance estimates or encounter rates have been published, and no resurveys have occurred since Hurricane Dorian in 2019 at the time of this writing.

*Activity and Trophic Ecology*

This species is active nocturnally and seeks prey by moving slowing along the ground poking its head in and out of crevices. It is thought to be largely terrestrial, and no published records of arboreal behavior or climbing exist (despite a Bahamian postage stamp portraying arboreal behavior). Nevertheless, some photos of boas resting in trees suggest that they do climb (S. Buckner, in litt. 19 May 2020). Potential prey items include curly-tail lizards (*Leiocephalus carinatus*), *Anolis* lizards, and possibly rodents (*Mus musculus.*, *Rattus* sp.). They might also take birds, and a likely candidate prey source (historically) would be nests of the ground-nesting Abaco parrot (*Amazona leucocephala bahamensis*). An adult was observed consuming a Cuban treefrog (*Osteopilus septentrionalis*) on Great Abaco (S. Buckner, in litt. 19 May 2020),

**Figure 5.23.** Young adult male *Chilabothrus exsul* crossing a road running through coppice habitat on Great Abaco, a behavior that probably results in the deaths of dozens to hundreds of boas per year. This species of boa is so small that motorists likely do not notice them. Photo by R. Graham Reynolds.

**Figure 5.24.** Range of *Chilabothrus exsul* on the Little Bahama Bank. The species is found on Great and Little Abaco islands, as well as several small islands east of Great Abaco. Boas are rarely found in the sandy and mangrove areas of western Great Abaco.

suggesting that the boas are not impacted by the frog's skin toxins. Like many *Chilabothrus* species, juveniles likely consume *Anolis* lizards.

*Predators and Defensive Behavior*

No natural predators are recorded, and external and internal parasites are unknown. It is likely that diurnal and nocturnal birds of prey would take boas. Introduced feral cats are known to consume boas, particularly around urban areas, and feral pigs might also eat them.

Because no predation attempts have been observed, there is limited information on defensive behaviors. This species does not move when approached at night and never strikes when handled, although like most boas they void the contents of the cloaca when first picked up. Juveniles will ball up tightly when handled. When approached during the day near a hide, the species will retreat into the hide, although this usually does not happen immediately.

*Reproduction*

Like the majority of boas, this species is live-bearing. Like other *Chilabothrus* species, mating takes place in the spring and parturition in the summer, with reproduction in wild females probably occurring biennially (Huff 1978). Male–male combat has been observed in captivity, consisting of "pinning" behavior, and courtship consists of tactile-chase and tail searching behavior (Tolson 1992b; Tolson and Henderson 1993). Boas in the genus *Chilabothrus* appear to follow two reproductive patterns, with smaller species like *C. exsul* having shorter gestation times and smaller litters relative to larger species, which have longer gestation times and larger litters. *Chilabothrus exsul* is thought to have a gestation time of between 147 and 159 days with a litter size of two to eleven neonates (Tolson and Henderson 1993). Dead neonates and infertile ova are known to be consumed by the female following parturition (Tolson and Henderson 1993). This species reproduces readily in captivity and has been recorded to live over 21 years in captivity (Snider and Bowler 1992). Schwartz and Henderson (1991) reported probable thermoregulation by gravid females beneath flotsam and debris at the high tide mark on Great Abaco.

*Conservation Status*

This species is listed on CITES Appendix II and is protected by the Bahamas Wild Animals Protection Act of 1968. It is considered Vulnerable by the IUCN Red List (Reynolds and Buckner 2016) owing to invasive vertebrate predators such as cats, human persecution, loss of preferred coppice habitat, and road mortality (Fig. 5.23). A study conducted in 2015 on Abaco Island determined that adult and juvenile boas were routinely killed by motorists, resulting in an average of 1.6 dead boas per week in an area with high boa densities and possibly representing a loss of hundreds of boas per year across Great Abaco Island (Reynolds et al. 2016b). No extirpations are known, although Netting and Goin (1944) suggested that boas were extirpated on Stranger's Cay by feral pigs; this appears to have been an extirpation of the racer *Cubophis vudii*, as boas were never conclusively reported from that island. A portion of the range is protected in the Abaco National Park and the Tilloo Cay Reserve. Figure 5.24 shows the distribution of *Chilabothrus exsul*.

## CROOKED-ACKLINS BOA

*Chilabothrus schwartzi* (Buden 1975)

*Taxonomy*

This little-known species was recently elevated from a subspecies of the Southern Bahamas boa (*C. chrysogaster*) complex. Previously designated *C. chrysogaster schwartzi*, the species had not been reported from a living specimen in the literature prior to 2017 and was only known from four museum specimens collected postmortem by Buden (1975) and from anecdotal reports. Both morphological and molecular data demonstrate that the Crooked-Acklins boa is not a subspecies of *C. chrysogaster* (Reynolds et al. 2018); instead, it is a distinct species endemic to the Crooked-Acklins Bank and is phylogenetically sister to *C. argentum*.

*Etymology*

The species epithet is a patronym honoring the late Albert Schwartz, a major contributor to Caribbean herpetology. Like most boas in the Lucayan Archipelago, this snake is known colloquially as the "fowl snake," although it is rarely encountered by locals.

*Description*

Crooked-Acklins boas are known from only eight documented specimens (only five complete specimens), so generalizing about body sizes and color variation is difficult. The largest specimen was reported to be "4–5 ft" (1.2–1.5 m) in Buden (1975), although only the head was retained for a museum specimen (KUH 260084). Two reliable measurements of sexually mature snakes (both females) are 774 mm SVL (mass 95 g) and 785 mm SVL (Louisiana State University Museum of Natural Science [LSUMZ] 27500, mass unknown) with tail lengths of 168 mm and 175 mm, respectively. Hence, we presently consider this species to be among the smaller West Indian boas, although anecdotal reports suggest that females can reach >1 m SVL. Only three juveniles have been observed, ranging from 523 to 576 mm SVL, with a tail length of 116–130 mm and mass of 30.5 to 40.5 g. This is a somewhat larger size than juveniles of other closely related species, such as *C. chrysogaster*, which undergoes ontogenetic color change prior to reaching 400 mm SVL. Like other members of the genus, the tail tapers grad-

**Figure 5.25.** *Top*: *Chilabothrus schwartzi* (LSUMZ 27500) from Crooked Island, the only intact museum specimen of this species. Note that the saddle-shaped bars on the dorsum have faded with preservation and that the rufous head coloration is also not well preserved. Photo by R. Graham Reynolds. *Bottom*: Adult female *C. schwartzi* from Crooked Island, exhibiting an unusual lack of dorsal patterning and little rufous coloring on the head. Photo by Joseph Burgess.

ually to a very fine point. The scales are infused with mottled pigment and taper to a relatively sharp point, especially on the flanks, but are more rounded on the dorsum. There are 1–2 loreal scales, and squamation is similar to other Lucayan boas; ventrals are 277, subcaudals are 95, supralabials 12–13 supralabials, and there are 9–10 circumorbitals (Table 5.1).

Adults and juveniles have a cream-colored venter and a faint gray mottling near the base of the ventral scales. Dorsal pattern is usually a series of irregularly shaped saddles of more darkly pigmented scales against a gray background color, transitioning from blotches anteriorly to distinct saddles reaching completely across the dorsum (contacting the ventral scales) at the midsection (Figs. 5.25, 5.26). These saddles are the most distinct pattern elements, with pattern fading toward the anterior and posterior; are dark brown in color; and contrast with the light gray dorsal background color. Unlike the dorsal patterns in other Lucayan species, these patterns are not bordered by darker scales but instead consist of uniformly colored scales. One adult largely lacked any dorsal pattern, suggesting some variation in patterning might be present (Fig. 5.25). The head has a distinct (likely diagnostic) rufus coloration on top in both juveniles and adults, and there is no postocular stripe or other head patterning as seen in other *Chilabothrus*. The pupil is black, and the iris is a mottled silver gray with brown streaks; in juveniles the iris has an orangish tint near the pupil. The tongue is black with white tips.

Like some other *Chilabothrus* species, juveniles

**Figure 5.26.** Juvenile *Chilabothrus schwartzi* from Crooked Island. Note the lighter coloration that will fade with age. The rufous head is apparent. Photos by Alberto R. Puente-Rolón.

**Figure 5.27.** Juvenile *Chilabothrus schwartzi* foraging in low vegetation on Crooked Island. This individual is showing transition to adult coloration. Photo by Alberto R. Puente-Rolón.

undergo ontogenetic color change, from dull burnt-orange to orangish-brown juvenile coloration to a light-gray adult coloration (Fig. 5.26). Juveniles observed in the field (SVL 523–576, $n$ = 3) displayed the juvenile coloration, while adults (SVL 774–785, $n$ = 2) were gray (Buden 1975; Reynolds et al. 2018). Thus, ontogenetic color change likely occurs around 600–700 mm SVL, although this could vary among populations.

*Distribution*

Crooked-Acklins boas are currently known from only two islands—Crooked Island (252 km$^2$) and Acklins Island (497 km$^2$) on the Crooked-Acklins Bank in the central Bahamas Archipelago. The Crooked-Acklins Bank supports a number of smaller islands, including Fish Cay (0.88 km$^2$), the Guana Cays (0.3 km$^2$), and Castle Cay (1.93 km$^2$), among others (Buckner et al. 2012). Surveys have concluded that the species does not occur on Fish Cay in the Bight of Acklins (Reynolds et al. 2018), although surveys of other islands and cays could potentially reveal additional populations.

*Habitat*

Very little is known about the Crooked-Acklins boa, but they likely prefer coppice forest or scrub upland habitat, also known as whiteland or blackland coppice, with a rocky substrate or an abundance of rocks lying on the surface. Recently described individuals on Crooked Island (Reynolds et al. 2018) occupied a hillside consisting of coppice scrub dominated by larger shrubs and small trees over abundant large, loose rocks. The hill was surrounded at the leeward base by taller-canopy coppice forest with thick vegetation and undergrowth. Other individuals have been found near human settlement on both Crooked and Acklins islands (Buden 1975), and we have heard (infrequent) reports of locals encountering larger snakes in gardens. This suggests that, like other Lucayan boas, the species might readily use rock walls that line gardens.

Juveniles of this species have been found only in arboreal situations (Fig. 5.27), actively foraging at night in shrubs and trees 1–2 m off the ground (Reynolds et al. 2018). The only live adult described in the literature was found on the ground, actively moving at night.

*Abundance*

No estimates of abundance are available, and this species is known from only a handful of specimens. Records of the species include six individuals from three localities on Crooked Island and individuals from two localities on Acklins Island. Only two bouts of records have been recorded in the literature separated by over four decades—four observations from 1972 (Buden 1975) and four observations from 2017 (Reynolds et al. 2018). It is likely a rare species, although it can apparently be readily found if ideal habitat is located during the summer months.

*Activity and Trophic Ecology*

All observations of live individuals of this species have occurred at night (Reynolds et al. 2018). The boas were active and likely foraging, indicating that it is probable that, like other members of the genus, they are nocturnal. In total, four live individuals have been recorded in the literature, although we are aware of several anecdotal reports and photographs of this species.

Little is known directly about the feeding ecology of this species. Given the observed arboreal foraging of juveniles it is likely that they, like other juvenile *Chilabothrus*, actively hunt sleeping *Anolis* lizards on branches. Reynolds et al. (2018) report that at least one juvenile had a recently ingested anole meal, although the species of anole was undetermined but guessed to be *A. sagrei* based on the numbers of that species of anole found at the site. Two species of anoles are found on the Bank, the larger and more arboreal *A. brun-*

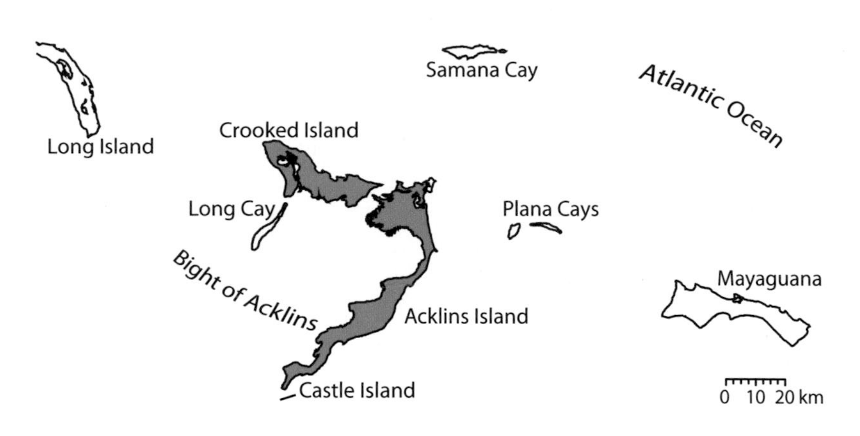


**Figure 5.28.** Range of *Chilabothrus schwartzi* on the Crooked and Acklins islands in the southern Bahamas. The species is not recorded from cays in the Bight of Acklins or from Castle Island.

*neus* (SVL 70–76 mm; Schwartz and Henderson 1991; Henderson and Powell 2009) and the relatively smaller *A. sagrei* (SVL 25–48 mm in the Crooked Island population [RGR, unpubl. data]). Adult diet is unknown, although it is presumably like that of the similarly sized terrestrial Lucayan species *C. exsul* and *C. chrysogaster*, which would be largely saurophagous.

*Predators and Defensive Behavior*

Natural predators are unknown, although likely include wading birds and raptors. Predation on members of the genus by native species is largely undocumented (Tolson and Henderson 1993; Henderson and Powell 2009). This species is almost certainly preyed upon by the numerous feral cats on the islands. Parasites are unknown in this species.

*Reproduction*

Nothing is known of the reproductive biology of this species, although it is presumably live-bearing and likely resembles the reproductive biology of other members of the genus (see *C. chrysogaster* account). We do not currently know size at parturition nor litter size. The most closely related species, *C. argentum*, is also lacking any information on reproductive biology.

*Conservation Status*

Only four live Crooked-Acklins boas have been reported in the literature (Reynolds et al. 2018), and an additional four were collected as freshly killed specimens and accessioned into museum collections (Buden 1975). Thus, few predictions can be made regarding the conservation status of the species. Reynolds et al. (2021) assessed the species as Endangered on the IUCN Red List. Given the infrequency of occurrence, coupled with interviews with locals, it is probable that the species is quite rare across its range, particularly in areas with human habitation. Nevertheless, in appropriate undisturbed habitat the species can apparently be occasionally found during the summer months (Reynolds et al. 2018). It is not clear how much boa habitat exists across the Crooked and Acklins islands, and it is even less clear what constitutes ideal habitat for this species. It is likely that the boas need to be insulated from human persecution, forest destruction, rock removal, and invasive predators such as cats to persist in relatively high densities. As this is based on research done at a single site on Crooked Island (Reynolds et al. 2018), this generalization might not be extensible to the rest of the range of the species. Other members of this genus that are sensitive to disturbance require similar conditions (e.g., Reynolds and Gerber 2012). This species was known to be present in the U.S. pet trade, although apparently it was represented in a single private collection by an adult male and adult female. No offspring appear to have circulated. Like other boa species, poaching for the pet trade is a mild concern for the longevity of wild populations of boas (Dodd 1986), although the Bahamian government takes a strong stance on prosecuting illegal wildlife trade (Isaacs 2014). At least one likely road-killed boa has been observed near Colonel Hill Airport on Crooked Island (J. Iverson, pers. comm.). This species is protected by the Bahamas Wild Animals Protection Act of 1968. Figure 5.28 shows the distribution of *Chilabothrus schwartzi*.

## BAHAMAS BOA, BAHAMIAN BOA

*Chilabothrus strigilatus* (Cope 1862)
Cat Island boa (*Chilabothrus strigilatus ailurus*) Sheplan and Schwartz 1974
Bimini boa (*Chilabothrus strigilatus fosteri*) Barbour 1941
Andros boa (*Chilabothrus strigilatus fowleri*) Sheplan and Schwartz 1974
Ragged Islands boa (*Chilabothrus strigilatus mccraniei*) Sheplan and Schwartz 1974
Bahamas boa (*Chilabothrus strigilatus strigilatus*) Cope 1862

### *Taxonomy*

This species was originally described as *Homalochilus strigilatus,* a distinct lineage of West Indian boas, by Cope (1862). Various populations were then placed into the Hispaniolan *Epicrates striatus* by Stull (1935) and by Sheplan and Schwartz (1974), the latter work subsuming all Bahamas populations into *Epicrates striatus.* Reynolds et al. (2013a) elevated the specific epithet *strigilatus* to refer to all boa populations on the Great Bahamas Bank, recognizing their uniqueness due to genetic, morphological, and geographic differences from Hispaniolan boas. This nomenclature was further codified by Pyron et al. (2014) and Reynolds and Henderson (2018). Bahamas boas are presently represented by five subspecies recognized by Sheplan and Schwartz (1974) owing to apparent differences in coloration and squamation, although genetic data might not be able to resolve differences among the subspecies (Reynolds et al. 2013a), suggesting limited divergence of these lineages. Molecular phylogenetic data suggest (although are not yet conclusive) that the species colonized the Great Bahama Bank in the Pleistocene, and the Silver, Mouchoir, and Navidad Banks (now submerged) might have facilitated colonization from Hispaniola. Earlier molecular phylogenies suggest that the species appears to have colonized the Bahamas separately from other Lucayan boas (Reynolds et al. 2013a), although recent analyses point to a single colonization of the Bahamas by an ancestral lineage that gave rise to all but *C. chrysogaster* (RGR, unpubl. data; *C. chrysogaster* definitely colonized the region separately).

### *Etymology*

Cope (1862) does not reveal the reason for selecting the species epithet, but it seems plausible that it is from a type of clay or stone decoration called strigilature, which resembles curved stripes or streaks. Hence we interpret the name to reference the diagnostic color pattern difference that distinguishes most Bahamas boas (*C. strigilatus*), which have a dorsal stripe on the anterior portion of the body (Fig. 5.29), from Hispaniolan boas (*C. striatus*), which usually do not. This species is colloquially referred to as "fowl snake" in the Bahamas owing to its proclivity for predation on domesticated birds. The subspecific epithet *ailurus* is from the ancient Greek for cat and refers to the population of boas endemic to Cat Island, Bahamas. The subspecific epithet *fosteri* is a patronym honoring R. W. Foster, who secured the holotype of the subspecies from North Bimini, Bahamas, in 1941. The subspecific epithet *fowleri* is a patronym honoring D. Fowler, who secured the holotype of the subspecies from Andros Island, Bahamas, in 1970. The subspecific epithet *mccraniei* is a patronym honoring J. R. McCranie, who secured the paratypes of the subspecies from Little Ragged Island, Bahamas, in 1970.

### *Description*

This species is a large-sized member of the genus *Chilabothrus.* Adult females can reach just over 2180 mm SVL and 2300 g in mass, although female sizes around 1000–1600 mm SVL and 330–1200 g in mass are far more typical. Males can reach 2330 mm SVL, but most males are far smaller. Sexual size dimorphism is minimal and not dramatic—males are capable of reaching a similar maximum SVL as females, but females tend to be much more massive (in weight). Neonates are born between 318 and 500 mm SVL and 12 and 20 g in mass (Tolson and Henderson 1993; Reynolds et al. 2016a). It does not appear that differences exist among the subspecies or populations in terms of their potential maximum body sizes. Indeed, the largest *C. strigilatus* recorded at 2330 mm SVL (*C. s. fosteri*; presumed to be Museum of Comparative Zoology, Harvard University [MCZ] R-46178; Sheplan and Schwartz 1974; measured as 2184 mm by RGR in 2012) was collected from either North or South Bimini. Nevertheless, the ability of a boa to live well into adulthood can be diminished in areas with higher predation or human distur-

**Figure 5.29.** Young adult male *Chilabothrus strigilatus fosteri* from mature blackland coppice forest on South Bimini. Photo by R. Graham Reynolds.

bance. It is generally thought that some of the largest boas presently occur on Long Island, although this is anecdotal and has yet to be demonstrated. Truly large boas (>2000 mm SVL) are regularly reported from Andros (*C. s. fowleri*; e.g., Knapp and Owens 2004) and Long Island (*C. s. strigilatus*; RGR, pers. obs.), and large females (1500 mm SVL) are commonly found on the heavily developed island of New Providence (*C. s. strigilatus*). A recent study on Eleuthera (S. Hoefer, pers. comm. to RGR) has documented some large individuals, including a 2180 mm total length [TL] (~2000 mm SVL) female found dead on the road.

The tail tapers to a fine tip, especially in juveniles, and is not prehensile in adults, although it is slightly capable of coiling in juveniles. Many adult snakes have lost the tip of the tail, and like other Lucayan *Chilabothrus*, larger individuals frequently have damaged or missing tail tips. Tail tips heal readily and do not appear to impede the animals. Males frequently have relatively large spurs, and recent focused work has revealed that spur size is a diagnostic trait (Hoefer et al. 2021). It is unknown whether male tail length can be used to diagnose sex.

Bahamas boas have some distinct squamation characteristics from other Lucayan and Hispaniolan boas, including a single loreal scale and modally lower counts for intersupraocular, pre-intersupraocular, supralabials, and circumorbitals. The ventral and subcaudal scale counts are highly variable and overlap with those of *C. striatus* (Sheplan and Schwartz 1974). The dorsal scales of adults are mildly tapering, with a more rounded distal tip than those of other Lucayan boas, while middorsal scales of juveniles can taper to a point.

The pupil is black and the iris is light gray with a variable mottling of black, brown, and reddish orange. The tongue is gray with white tips. Neonates are born with similar coloration to the adults, although changes in the relative amounts of color elements can shift the appearance of adults and juveniles. For example,

**Figure 5.30.** Young *Chilabothrus s. strigilatus* showing variation in juvenile coloration. *Top*: A juvenile from Staniel Cay in the Exumas. Photo by Anthony Geneva. *Bottom*: A juvenile *C. s. strigilatus* from Long Island. Note the more reddish coloration. Photo by R. Graham Reynolds.

**Figure 5.31.** Adult female *Chilabothrus s. strigilatus* (~1.5 m SVL) foraging in blackland coppice on a hillside on Long Island. Photo by Joseph Burgess.

juvenile Bimini boas can appear almost black from a distance. Although the species is not thought to display ontogenetic color change, juvenile *C. s. fowleri* on Andros and *C. s. strigilatus* on Long Island can be a deep brick-red color that will fade to a grayer coloration as they age, while other populations do not display such color change (Fig. 5.30).

Generally, coloration in the Bahamas boa consists of a cream- to gray-colored venter with some mottling, a light- to dark-gray dorsal background color that transitions from light to dark from the venter to the middorsum, and a series of large darker blotches or stripes laterally and dorsally, usually containing lighter scales in the blotches (Fig. 5.31). Coloration is highly variable, although between-population color variation is more apparent than most within-population variation. Ventral colors might also vary by age, with older individuals seeming to have more freckling or mottling on the venter (Fig. 5.32), although not nearly as much as some *C. striatus*. Because of their iridescence, even darkly colored boas will appear lighter during nighttime flash photography or under an LED lamp, so viewing the animals in life during the day is the best way to see the color pattern differences. Similarly, boa color patterns can start to fade not long after death. Subspecies of Bahamas boa are mostly defined based on geography and dorsal color and color pattern (Sheplan and Schwartz 1974). We discuss subspecific variation below.

*Chilabothrus strigilatus ailurus* (Fig. 5.33) is generally characterized by having a strongly accented lateral neck stripe composed of two to three dark scales surrounded by light-gray scales. The dark scales of the middorsal blotches are often fused, while toward the

**Figure 5.32.** Ventral color variation in *Chilabothrus strigilatus*. *Left*: Venter of an adult female *C. s. ailurus* from Cat Island. Photo by R. Graham Reynolds. *Right*: Venter of a juvenile *C. s. strigilatus* from Staniel Cay, Exumas. Photo by Anthony Geneva.

**Figure 5.33.** Adult female *Chilabothrus strigilatus ailurus* from Cat Island, profile and close up of scalation and lateral color pattern. Photo by R. Graham Reynolds courtesy of Mark Keasler.

tail the light scales of the blotches fuse, producing the appearance of a dorsal stipe. No other Bahamas boa population has individuals with a light-colored posterior dorsal stripe extending onto the tail. Cat Island boas also frequently appear lighter in overall color—the dorsal scales are browner to brownish gray and the lateral scales are very light gray. These boas are distinctly lighter in overall color, particularly when compared with boas from Bimini, which represent the other end of the spectrum and are noticeably dark. The venter is frequently more darkly mottled than other subspecies.

*Chilabothrus strigilatus fosteri* (Fig. 5.34) is well known as a darkly colored subspecies, owing to the near fusion of blotches composed of darkly pigmented scales on the dorsum. The effect is to produce what looks like a dark, almost black, wide stripe on the middorsum, particularly from a distance. Up close the pattern is much more variable, and the scales forming this pattern are dark brown to dark gray, rarely black. The lateral neck stripe can be less pronounced than in *C. s. ailurus* and disintegrates relatively quickly into lateral blotches or may also fuse in places with the dorsal dark patterns. Dorsal blotches can consist of dark scales edged by lighter-colored scales or spots of light scales surrounded by dark scales, as is more typical in other subspecies. The lateral flanks have large dark blotches or patterning, and there is much less area covered by lighter scales as in other subspecies. Nevertheless, some individuals have very light-colored scales on the flanks, creating an almost black-and-white pattern that is particularly striking (see illustration 47 in Tolson and Henderson 1993).

*Chilabothrus strigilatus fowleri* (Fig. 5.35) coloration is similar to that of *C. s. fosteri*, although appearing more brown than black, and generally consists of a very dark overall appearance due to the fusion of the middorsal dark blotches. The main diagnostic characteristic, according to the describers (Sheplan and Schwartz 1974), is that some faint separation in the form of light scales exists between middorsal blotches and that the lateral neck stripe is extremely short or even not apparent, as it rapidly degrades into mixed pattern elements. Because of the variation present in these populations, it is our opinion (and experience) that distinguishing between *C. s. fowleri* and other *C. strigilatus* subspecies without knowing their origin can be difficult.

**Figure 5.34.** Adult female *Chilabothrus strigilatus fosteri* from South Bimini. Photo by R. Graham Reynolds.

**Figure 5.35.** Ontogenetic color change in *Chilabothrus strigilatus fowleri* from North Andros Island. *Left*: Profile and head portrait of a juvenile. *Right*: Profile and head portrait of an adult. Photos by Jon Suh.

**Figure 5.36.** *Left*: Adult female; *right*: adult male *Chilabothrus strigilatus mccraniei* from Ragged Island near Duncan Town. Note the difference in color that is likely due to variation in the population, not specific to sex. Photos by Shannan Yates.

*Chilabothrus strigilatus mccraniei* (Fig. 5.36) also has a darker overall appearance, although the middorsal dark blotches do not fuse as in *C. s. fosteri* or *C. s. fowleri*. Instead, the blotches are distinct, with light scales surrounded by very dark scales. The overall dark appearance comes from the very dark gray to dark brown scales interspersed between the middorsal blotches. Nevertheless, as in all populations of *C. strigilatus*, variation in coloration exists and light-colored individuals are also known.

*Chilabothrus strigilatus strigilatus* (Fig. 5.37) has a recognizable color pattern consisting of a distinct shift of light-gray scales mixed with darker pattern elements on the flanks to increasingly darker gray to grayish-brown scales on the dorsum—also mixed with yet darker pattern elements. The middorsum is often very dark gray. Starting at the anterior end, the flanks have dark stripes, about three scales wide, starting behind the eyes and running between one-fourth and one-third of the distance to the vent, at which point they disintegrate into either angular blotches or mixed pattern elements. A series of light blotches composed of two to eight light-gray scales, surrounded by nearly black scales, starts at the base of the head and continues posteriorly to the tail. These blotches, composed of highly contrasting light and dark scales, are separated by dark-gray to grayish-brown scales. Despite these generalities, most snakes differ in some respects in their dorsal color patterns, and there is a great deal of variation in this subspecies. Potential variation among islands has not been well documented.

*Distribution*

Bahamas boas are widely distributed across the Great Bahamas Bank, occurring on the largest islands, such as North Andros (5959 km$^2$), down to some very small islands, such as some of the smallest Exuma cays (0.992 km$^2$). The subspecies are in part geographically

**Figure 5.37.** Adult female *Chilabothrus strigilatus strigilatus* from Long Island, profile view and close up of scalation and dorsal pattern. Photo by Joseph Burgess.

defined, with two subspecies occurring on the western end of the Bank (west of the Tongue of the Ocean), two subspecies occurring on the eastern end of the Bank, and one subspecies occurring on the very southern end of the Bank.

*Chilabothrus strigilatus ailurus* is endemic to Cat Island, located south of Eleuthera and east of the Exumas across the Exuma Sound. The subspecies also occurs on Alligator Cay, which is an island nearly attached at low tide to the northwestern edge of Cat Island near Bennet's Harbour Settlement. No boas are recorded from Little San Salvador to the west of Cat Island.

*Chilabothrus strigilatus fosteri* is endemic to the Bimini Islands on the northwestern edge of the Great Bahamas Bank. They are recorded from three islands: East Bimini, North Bimini, and South Bimini. A fourth record exists from "Easter Cay," probably one of the marl islands adjacent to East Bimini. Boas do not occur elsewhere in the Biminis—they are not recorded from Gun Cay or the Cat Cays to the south, nor from Great Isaac Cay to the north.

*Chilabothrus strigilatus fowleri* is found on the Andros islands, which include the major islands known as Andros, South Andros, and Mangrove Cay. The Androsian Archipelago comprises dozens of smaller islands, and no systematic survey has determined the exact range of boas on these islands. Assuming appropriate habitat exists (see below), it is likely that they are widespread on these smaller cays. They are also known from two of the Berry Islands to the north of Andros: Great Harbour Cay and Chubb Cay. Boas might occur elsewhere in the Berry Islands, although some islands are privately owned and access is restricted.

*Chilabothrus strigilatus mccraniei* is endemic to the lower Ragged Islands range, the southern edge of a sweeping arc of remote islands that approaches Long Island to the northeast via the Flamingo and Jumentos Cays. The subspecies is recorded from the privately owned Little Ragged Island, as well as from Great Ragged Island and Margaret Cay (Reynolds and Puente-Rolón 2016). It is possible that boas occur on other cays to the north, south of Nurse Channel.

*Chilabothrus strigilatus strigilatus* is a poorly defined and variable subspecies thought to occur across the central portion of the Great Bahamas Bank, east of the Tongue of the Ocean. This subspecies is known from New Providence and Rose Island, Eleuthera Island and Windermere Island, the Exuma Cays including Bell Island, Hawksbill Cay, Compass Cay, Lee Stocking Island, Staniel Cay, Great Exuma, additional unnamed small cays, and Long Island. Boas are widespread on the larger islands, although often localized or with variable abundance in different areas.

*Habitat*

Bahamas boas are habitat generalists and can be found from the densest blackland mahogany forests, to mangrove forests, to pineland, to open coastal scrub. In these environments boas will use nearly all available habitats: arboreal habitats from the tops of large trees to smaller shrubs, vines, and branches; terrestrial habitats such as rock walls and debris piles; and even subterranean habitats such as caves (Fig. 5.38). Rock walls are especially favored, particularly when surrounded by coppice forest. This species is an excellent climber (Fig. 5.38), and daytime retreats include tree canopies, mangrove branches, and roof rafters. Indeed, boas are most frequently seen during the day when found resting in the rafters of thatch roofed structures. Boas are

**Figure 5.38.** *Chilabothrus strigilatus* are excellent climbers. *Top left*: *C. s. strigilatus* climbing across vegetation on Long Island. *Top right*: *C. s. fosteri* from South Bimini retreating into a tree hole. *Bottom*: Cave habitat on Long Island where boas can occasionally be found. They do not occupy caves in the numbers that other species in the genus do, but cave habitat does provide shelter and feeding opportunities on some of the larger islands in the Lucayan Archipelago. Note the vegetation that facilitates access to and from the cave. Photos by R. Graham Reynolds.

occasionally found during the day under debris, including plywood and other artificial cover, or in rock walls (where a portion of the body might be visible). Boas are also found in caves on Long Island and Eleuthera (RGR, pers. obs.). More so than other *Chilabothrus* species, Bahamas boas will frequent mangrove forest, probably hunting birds (RGR, pers. obs.). Nocturnally, boas can be found in almost any habitat or situation, from the ground to the canopy and from forest to patches of urban greenspace or gardens. Bimini boas are readily found in mangrove forest as well as near the shoreline. This seems to be a less-preferred habitat among other populations and subspecies. Androsian boas are locally common in coppice forest and mangrove forest, are less common in pineland, and are occasionally found in mangrove marl habitat. Boas can be found in the most remote portions of these islands, as well as close to human habitation. In the South Andros area, refugia include under leaf litter, in logs, in limestone holes, underground, and in water (Knapp and Owens 2004). On New Providence boas largely avoid pineland but are readily found near human habitation and in relatively small patches of hardwood coppice forest. For example, there are probably dozens of boas living in a 2.5-ha patch of greenspace near downtown Nassau. On Long Island, boas can be found island-wide, as well as on small islands in brackish lakes with coppice forest (Herrmann et al. 2018). In the Exumas, boas can apparently survive on very small cays, such as around Staniel Cay, although whether they move between these cays is unknown (although likely). On Eleuthera, boas are locally common in coppice forest near Cape Eleuthera, Governors Harbour, and Rock Sound but become scarcer near the modified agricultural lands of Hatchet Bay. This is an apparent shift, as boas used to be readily found near poultry farms in this region despite the reduction of intact coppice forest. On Cat Island boas are found island-wide, including in the mangrove forests but, like elsewhere, are most common in coppice forest developed over limestone. One individual was found in a large garbage dump near Bennet's Harbour Settlement (RGR, pers. obs.). On Ragged Island, boas are found in and around Duncan Town, near goat pastures, and in scrubby coastal coppice (RGR and ARPR, pers. obs.).

*Abundance*

Bahamas boas are generally considered rare to occasional across most of the range, with a few locations where they can be locally common during certain times of the year (usually June through September). Bahamas boa abundance is highly variable depending on the island, habitat, and time of year. For example, it is possible to conduct focused surveys for weeks on Long Island in November without seeing any animals, followed by finding a dozen in a couple of days in June. Bimini boas are relatively frequently found from March to August in coppice habitats on South Bimini, and road kills are more frequent across all islands during the summer months. Despite having so many people, New Providence retains a healthy boa population on parts of the island. Small forested areas to where boas are relocated after being removed from people's houses are excellent places to observe several individuals per night. Most abundance observations for this species are anecdotal—no focused studies of Bahamas boa abundance or activity have been published.

*Activity and Trophic Ecology*

Boas are most active during the summer, with an apparent peak of detectability during June and July (RGR, pers. obs.). Nevertheless, boas can be found active at any other time of the year. Activity seems to depend on diurnal and nocturnal temperatures, rainfall, and season. Warm, rainy seasons appear to trigger the most surface activity in boas on most islands. Like other members of the genus, the species is nocturnal. In the South Andros area, *C. s. fowleri* moved 0–159 m/d; convex polygon home ranges were 0.03 to 2.0 ha (Knapp and Owens 2004).

There are apparent age effects related to trophic ecology in this species. Neonates are very small, arboreal, and likely feed on young *Anolis* lizards, although they have been documented to eat *Osteopilus septentrionalis* in an artificial well on New Providence (Franz et al. 1993). Juveniles under 600 mm SVL feed on mostly *Anolis*, with a trophic switch to additional prey taking place between 600 and 800 mm SVL (Ottenwalder 1980; Tolson and Henderson 1993). Males and females are capable of growing large enough to take small *Cyclura* iguanas and readily consume birds. Indeed, adults are generalists, consuming a variety of endo- and ectothermic prey. On Andros Island, boas will consume hatchling *Cyclura* iguanas, sometimes eating several individuals from a single clutch (Knapp and Owens 2004), and *Cyclura* were probably consumed by other populations throughout the range prior to the extirpation of the iguanas from most islands. Iguanas and hutias, once widespread throughout the Bahamas in the recent past, would likely have been major prey items for adult snakes. Boas will also consume frogs (*Osteopilus*; Franz et al. 1993), rats, domestic fowl, and feral cats (Schwartz and Henderson 1991). Boas were known to congregate around domestic fowl production facilities, such as those that once existed at Hatchet Bay, Eleuthera. Anecdotally, boas have been reported to consume entire litters of feral kittens. Other possible prey items include curly-tail lizards (*Leiocephalus*), geckos (*Aristelliger* and *Tarentola*), mice, and other small native and introduced vertebrates, such as bats in caves.

In captivity, boas have been known to consume their own offspring (Hanlon 1964), and one individual consumed numerous racers (*Alsophis vudii*) and a smaller conspecific placed into its enclosure as cage mates (Mittermeier 2011). No wild boas have been observed to exhibit ophidiophagy.

*Predators and Defensive Behavior*

Predators of juvenile boas likely include night herons, which frequently hunt in the same areas where the boas forage. Other natural predators are unknown but likely include wading birds and raptors. External parasites are rare, and internal parasites are unknown. Many adults are missing the tip of the tail, though the reasons for this are unclear and could be related to either predation attempts, injury by shifting rocks, or possibly filarial worm infection, although none of these have been demonstrated causes. Introduced vertebrate predators include cats, which can consume young to medium-sized boas and have been observed doing so. Rats might eat neonates, although this has not been observed.

Defensive behaviors include occasionally attempting to flee when approached, particularly when near a hole or crevice. Adults of this species rarely defensively strike when handled, although when they do, they usually only strike once and then resume attempts to flee. When handled the species will void the contents of the cloaca.

*Reproduction*

Like most *Chilabothrus* species, female Bahamas boas are thought to reproduce every other year, likely owing to the necessity to accumulate sufficient energetic reserves for reproduction. Breeding takes place in the spring from February to May, and while courtship is unknown from the wild, in captivity it consists of "tail-searching" and physical contact between potential mates. Male–male combat is known (Tolson and Teubner 1987), and courtship can occasionally last several weeks (Tolson and Henderson 1993). Gestation likely requires between 192 and 225 days, and neonates are most frequently observed in the late fall (Tolson 1980; RGR, unpubl. data). Litter sizes and neonate measurements are not well characterized, but in general litters can consist of six to fifty-one young, born with an SVL of around 450 mm and mass of around 15 g. A New Providence female (1676 mm total length) produced fifty-one babies on 19 June (Hanlon 1964). Neonates and young-of-year are very rarely observed in the wild—like many other *Chilabothrus* they are likely quite cryptic. This species reproduces readily in captivity and many individuals exist in Canadian, U.S., and European private collections. This species can likely reach 22 years of age in captivity (Bowler 1977; although this record could refer to Hispaniolan *C. striatus*).

*Conservation Status*

Bahamas boas are listed on CITES Appendix II and are assessed as Least Concern on the IUCN Red List (Reynolds and Buckner 2019). Boas can be abundant where the habitat is good and the number of predators, such as cats, and persecutors, such as humans, are low. Bahamas boas are adaptable and can persist in small forest patches near development as long as they are not subject to strong predation pressure or road mortality.

Feral cats hunt juvenile and young boas successfully and likely create a large impact on recruitment in areas where cats are abundant. Cats occur on almost every island in the Bahamas and thus represent a persistent and widespread threat. Nevertheless, boas are capable of growing large enough to consume cats and will eat kittens.

Development and the associated habitat disturbance and human activity are likely the biggest threat to boas beyond feral cats. Developments on North and

**Figure 5.39.** Range of *Chilabothrus strigilatus*. The ranges of recognized subspecies are shown in different colors. Note that this species is entirely restricted to the extent of the Great Bahama Bank. Boas are found on some, but not all, of the Exuma Cays.

South Bimini are likely impacting boa populations. North Bimini is the site of a large hotel complex and cruise ship port, and little habitat remains between the southern tip of North Bimini through Alice Town to the end of the resort development. Some habitat exists on the eastern side of North Bimini before the channel separating it from East Bimini, although there is little coppice habitat there. On South Bimini, an airport expansion project resulted in the clearing of a large patch of forest near the tarmac, and traffic between the airport and the developed areas to the west passes through excellent boa habitat and results in frequent road kills. On Ragged Island, construction on the north end of the island has resulted in a paved road being built between the north end and Duncan Town to the south. This allows travel at a relatively high speed, and the road passes next to some of the best boa habitat on the island, resulting in significant mortality for this small population. Boas are regularly found dead on the main roads ("Queens Highway") on Long Island and Eleuthera, particularly in the summer months. An ongoing study has found an alarming number of road kills reported from near Cape Eleuthera (S. Hoefer, pers. comm.).

Human persecution is a tremendous threat to boas living in developed or rural areas. Most boas that encounter a human end up dead—people fear them and kill them on sight. Happily, a major effort is underway on New Providence, led by the Bahamas National Trust, to educate people about the benefits of the boas. A boa rescue team, composed of volunteers, responds to calls or Facebook posts about boas in peoples' houses or yards, and they safely relocate the animals. This sort of citizen action, on a large scale, is what is needed to change attitudes about snakes.

Poaching is also an ongoing concern, and a large number of boas were poached from the Biminis (Dodd 1986), resulting in a likely population crash that is only recently starting to recover. The increased enforcement of poaching regulations has probably meant that it is harder to poach this species; nevertheless, it likely still happens albeit at a much lower scale. A Bimini boa discovered in a Miami parking lot could have been a poached animal (see Chapter 3).

The Commonwealth of the Bahamas has a significant number of protected areas across the Great Bahamas Bank, managed as national parks by the Bahamas National Trust. Many of these parks have wardens who serve to enforce wildlife laws. Parks like West End on Andros and the Retreat and Primeval Forest on New Providence protect large areas of boa habitat. But some of the most abundant boa populations occur in areas or tracts of private landownership, so-called freeholds or generational land. These lands are always vulnerable to clearing or development, and so additional protection or incentives for protection would go a long way in helping to preserve some important boa populations on islands like Cat Island, Eleuthera, and Long Island. Figure 5.39 shows the distribution of *Chilabothrus strigilatus*.

## CUBA

The Cuban archipelago stretches for over 1100 km in the west–east axis, from about 210 km eastward of the Yucatán Peninsula to about 78 km west of Haiti. It comprises the main island of Cuba (the largest of the Greater Antilles, 104,338 km$^2$), Isla de la Juventud (formerly Isla de Pinos, the second largest island of this archipelago: 2204 km$^2$), and over 1600 smaller islets and cays grouped into four main micro-archipelagos (i.e., Los Jardines de la Reina, Sabana-Camagüey, Los Canarreos, and Los Colorados), which sum all together to a land surface of 106,542 km$^2$. The topography of the archipelago is complex, characterized by extensive lowland plains that alternate with undulating hills and four main mountain systems that occur on the largest island: Cordillera de Guaniguanico (maximum elevation [max. elev.] 700 m at Pan de Guajaibón) in the western region, Macizo de Guamuhaya or Escambray (max. elev. 1140 m at Pico San Juan) in the central region, and Sierra Maestra (max. elev. 1974 m at Pico Real del Turquino) and the mountains of Nipe-Sagua-Baracoa (max. elev. 1231 m at Pico Cristal) in the eastern region. These mountain ranges, together with several important wetlands and some areas of serpentine soil, contain the best-preserved ecosystems in the Cuban archipelago and harbor the highest number of endemic species. In part, the high endemism results from these higher terrains functioning as paleo islands during the interglacial periods when the sea level reached its maximum and flooded most lowlands. The first emergent lands of the Cuban archipelago capable of sustaining a permanent terrestrial biota likely appeared only after the Middle Eocene, about 40 MY. About 90% of Cuba's forests were cut down for agriculture, stockbreeding, and the development of human settlements during the last few centuries; natural ecosystems have persisted only in the most difficult-to-access areas such as mountains and swamps.

Geologically, two-thirds of Cuba's surface is formed by carbonated rocks, which translates to an extraordinary abundance of caves (estimated number more than 10,000) that serve as refuge for many species. At least one hundred hot caves have been reported across the archipelago, which are characterized by having air temperature and relative humidity approaching 40 °C and 100%, respectively. These caves shelter bat colonies of thousands of individuals, as well as a rich invertebrate fauna that relies on the bat guano and the microclimatic conditions generated by these mammals. Only one species of boa occurs in the Cuban archipelago, *Chilabothrus angulifer*, which frequently uses caves both as shelter and as a source for bat prey.

## CUBAN BOA

*Chilabothrus angulifer* (Bibron 1843)

### *Taxonomy*

The species was thought to represent the earliest-divergent lineage of the *Chilabothrus* radiation in the West Indies, which was estimated to have split off from other Caribbean taxa during the Late Oligocene–Early Miocene, between 16.9 and 26.0 MY (Tolson 1987; Reynolds et al. 2013a). However, further molecular studies placed it in a clade shared with *Chilabothrus* species from the Puerto Rico Bank and Isla de Mona (i.e., *C. inornatus*, *C. granti*, and *C. monensis*; Rivera et al. 2011; Reynolds et al. 2014a, 2015, 2016a, 2018), hence it remains unclear what is the earliest diverging lineage of *Chilabothrus* (although current work by RGR is likely to resolve this). There are no recognized subspecies, but a clinal variation in color pattern can be observed from the western to the eastern populations (see Description for more details). Mertens (1939) combined this species with *Epicrates striatus*, *E. strigilatus*, and *E. chrysogaster*, but this arrangement was not followed by subsequent authors (Sheplan and Schwartz 1974).

### *Etymology*

The specific epithet *angulifer* referrers to the dark-bordered angulate markings on the dorsum (Fig. 5.40). This species is known locally as "majá" or "majá de Santa María."

### *Description*

This is by far the largest and stoutest species of *Chilabothrus*. The large size that this boa can attain impressed the first Europeans who arrived in Cuba during the fifteenth and sixteenth centuries. The Spanish captain Gonzalo Fernández de Oviedo y Valdés (1535, officially published in 1851a) mentioned snakes of up to 25–30 feet (7.62–9.14 m) TL and as stout as a man's thigh (see also Rodríguez-Cabrera et al. 2016a for a review). However, more realistic approaches were given during the nineteenth century. Rodríguez Ferrer (1876) recorded specimens of "6 varas" TL (5.49 m). Gundlach (1880) mentioned having observed specimens of about "7 varas" TL (6.4 m) and kept a wild-caught one himself in captivity of "5 varas" TL (4.6 m). Barbour (1914) mentioned that the Cuban boa "grows commonly to be twelve feet long [3.66 m] and to be as large in circumference as any boa, or as a Python of far greater length." Barbour and Ramsden (1919) cited Gundlach's "seven yards" record and mentioned that "a specimen of almost this size was received some years ago alive at the New York Zoological Park and was a most extraordinarily bulky reptile." However, Kluge (1989:19) regarded this measurement as probably erroneous. A very large female measuring 5.65 m TL and almost certainly exceeding 40 kg was killed on a farm in the mountains of Guamuhaya, central Cuba, in 1987 (J. D. León in Rodríguez-Cabrera et al. 2020b; see also Activity and Trophic Ecology for additional details). Another very large snake was killed on a road on the U.S. Naval Station Guantanamo Bay in 1989 and measured "15 ft., 11 in." TL (4.85 m; Tolson and Henderson 1993). Female boas up to 3.65 m TL and 20.4 kg have been captured at the Station in recent years (Petersen et al. 2015; P. J. Tolson in Rodríguez-Cabrera et al. 2016a). One female 3.25 m TL and 33.3 kg was collected at the Guanahacabibes Peninsula in the westernmost region of Cuba and kept in the Parque Zoológico Nacional de Cuba for some time (A. Arango, in litt. to TMRC, 28 May 2020). On 28 April 2016 a large individual approaching 4.0 m TL was observed and photographed on the karstic hills northeast of Platero, Yaguajay, Sancti Spiritus Province (H. Vela, pers. comm. to TMRC, 30 May 2020). The latest truly large boa was killed by a local and photographed on 15 April 2020 in a forest patch surrounded by degraded areas northwest of Vegas, Mayabeque Province; it approached 4.0 m TL (3.8 m with the head cut off) and weighed 19.5 kg (Y. C. Pérez, in litt. to TMRC, 17 April 2020). Large boas approaching 3.0 m TL have become increasingly rare during the last century (Fig. 5.41), perhaps because of strong persecution by humans (directional selection toward smaller sizes), accelerated habitat loss, and the introduction of mammalian predators (interference competition) (Gundlach 1880; Barbour and Ramsden 1919; Sheplan and Schwartz 1974; Díaz 2006; Tolson and Henderson 2006; Rodríguez-Cabrera et al. 2016a; see Conservation for more information). Neonates are the largest in the genus and among the largest in the family Boidae (Šimková 2007), frequently exceeding 600 mm SVL and 180 g (versus 318–530 mm SVL and 12.0–21.7 g in other large *Chilabothrus* species;

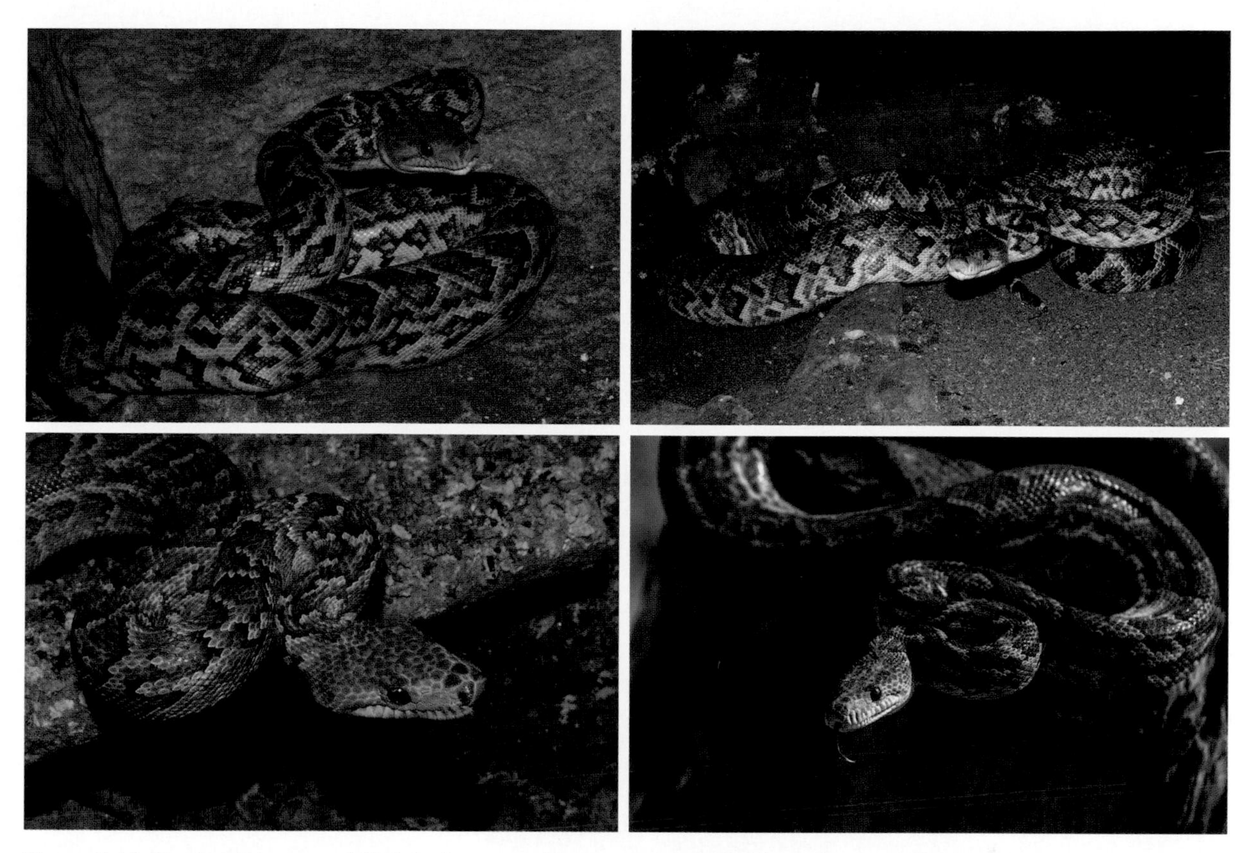

**Figure 5.40.** The dorsal pattern of *Chilabothrus angulifer* is composed of dark-bordered angulate markings. Like other members of the genus, the skin produces notable iridescent reflection, especially under sunlight (*lower right*). Photographs by Tomás M. Rodríguez-Cabrera (upper right and lower left) and Raimundo López-Silvero (upper left and lower right).

**Figure 5.41.** TMRC (*left*) and his colleague Hansel Caballero holding a female *Chilabothrus angulifer* approaching 3.0 m total length and 12 kg. Boas of this size and larger have become increasingly rare in most parts of the country owing to the habitat loss and the strong persecution by humans. Photograph by Raimundo López-Silvero.

**Figure 5.42.** *Chilabothrus angulifer* is the only species in the genus with infraoculars separating the eye from all supralabials and heat-sensing labial pits (*above*), although occasionally one supralabial may enter the eye (*below*). Photographs by Tomás M. Rodríguez-Cabrera (above) and Luis M. Díaz (below).

Bloxam and Tonge 1981; Tolson 1992b; Tolson and Henderson 1993; Polo Leal and Moreno 2007; Morell Savall 2009; Rodríguez-Cabrera et al. 2016b; see Reproduction). Only neonatal green anacondas (*Eunectes murinus*) and sometimes those of *Boa constrictor* have comparable sizes at birth to those of *C. angulifer* (e.g., Pizzato and Marques 2007; Reed and Rodda 2009; Rivas 2020).

The Cuban boa can be readily distinguished from other *Chilabothrus* species on the basis of modally higher counts of dorsal scale rows at midbody (53–69 versus 35–56 in other species); infraoculars separating the eye from all supralabials (versus eye in contact with some supralabials in other species); occasionally the seventh or eighth supralabials entering the eye in *C. angulifer*, apparently because one supralabial and one infraocular fuse together (Fig. 5.42); a shorter and blunter tail (averages 10% of snake's TL) (Fig. 5.43); lower counts of subcaudals (45–55 versus 66–93 in other species); presence of heat-sensing labial pits (versus absent in other species), and color pattern (Sheplan and Schwartz 1974; Tolson 1987; Schwartz and Henderson 1991; Tolson and Henderson 1993; Henderson and Arias Barreto 2001; Reynolds et al. 2013a; Rodríguez-Cabrera et al. 2015, 2016a). The species is characterized also by the following meristic values: dorsal scale rows on the neck 44–54 and anterior to the vent 28–38; ventrals 272–292 in males and 268–290 in females; subcaudals 45–55 in males and 46–54 in females; supralabials 12–16 (modally 14), infralabials 13–19 (modally 17), circumorbitals 8–9 (occasionally 10), and loreal usually 1; head scale formula 3–3–4

**Figure 5.43.** The tail of *Chilabothrus angulifer* is much shorter relative to body length than that of any other species in the genus (averages 10% of snake's TL) and is blunt at the tip. Photograph by Rosario Domínguez.

(Sheplan and Schwartz 1974; Schwartz and Henderson 1991; Tolson and Henderson 1993; Henderson and Arias Barreto 2001).

Da Silva et al. (2019) found that *C. angulifer* has one of the thickest corneas (833 ± 212 μm) and largest spectacle diameters (6.4 ± 0.7 mm) and spectacle thickness (226 ± 34 μm) among forty-four snake species analyzed belonging to fourteen families. These authors categorized *C. angulifer* as an arboreal and nocturnal snake and related such eye characteristics mainly to arboreality. However, this species is a generalist regarding both habitat use and activity, not a nocturnal, arboreal specialist such as *C. gracilis*, *C. granti*, *C. ampelophis*, *C. monensis*, and *C. argentum* (see the respective accounts). Since Da Silva et al. (2019) used multiple published sources to categorize the species studied, some of them not based on extensive studies, the reasons for such eye features in *C. angulifer* remain uncertain and deserve further investigations.

The dorsal ground coloration is usually yellowish tan, light to dark brown, or grayish brown. Ventral coloration is usually yellowish, cream, or grayish white, with the chin, throat, and anterior part of the neck usually paler and becoming heavily suffused or even spotted with pale to dark gray posteriorly, sometimes producing tan to brown scales with pale edges. Dorsal pattern consists of a series of more or less defined darker angulate markings with a dorsolateral row of irregularly shaped secondary blotches, which can form almost perfect rhombuses. The pattern color varies from light brown to dark brown or even black and the blotches are frequently outlined by paler scales (Figs. 5.40, 5.44). The blotches may be solid dark or centrally pale. For those snakes with defined dorsal patterns that allow blotch count, it varies from 42–65 on the body and from 5–12 on the tail. There are not strong ontogenetic shifts in body color and pattern; juveniles retain essentially the same pattern as adults, although usually more contrasting and may undergo only a slight shift from a lighter to darker ground color (Fig. 5.44). Juveniles may occasionally show a faint, dark postocular stripe one-scale wide immediately above the supralabials (Fig. 5.44). The iris is usually dark brown, with the pupil barely distinguishable, giving the eye a uniform dark appearance (Fig. 5.42). Young individuals tend to have a light brown to tan iris mottled and streaked dark brown and solid dark brown around the pupil (Fig. 5.44), which usually shift to completely solid dark brown as they age; only a few snakes retain a light iris when adults. Like all members of the genus, the skin produces notable iridescent reflection, especially under sunlight (Fig. 5.42). The tongue is dark gray or dark grayish blue with whitish tips starting at the fork (Figs 5.42, 5.44).

There is a clinal variation in color pattern from west to east (Sheplan and Schwartz 1974; Tolson and Henderson 1993; Henderson and Arias Barreto 2001; Rodríguez-Cabrera et al. 2016b) (Fig. 5.45). Snakes from Pinar del Río and Artemisa Provinces in the west usually show faded and paler dorsal markings, which makes blotch counts sometimes difficult. This pattern is gradually replaced by a pattern with darker and better-defined dorsal markings and secondary side blotches, starting in the region of La Habana and almost completely replaced in central Cuba. Nonetheless, exceptions exist on both sides of the cline. Consistent with such a phenotypic clinal variation, Starostová et al. (2006) sequenced a 554 base pair fragment of cytochrome *b* of eighty-nine individuals from most of the European zoos and found twenty-two haplotypes corresponding to two clearly distinguished lineages, western and central-eastern, that possess up to 6.1% sequence difference. Such east–west divisions in mitochondrial DNA are seen in a number of terrestrial Cuban reptiles (e.g., Reynolds et al. 2020c).

Sexual dimorphism is more evident in mature

**Figure 5.44.** Contrary to some other species in the genus, neonate *Chilabothrus angulifer* have essentially the same color and pattern as adults, only the color of the iris darkens as the snakes age. Occasionally they may have a faint, dark postocular stripe (*lower left*). Photographs by Raimundo López-Silvero (*above*) and Tomás M. Rodríguez-Cabrera (*below*).

individuals, with females usually longer and heavier than males. As they grow, sexual size dimorphism becomes stronger, with males being relatively slender and females relatively stout; adult females can be twice as heavy as males of similar SVL. Some very large males have been recorded: one male 3.15 m TL and 27.6 kg was collected at the Guanahacabibes Peninsula, and another male reached a similar TL and 10 kg in captivity; both were kept in the Parque Zoológico Nacional de Cuba for some time (A. Arango, in litt. to TMRC, 28 May 2020). Subadult and mature males have larger and more pointed spurs (Rodríguez-Cabrera et al. 2016b) (Fig. 5.46), and the tail averages 10.4% of snake's TL (range: 8.6–12.3%; $n$ = 40 snakes) versus 9.7% of snake's TL in females (range: 7.8–11.1%; $n$ = 36 snakes), although a considerable overlap exists in the latter character. These sexually dimorphic characters are more ambiguous in juveniles. Neonates show no sexual dimorphism in size and body shape, and probing is necessary for a definitive sex diagnosis (Frynta et al. 2016). A clearly bimodal distribution in the distance of penetration using probes inserted into the cloaca is observed even at this age (i.e., females: <6 subcaudals; males: >7 subcaudals; Frynta et al. 2016).

The retracted hemipenis was first illustrated by Cope (1895), without any description; a reference to external characters and an illustration of a partially everted organ is in Dowling and Savage (1960); a detailed description (not illustrated) was provided by Sheplan and Schwartz (1974); redescription and comparisons with other boas are in Branch (1981). The hemipenis (Fig. 5.46) is longer than it is wide, bilobed over the distal three-quarters of its length, and the proximal half mostly nude. The asulcate surface

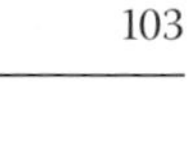

has a prominent stout papilla in a submedial position (Fig. 5.46), followed by six to eight deep, transversal flounces with irregular edges, directed downward. The flounces become gradually fragmented in distal progression, giving place to relatively large, irregular, and spaced papillae on lobes. Much smaller papillae may be interspersed among flounces and the larger papillae. The sulcate surface has flounces medially interrupted by the sulcus spermaticus, which bifurcates below the division of the lobes, each branch extending to near the tip of its respective lobe (Fig. 5.46). The margins of the sulcus spermaticus are much stouter on the lobes, and the lobe papillae are comparatively smaller, digitiform, and tighter on the sulcate surface than on the asulcate surface.

Jan (1864) illustrated *C. angulifer*, among other species, including its skull in dorsal, lateral, and ventral views. Kluge (1989) listed the states of numerous cranial and axial osteological characters in several species of *Chilabothrus*, including *C. angulifer.*

*Distribution*

This species is widely distributed over the main island of Cuba and Isla de la Juventud (= Isla de Pinos), but it has been reported only on about thirty satellite cays, which constitute approximately 2% of the more than 1600 cays of the Cuban archipelago (Estrada 2012; Rodríguez Schettino et al. 2013; Martínez Reyes and Arias Barreto 2014; Marichal Arbona 2016). The species is not recorded from vast areas of Pinar del Río, Artemisa, Mayabeque, Matanzas, Ciego de Ávila, Las Tunas, and Holguín Provinces (see Rodríguez Schettino et al. 2013). What all these regions have in common are the predominantly degraded areas devoted to agriculture and stockbreeding, which might affect the species occurrence. Another possible reason might be that boas occurring in human-altered areas not frequented by researchers just go unnoticed. Boas have not been recorded from the mountain peaks above 1200 m at Sierra Maestra in eastern Cuba.

*Fossil Records*

Brattstrom (1958) reported semifossilized vertebrae, ribs, and a maxilla of *C. angulifer* from Late Pleistocene deposits at three localities in Cuba. This author commented on massive vertebrae in which the height, width, and length of the centrum of the largest was

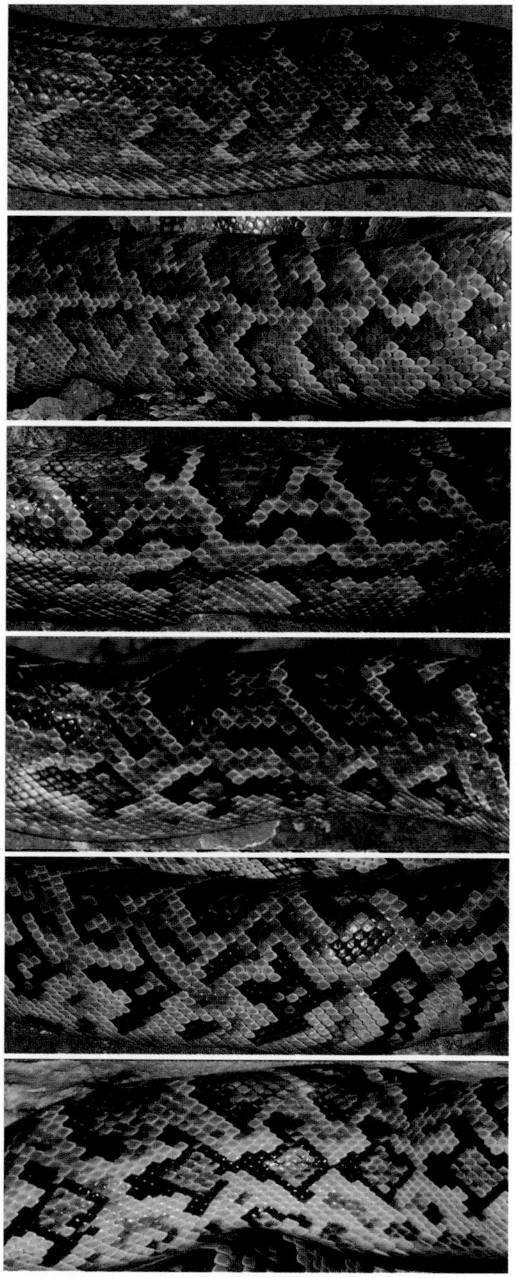

**Figure 5.45.** There is a clinal variation in dorsal color pattern in *Chilabothrus angulifer* from east to west. *Top images*: Snakes from Pinar del Río and Artemisa Provinces in the west usually show faded and paler dorsal markings, which makes blotch counts very difficult. This pattern is gradually replaced by a pattern with darker and better-defined dorsal markings and secondary side blotches, starting in the region of La Habana and almost completely replaced in central Cuba. Photographs by Tomás M. Rodríguez-Cabrera.

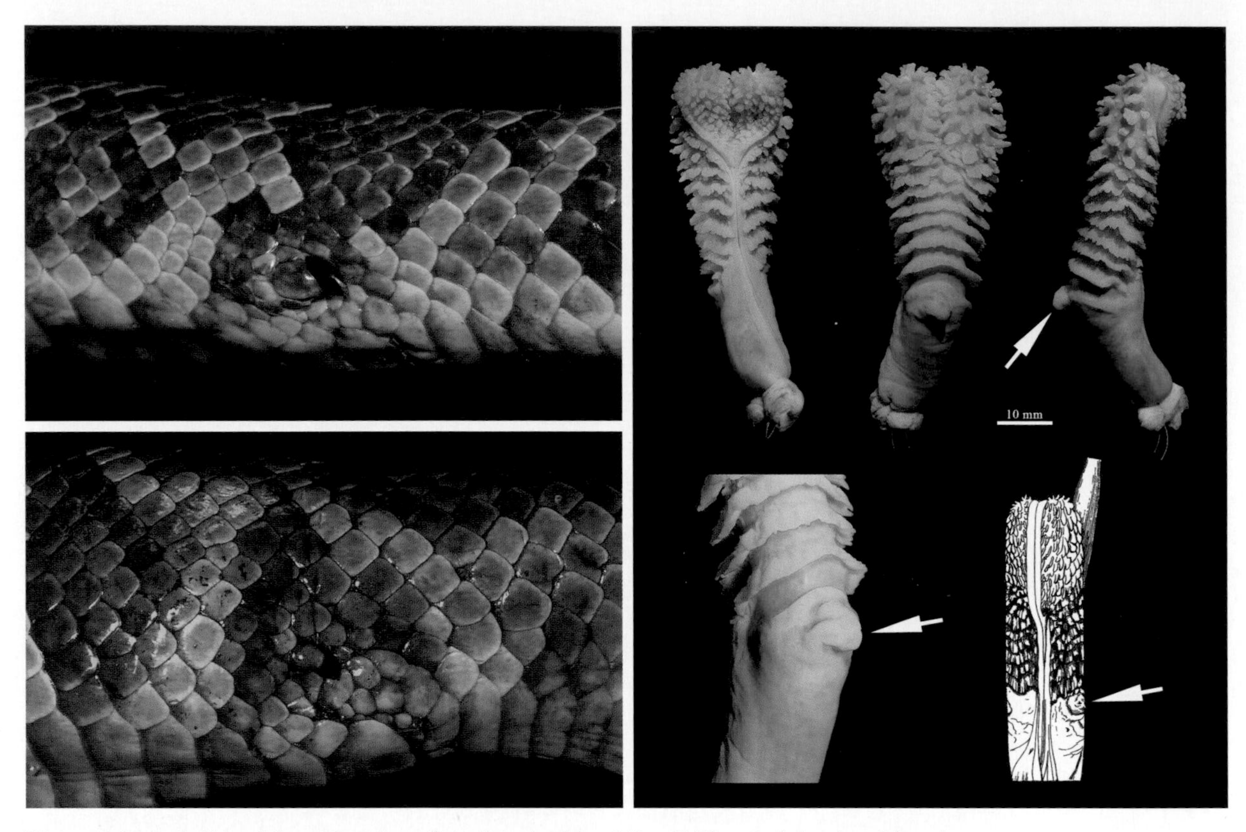


**Figure 5.46.** *Left*: Side of the vent region of similar-sized (ca. 1.5 m SVL) male (*above*) and female (*below*) *Chilabothrus angulifer*, showing sexual dimorphism in spur size. *Right*: Hemipenis of *C. angulifer* in (*clockwise from top left*) sulcate, asulcate, lateral view, an illustration in Cope 1895 shown for comparison, and detail of the medial region. The arrows indicate the prominent stout papilla in submedial position. Hemipenis mounting and photographs by Tomás M. Rodríguez-Cabrera.

22.7, 15.6, and 11.0 mm, respectively. Arredondo (1971) mentioned *C. angulifer* as part of the associated fossil fauna of Cueva del Agua, Quivicán, Artemisa Province in the description of extinct Quaternarian vultures. Arredondo (1976) also reported fossil material of this boa among the associated fauna of two type localities of extinct birds of prey at Cueva del Túnel and Cueva de Paredones, Artemisa Province. Varona and Arredondo (1979) reported *C. angulifer* as part of the Pleistocene–Holocene-associated fossil fauna of eight localities (mostly caves) in western, central, and eastern Cuba in the description of several hutia species (see also Acevedo González 1983; Henderson and Arias Barreto 2001). Jaimez Salgado et al. (1992) reported unidentified material of a fossil boid from Cueva del Mono Fósil and Cueva Alta, in western Cuba. Arredondo Antúnez (1997) listed *C. angulifer* from the Pleistocene of Sierra de Cubitas, Camagüey Province, and cited a record by Koopman and Ruibal (1955); however, the late authors mentioned *Alsophis angulifer* (= *Cubophis cantherigerus*). Iturralde-Vinent et al. (2000) reported fossil material of unidentified boids from Las Breas de San Felipe, a presumably Late Pleistocene–Early Holocene asphalt seep in Matanzas Province. Suárez (2000) reported *C. angulifer* among the fossil vertebrate taxa recovered from Cueva de Sandoval, Artemisa Province. The Cuban boa is also present in Quaternary fossil deposits at Solapa del Megalocnus, northeast Villa Clara Province (Arredondo Antúnez and Villavicencio 2004). Jiménez Vázquez et al. (2005) reported one vertebra from Cuevas Blancas, Quivicán, Mayabeque Province, dating it from Early Middle Ho-

**Figure 5.47.** *Chilabothrus angulifer* seems to show a particular preference for habitats consisting of forest developed over karstic limestone (*upper left*) with abundance of caves (*upper right*), but it also is found in other habitat types such as grasslands (*lower left*) and flooded rice fields (*lower right*). Photographs by Tomás M. Rodríguez-Cabrera and Raimundo López-Silvero (lower left).

locene. Rojas Consuegra et al. (2012) and Viñola López et al. (2018) reported fossil remains dating from the transition Pleistocene–Holocene (ca. 10,000 years ago) from Cueva del Indio, Mayabeque Province. Aranda et al. (2017) listed six fossil vertebrae from Quaternary deposits at Las Llanadas, Sancti Spíritus Province. Aranda Pedroso (2019) reviewed the Squamata fossil record in the Antilles and Bahamas and reported Pleistocene fossil material of *Chilabothrus* from various localities in western and central Cuba, including Isla de la Juventud; he commented on the possible finding of a new species of *Chilabothrus* from Cuba, but a more thorough review of the latter fossil material revealed that it belonged to a very large specimen of *C. angulifer* (L.W. Viñola López, in litt. to TMRC, 8 April 2020). Aranda et al. (2020) reported and illustrated fossil remains of *C. angulifer* from Isla de la Juventud, presumably from the Late Pleistocene–Early Holocene.

*Habitat*

*Chilabothrus angulifer* exhibits great ecological plasticity and tolerance to human disturbance. It has been found in virtually every type of terrestrial ecosystem present in Cuba and even in wetlands, from sea level to 1214 m asl (Schwartz and Henderson 1991; Tolson and Henderson 1993; Henderson and Powell 2009; Rodríguez Schettino et al. 2010). However, it seems to show a particular preference for habitats consisting of forest developed over karstic limestone with an abundance of caves (see Abundance and Activity and Trophic Ecology) (Fig. 5.47). Several authors have reported that the species is frequently associated with caves, where the

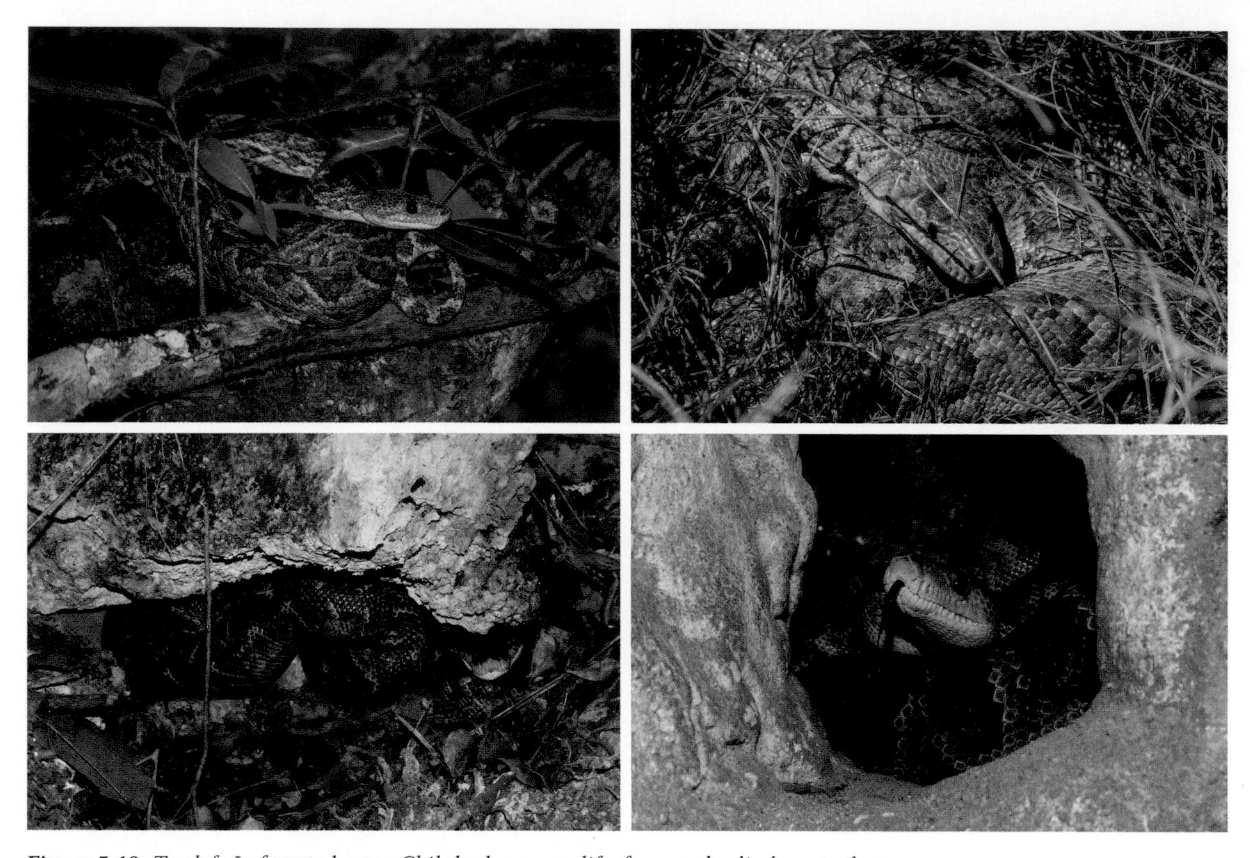

**Figure 5.48.** *Top left*: In forested areas *Chilabothrus angulifer* frequently climb up to the trees searching for prey, to bask, or simply to rest coiled on the branches or in tree holes. *Top right*: In xeric open areas, boas frequently hide on the ground in grasslands. *Bottom photos*: In karstic situations they often use limestone holes and crevices and caves. Photographs from top left clockwise by Rosario Domínguez, Wayne Fidler, Rolando Teruel, and Tomás M. Rodríguez-Cabrera.

highest population densities have been recorded (e.g., Hardy 1957; Silva-Taboada 1979; Berovides Álvarez and Carbonell Paneque 1998; Linares Rodríguez et al. 2011; Rodríguez-Cabrera et al. 2015; Dinets 2017). Cruz (1992) noted that when associated with hot caves, Cuban boas never occupy the hottest chambers but may be common outside. This agrees with Berovides Álvarez and Carbonell Paneque (1998), who observed that the highest numbers (nineteen to twenty individuals) of a population they surveyed in a hot cave were found in passages where the temperature was about 28 °C (versus nearly 40 °C in the hottest chamber). At Sierra de Cubitas, Camagüey Province, *C. angulifer* has been repeatedly observed resting by day coiled on tree branches, sometimes more than 9.0 m above the ground (Sheplan and Schwartz 1974; Vergner 1989; Díaz 2006). They can also be seen resting by day on tree branches, under rocks, on rock cornices, and in rock or tree holes and crevices near the cave entrances they regularly use to forage for bats (Hardy 1957; Dinets 2017). In a sample of thirty-two individuals from Mogotes de Jumagua Ecological Reserve, north of Villa Clara Province, 68.8% were found either in hot caves or in regular ("cold") caves, 12.5% were found on trees, 9.4% on the ground, 6.2% in fallen palm trunks, and 3.1% in human buildings (Morell Savall et al. 1998). In xeric areas such as the U.S. Naval Station Guantanamo Bay, Cuban boas are frequently found in grasslands (Petersen et al. 2007, 2015) (Figs. 5.47, 5.48).

Cave-associated boas seem to have a high fidelity

for these sites with abundance of trophic resources. Rodríguez-Cabrera et al. (2016b) recaptured eight previously marked individuals associated with a hot cave in central Cuba three ($n$ = 2), six ($n$ = 3), nine ($n$ = 1), and 12 ($n$ = 1) months after being marked, and one was recaptured twice (on the third and ninth month). In January 2012, fourteen boas were marked in a bat cave northwest of Yaguajay, Sancti Spíritus Province. Four of those boas were recaptured in the same cave in July 2012 (ca. 5.5 months later), and one was recaptured in January 2019 (ca. 7 years later; Rodríguez-Cabrera et al. 2020b). Three more boas marked in a bat cave at Cariblanca, Sancti Spíritus Province, in January and April 2013, were recaptured in December 2017 (ca. 5 years later; Rodríguez-Cabrera et al. 2020b). This relationship between hot caves and Cuban boas is so well known that many of these caves all over the country are named "Cueva del Majá" or "Cueva de los Majaes" (i.e., Cave of the Boas) (e.g., Silva-Taboada 1979, 1988; Armas et al. 1990; Longueira 2006).

Rodríguez Batista et al. (2014) listed the populations inhabiting the cays north of Cuba as being associated with xeromorph coastal and subcoastal scrubland on sand, semideciduous forest, microphyllous evergreen forest, freshwater vegetation, and secondary vegetation. Similarly, the species was recently recorded from Jardines de la Reina Archipelago, south of central-eastern Cuba, where vegetation types are mangrove, scrubland on sand, and sandy beach vegetation (Marichal Arbona 2016). It has also been observed in mangrove forests sustaining large bird nesting colonies (mostly herons). At "Wiso Colony," near Cauto River Delta (Granma Province), a boa approaching 1.5 m TL was found coiled on a trunk fork with a stomach bulge, presumably containing a medium-sized chick of the Cattle Egret (*Bubulcus ibis*), the most common species nesting in the area; the soil was flooded with brackish water and mangrove pneumatophores (*Avicennia germinans*; D. Denis in Rodríguez-Cabrera et al. 2020b; see also Denis 2006). Cuban boas have been repeatedly observed also in mangroves (*Rhizophora mangle*) south of Cayo Santa María, north of Villa Clara Province, with the soil entirely flooded with salt water (ca. 1.5 m depth), about 400–500 m from the cay's mainland (A. Arias Barreto, pers. comm. to TMRC, 19 April 2020). Boas with stomach bulges have also been observed coiled on branches or in hollow dead trunks in mangroves south of Las Tunas Province, always where the vegetation structure provides a continuous route for the snakes to move throughout the branches and aerial roots, since the soil is flooded with salt water (M. Alonso in Rodríguez-Cabrera et al. 2020b). Mangroves are known to support the largest populations of hutias (*Capromys pilorides*) in the Cuban archipelago, with densities sometimes exceeding 150 hutias/ha (Comas González and Berovides 1990; Berovides and Comas 1997a, 1997b; Borroto-Páez and Mancina 2006). Since hutias represent the most frequent prey item of adult Cuban boas (see Activity and Trophic Ecology), it is not surprising to find these snakes associated with this kind of ecosystem.

This snake can survive in anthropogenic habitats such as farms and small villages for some time, feeding mostly on domestic animals (Estrada 1994; Rodríguez-Cabrera et al. 2020b; see Activity and Trophic Ecology), but at some point they are detected by local people and almost always such encounters end up in the snake's death (see Conservation). Agroecosystems such as sugarcane plantations and flooded rice fields (Fig. 5.47) seem to harbor relatively large populations as well, but no studies exist in such habitats. Fishers and American bullfrog (*Lithobates catesbeianus*) hunters repeatedly observed large boas (>2.0 m TL) on dikes in paddy rice cultivations at night (R. A. Fuentes, pers. comm. to TMRC, 2014). These cultivated areas are particularly rich in amphibians, birds, and introduced murid rodents that must represent a major portion of the diet for boas associated with these ecosystems (e.g., Borroto-Páez et al. 1990; Acosta and Mugica 2006; Borroto-Páez 2011, 2013). Several aquatic bird species nest in these flooded rice cultivations (e.g., Mugica et al. 1989; Jiménez et al. 2002; Acosta and Mugica 2006), providing an abundant food supply for any medium-sized to large snakes all year round. However, after the introduction of two Old World invasive catfishes (*Clarias gariepinus* and *C. macrocephalus*) in the late 1990s to early 2000s, the boa populations in these agroecosystems seem to have declined, possibly as a consequence of the negative impact by these catfishes on the local prey base. Boas have been observed (apparently foraging in ambush for aquatic birds) among the riparian vegetation on riverbanks, with the body completely submerged and only the head out of the water (T. M. Rodríguez Águila, pers. comm. to TMRC, 2016). During the last few decades of the twentieth century, fishers routinely extracted boas from partially flooded holes on the riverbanks in central Cuba (TMRC, pers. obs.).

*Abundance*

Few demographic studies other than in cave habitats have been conducted for the Cuban boa, with the exception of those by Alfonso Álvarez et al. (1998) and Morell Savall et al. (1998) at the Mogotes de Jumagua Ecological Reserve, north-central Cuba. These authors observed mean relative densities from 0.67 to 6.0 boas/h. An unpublished long-term field research project led by Peter J. Tolson on the U.S. Naval Station Guantanamo Bay will eventually provide some abundance data (e.g., Tolson and Henderson 2006; P. J. Tolson in Henderson and Powell 2009; Petersen et al. 2015). During fifteen different surveys in a hot cave at Guanayara, Trinidad, south-central Cuba, Hardy (1957) recorded a total of forty-one boas, and the number of individuals observed per survey varied from zero to seven. Ševčík and Ševčík (1988) found nine individuals together in a cave near Higuanojo, Sancti Spíritus Province. Linares Rodríguez et al. (2011) reported from three to seventeen individuals ($n$ = 74) in ten different caves from Guanahacabibes Peninsula in the westernmost tip of Cuba, Pinar del Río Province. They also correlated this abundance with the observation of potential prey species in such caves, including birds, bats, and hutias, and commented on testimonies from local people reporting up to thirty individuals in a single cave at the same time during the breeding period. This coincides with monthly surveys in a cave near Galalón, La Palma, Pinar del Río Province, where densities from nine to twenty-eight individuals per survey were reported, with more than twenty individuals observed in seven of the thirteen surveys (Berovides Álvarez and Carbonell Paneque 1998). Across the year, the highest number (i.e., twenty-eight) was observed in June, whereas the lowest numbers (i.e., nine, thirteen, fourteen, seventeen, and nineteen) were observed mostly from January to April. Rodríguez-Cabrera et al. (2015) reported seven and fourteen individuals in bat caves (not hot caves) from central Cuba in March and January, respectively; the lowest numbers observed were one individual in April and five individuals in July. Populations associated with habitats different from caves seem to be more dispersed and less dense, but additional studies are required outside of cave aggregations. In general, it is difficult and hazardous to find Cuban boas in surface habitats during both daytime and nighttime surveys, even in the most appropriate habitats (TMRC, pers. obs.).

*Activity and Trophic Ecology*

The Cuban boa is largely nocturnal, although it can be observed basking by day, particularly after a meal or by gravid females during the summer (see Reproduction). Males are also active by day at the onset of the reproductive season, apparently searching for receptive females (Tolson 1992b, 1994; Rodríguez-Cabrera et al. 2016b). Juveniles can also occasionally be observed active by day (see also Lando and Williams 1969). Predation events have been observed in natural habitats, and 60% (twenty-seven of forty-five) of these observations were during the day. In contrast, about fifty predation events were observed in anthropogenic habitats, and forty-seven (96%) occurred at night (Rodríguez-Cabrera et al. 2020b).

*Chilabothrus angulifer* will employ both active foraging and ambush strategies in similar proportions in natural habitats (Tolson and Henderson 1993; Rodríguez-Cabrera et al. 2020b). However, in anthropogenic habitats it seems to use mostly an active foraging mode. In natural habitats they use an active foraging strategy involving predation on native species such as treefrogs, anoles, bird eggs and nestlings, and rats (Rodríguez-Cabrera et al. 2020b). They are occasionally observed using a combination of both ambush and active foraging modes involving predation of introduced murid rodents in natural habitats. A 1.5-m-long individual was observed ambushing a rat (*Rattus rattus*) near Cueva del Indio, Viñales, Pinar del Río Province in May 2009. The snake remained immobile on a tree fork about 4 m high, inspecting the air and bark with its tongue while the rodent moved nervously along one of the branches trying to find alternative ways to escape, but the only option was returning back to the place where snake was waiting (LMD, pers. obs.). An ambush strategy is usually seen at night in those individuals associated with caves harboring large bat colonies and by day on the forest floor where native birds more frequently serve as prey (Rodríguez-Cabrera et al. 2020b). Occasionally boas have been reported using an active foraging mode when searching

**Figure 5.49.** *Chilabothrus angulifer* associated with human-altered habitats will employ an active foraging mode to reach poultry roosting in trees. Photographs by Tomás M. Rodríguez-Cabrera.

for juvenile bats (reduced or absent flight capacity) in caves (Rodríguez-Cabrera et al. 2015). Nonetheless, the use of this strategy in caves must increase during the postbirth period of the bats (Silva Taboada 1979), when there are many suckling bats on the walls, roof, and even on the cave floor, rendering them much easier to capture and more energetically rewarding than flying adults. An adult Cuban flower bat (*Phyllonycteris poeyi*) first seen roosting on a cave wall, apparently ill since it did not react under intense lighting, was observed being captured by a Cuban boa a few minutes later (LMD, pers. obs.). Domestic animals are actively searched for at night on farms, particularly poultry roosting in trees (Fig. 5.49). The boas apparently seek out concentrations of food resources, whether natural or domestic. The most frequent wild prey species of *C. angulifer* exhibit high gregariousness or population densities, or both (e.g., bats, hutias, rats, or colonial nesting birds; Rodríguez-Cabrera et al. 2020b). Domestic animals concentrated on farms are not very different in terms of organic waste production (chemical detection from a distance) and food supply from any bat colony, hutia family group, or bird nesting colony (e.g., pigeons, herons). Apparently, the boas make little distinction between natural and domestic prey.

The Cuban boa is a generalist and opportunistic predator that take advantage of the most abundant prey species in every kind of habitat. Rodríguez-Cabrera et al. (2020b) analyzed the diet of 218 Cuban boas. Forty-nine taxa were identified from a total 351 prey items (Table 5.3). Endothermic prey constituted a major portion in the diet of *C. angulifer* (96%), whereas ectotherms (anurans, hatchling freshwater turtles, lizards, and other snakes) represented only a small portion (4%). Domestic fowl (*Gallus gallus*) was the prey species most frequently consumed by *C. angulifer* (24.8%), followed by Desmarest's hutias (*Capromys pilorodes*; 14.5%), Jamaican fruit-eating bats (*Artibeus jamaicensis*; 8.8%), and rats (7.7%). Of twenty-five snakes with stomach contents, two snakes were feeding in situ and seven snakes contained prey items at the time that they were constricting or swallowing others (Hardy 1957; Sheplan and Schwartz 1974; Rodríguez-Cabrera et al. 2020b). Multiple prey taxa have been reported in a single snake only once, and they consisted on three bat species (one *Brachyphylla nana*, two *Mormoops blainvillii*, two *P. poeyi*) and one unidentifiable bat (Sheplan and Schwartz 1974).

*Chilabothrus angulifer* undergoes a dramatic shift in its diet and foraging strategies from natural to anthropogenic habitats (Rodríguez-Cabrera et al. 2020b). Native mammals dominate the diet of boas in natural habitats (82.1%), whereas in anthropogenic habitats domestic birds are the most frequent prey items

**Table 5.3.** Prey taxa confirmed for free-ranging *Chilabothrus angulifer* in natural (NAT) and anthropogenic (ANT) habitats.

The percentage each species represents of the total prey items is in parenthesis following the species name. The number of prey items is followed (in parenthesis) by the number of snakes involved; question mark (?) represents unknown data. When a prey species is reported by more than one author, numerical superscripts indicate the number of items referred in each reference.

| Prey | Prey items (Snakes) | | Source |
|---|---|---|---|
| | NAT | ANT | |
| **Amphibia** | | | |
| Anura: Hylidae | | | |
| *Osteopilus septentrionalis* (1.4%) | 5 (5) | — | 1 |
| Anura: Ranidae | | | |
| *Lithobates catesbeianus* (0.3%) | — | 1 (1) | 1 |
| Anura: Bufonidae | | | |
| *Peltophryne fustiger* (0.3%) | 1 (1) | — | 1 |
| **Reptilia** | | | |
| Squamata: Dactyloidae | | | |
| *Anolis smallwoodi* (0.3%) | 1 (1) | — | 2 |
| *Anolis* sp.[a] (0.3%) | 1 (1) | — | 3 |
| Squamata: Leiocephalidae | | | |
| *Leiocephalus carinatus* (0.6%) | 2 (2) | — | 1 |
| Squamata: Iguanidae | | | |
| *Cyclura nubila* (0.3%) | — | 1 (1) | 1 |
| Squamata: Tropidophiidae | | | |
| *Tropidophis melanurus* (0.3%) | 1 (1) | — | 4 |
| Testudines: Emydidae | | | |
| *Trachemys decussata*[b] (0.3%) | 1 (1) | — | 5, 6 |
| **Aves** | | | |
| Accipitriformes: Accipitridae | | | |
| *Buteo jamaicensis* (0.3%) | 1 (1) | — | 1 |
| Cathartiformes: Cathartidae | | | |
| *Cathartes aura* (0.6%) | 2 (1) | — | 1 |
| Anseriformes: Anatidae | | | |
| *Cairina moschata* (0.9%) | — | 3 (1) | 1 |

*continues*

**Table 5.3 continued**

| Prey | Prey items (Snakes) | | Source |
|---|---|---|---|
| | NAT | ANT | |
| Columbiformes: Columbidae | | | |
| *Columba livia* (1.1%) (caged) | — | 4 (3) | 1 |
| *Geotrygon montana* (0.3%) (caged) | — | 1 (1) | 1 |
| *Patagioenas leucocephala*[c] (?) | ? | — | 7, 8, 1 |
| *Zenaida aurita* (0.3%) | 1 (1) | | 1 |
| *Zenaida macroura* (1.7%) | 6 (5) | — | 1 |
| Cuculiformes: Cuculidae | | | |
| *Coccyzus merlini* (0.3%) | 1 (1) | — | 1 |
| Falconiformes: Falconidae | | | |
| *Caracara cheriway* (0.3%) (caged) | — | 1 (1) | 1 |
| Galliformes: Phasianidae | | | |
| *Coturnix japónica* (2.6%) (caged) | — | 9 (1) | 1 |
| *Gallus gallus* (24.8%) (10 caged) | 2 (2) | 85 (61) | 1 |
| *Meleagris gallopavo* (1.1%) | — | 4 (2) | 1 |
| Galliformes: Numididae | | | |
| *Numida meleagris* (2.0%) | — | 7 (6) | 1 |
| Gruiformes: Rallidae | | | |
| *Porphyrio martinicus* (0.3%) | 1 (1) | — | 1 |
| Passeriformes: Icteridae | | | |
| *Quiscalus niger* (0.3%) | 1 (1) | — | 1 |
| Black bird indet. (0.3%) | 1 (1) | — | 1 |
| Passeriformes: Tyrannidae | | | |
| *Tyrannus dominicensis* (0.3%) | 1 (1) | — | 9 |
| Passeriformes: Estrildidae | | | |
| *Lonchura malacca* (1.7%) (caged) | — | 6 (1) | 1 |
| Passeriformes: Hirundinidae | | | |
| *Petrochelidon fulva*[d] (0.6%) | 2 (1) | — | 10 |
| Passeriformes: Thraupidae | | | |
| *Spindalis zena* (0.3%) (caged) | — | 1 (1) | 1 |
| Passeriformes: Turdidae | | | |
| *Turdus plumbeus* (0.6%) | 2 (2) | — | 1 |

*continues*

**Table 5.3 continued**

| Prey | Prey items (Snakes) | | Source |
|---|---|---|---|
| | NAT | ANT | |
| Psittaciformes: Psittacidae | | | |
| *Agapornis roseicollis* (0.3%) (caged) | — | 1 (1) | 1 |
| *Melopsittacus undulatus* (0.3%) (caged) | — | 1 (1) | 1 |
| *Psittacara euops* (0.3%) (caged) | — | 1 (1) | 1 |
| **Mammalia** | | | |
| Artiodactyla: Bovidae | | | |
| *Capra hircus* (0.6%) | — | 2 (2) | 11[1], 1[1] |
| *Ovis aries* (0.3%) | — | 1 (1) | 1 |
| Artiodactyla: Suidae | | | |
| *Sus scrofa* (3.7%) | 5 (1) | 8 (3) | 1 |
| Carnivora: Canidae | | | |
| *Canis lupus familiaris* (1.1%) | — | 4 (2) | 1 |
| Carnivora: Felidae | | | |
| *Felis catus* (1.1%) | 1(1) | 3 (3) | 1 |
| Chiroptera: Phyllostomidae | | | |
| *Artibeus jamaicensis* (8.8%) | 31 (9) | — | 12 |
| *Brachyphylla nana*[e] (0.6%) | 2 (2) | — | 13[1], 14[1] |
| *Erophylla sezekorni* (1.7%) | 6 (3) | — | 15[3], 1[3] |
| *Phyllonycteris poeyi*[e] (6.0) | 21 (10) | — | 1[1], 14[2], 15[3], 16[15] |
| Chiroptera: Mormoopidae | | | |
| *Mormoops blainvillei*[e] (1.1%) | 4 (2) | — | 14[2], 1[2] |
| Bats indet.[e] (1.7%) | 6 (3) | — | 14[1], 1[5] |
| Lagomorpha: Leporidae | | | |
| *Oryctolagus cuniculus* (4.3%) (caged) | — | 15 (2) | 1 |
| Rodentia: Capromyidae | | | |
| *Capromys pilorides* (14.5%) | 51 (37) | — | 17[1], 18[1], 1[49] |
| *Mysateles prehensilis* (0.3%) | — | 1 (1) | 1 |
| *Mesocapromys melanurus* (0.3%) | 1 (1) | — | 1 |
| Rodentia: Muridae | | | |
| *Rattus rattus* (6.3%) | 22 (17) | 5 (5) | 19[1], 20[1],1[25] |
| *Rattus norvegicus* (0.3%) | — | 1 (1) | 1 |
| *Rattus* sp. (0.6%) | 1 (1) | 1 (1) | 1 |

*continues*

**Table 5.3 continued**

| Prey | Prey items (Snakes) | | Source |
|---|---|---|---|
| | NAT | ANT | |
| **Total prey items (351)** | 184 | 167 | |
| **Total snakes (218)** | 114 | 104 | |
| **Total prey species (49)**[f] | 29 | 24 | |

*Sources*:

1. Rodríguez-Cabrera et al. 2020b.
2. Tolson 2012.
3. Holanova and Hribal 2004; Holanova in Rodríguez-Cabrera et al. 2020b.
4. Viña Dávila and Armas 1989.
5. Sampedro Marin and Montañez Huguez 1989.
6. Sampedro Marin 1998; Sampedro in Rodríguez-Cabrera et al. 2020b.
7. Vázquez Milian and Nieves Lorenzo 1980.
8. Godínez et al. 1987.
9. Segovia Vega et al. 2013.
10. Mancina and Llanes Sosa 1997; Mancina in Rodríguez-Cabrera et al. 2020b.
11. Buide 1966.
12. Dinets 2017.
13. Mancina 2011; Mancina in Rodríguez-Cabrera et al. 2020b.
14. Sheplan and Schwartz 1974.
15. Rodríguez-Cabrera et al. 2015.
16. Hardy 1957.
17. Hernández Matínez and Pimentel 2005.
18. Tolson and Petersen 2008; Tolson in Rodríguez-Cabrera et al. 2020b.
19. Borroto-Páez 2011.
20. Tolson and Henderson 1993; Tolson in Rodríguez-Cabrera et al. 2020b.

[a] Possibly *Anolis bartschi* (which certainly represents a different species to *A. smallwoodi*).

[b] *Pseudemys decussata*.

[c] *Columba leucocephala*.

[d] *Hirundo fulva*.

[e] The six bats (i.e., "two *Mormoops blainvillei*, two *Phyllonycteris poeyi*, one *Brachyphylla nana*, and one small unidentifiable bat") reported by Sheplan and Schwartz (1974), were found in a single snake.

[f] Four prey species were consumed in both habitat types.

**Figure 5.50.** *Chilabothrus angulifer* constricting a freshly captured buffy flower bat (*Erophylla sezekorni*) on a cave floor in central Cuba. In natural habitats, bats are more frequently consumed by boas smaller than 2.0 m SVL. Photograph by Raimundo López-Silvero.

(74.3%). Bats are the prey items most frequently consumed in natural habitats (seventy bats in twenty-six snakes) (Fig. 5.50), followed by hutias (fifty-two hutias in thirty-eight snakes), and rats (twenty-three rats in eighteen snakes). *Capromys pilorides*, the largest native terrestrial mammal in Cuba (to ca. 7.0 kg), are also frequently consumed in natural habitats (Fig. 5.51). Predation on *Ca. pilorides* by this boa is also very common on the U.S. Naval Station Guantanamo Bay, where the feces of every adult boa examined contained hutia hairs (Tolson and Henderson 2006; P. J. Tolson in Henderson and Powell 2009). Domestic fowl represented over a half (50.9%) of the prey items consumed in anthropogenic habitats and 68.6% of all birds taken in this habitat type. Only four taxa were consumed in both types of habitat (i.e., domestic fowl, pigs, cats, and rats), although in different proportions. Domestic species that were predated in natural habitats were feral or semiferal individuals. Although in smaller proportion, pets such as cats and dogs were also taken in anthropogenic habitats (seven items, five snakes). Roberto Martínez (in Rodríguez-Cabrera et al. 2020b) commented on a boa (ca. 2.0 m SVL) found swallowing a feral cat (ca. 4.0 kg) at the entrance of a bat cave, which presumably was preying on bats for some time before, as evidenced by the frequently encountered bat remains (wings) that were no longer seen after the cat was predated by the boa.

Neonates of this species are large enough to fall in the size range in which ontogenetic shift in diet from ectothermic to endothermic prey begins in other boa species with much smaller sizes at birth (see Description and Reproduction). Neonates of other *Chilabothrus* species are relatively small, slender, and

**Figure 5.51.** *Top*: *Chilabothrus angulifer* over 3.0 m TL digesting a Desmarest's hutia (*Capromys pilorides*) at the U.S. Naval Station Guantanamo Bay. Photo by Jeff Lemm. *Bottom*: The hutia is the largest native nonvolant mammal in Cuba (to ca. 7.0 kg) and in natural habitats is the most common prey species of boas in excess of 2.0 m SVL. Photo by Tomás M. Rodríguez-Cabrera.

highly arboreal and feed predominantly on sleeping anoles (see the other species accounts in this book), but that niche is occupied by semi-arboreal species of *Tropidophis* in Cuba, which are present in virtually every terrestrial ecosystem across the archipelago (Rodríguez-Cabrera et al. 2016a, 2020a). Rodríguez-Cabrera et al. (2016a) commented on the possibility that the early establishment and impressive adaptive radiation of *Tropidophis* in Cuba possibly prevented the in situ evolution and the later colonization of lineages of smaller specialist *Chilabothrus*. The latter probably also applies for the occurrence of small, arboreal, anole-eating neonate *Chilabothrus*. Consistent with that, Cuban boas are usually born with a minimum SVL and body mass larger than the maximum SVL of any semi-arboreal species of *Tropidophis* in Cuba (to a maximum of 488 mm SVL in *T. wrighti*; for reviews see Henderson and Powell 2009; Rodríguez-Cabrera et al. 2020a). Consequently, juvenile *C. angulifer* will prey upon ectotherms as well as on endotherms (particularly bats, passerine birds, and possibly introduced murid rodents) shortly after birth (Rodríguez-Cabrera et al. 2015, 2020b). Therefore, the sharp ontogenetic shift in diet from endothermic to ectothermic prey observed in most *Chilabothrus* species is not as evident in the Cuban boa (see Rodríguez-Cabrera et al. 2015, 2020b). Contrary to other *Chilabothrus* species, heat-sensing labial pits in the Cuban boa are present (and presumably functional) at birth (Rodríguez-Cabrera et al. 2015). Ectotherms are sporadically taken by small to medium-sized adults, although adults seem to drop completely this kind of prey from their diets as they approach 2.0 m SVL (Rodríguez-Cabrera et al. 2020b). The exception might be the Cuban iguana (*Cyclura nubila*), which can occasionally reach over 1.0 m TL and 5.5 kg (Beovides-Casas and Mancina 2006) and certainly represents an important dietary item for *C. angulifer* in some areas (P. J. Tolson in Rodríguez-Cabrera et al. 2020b). Nevertheless, a different shift in diet in endothermic prey occurs in this boa, mostly related to snake body size and prey size. For example, bats are taken only by juvenile snakes and small adults up to about 2.0 m SVL (see Historical Overview of Bat Predation). Beyond that size, the most frequent prey species consumed in natural habitats is *Ca. pilorides* (Fig. 5.51). Only a few boas smaller than 2.0 m SVL have been reported consuming hutias, and always young individuals <2.0 kg (Hernández Martinez and Pimentel 2005; Rodríguez-Cabrera et al. 2020b). Multiple hutias have been found in the stomachs of snakes larger than 2.0 m SVL, including up to three hutias per stomach in snakes exceeding 3.0 m SVL. Juvenile boas (<1.0 m SVL) in anthropogenic habitats consumed exclusively small caged birds. Snakes from 1.0 to 3.0 m SVL associated with human-altered habitats include a wide range of bird sizes in their diet. Large poultry species such as adult domestic fowl and young turkeys are usually taken by snakes >2.0 m SVL. Ungulates (goats, sheep, pigs) are also consumed by snakes larger >2.0 m SVL. A large boa (3.35 m TL) containing a young domestic goat (*Capra hircus*) of about 3.2 kg was killed near a fisherman's house at Hicacos Peninsula, Matanzas Province (Buide 1966). In 1987 on a farm in the mountains of Guamuhaya, another very large boa (5.65 m TL) was killed and contained a young goat ~6.0 kg (J. D. León in Rodríguez-Cabrera et al. 2020b).

Ophidiophagy seems quite rare in this species. Viña Dávila and Armas (1989) reported a wild Cuban boa slightly >1.0 m TL swallowing (tail-first) a giant trope (*Tropidophis melanurus*) 0.42 m TL in eastern Cuba. A captive-born yearling ingested a sibling only about 1.0 cm smaller after coming into conflict for a lizard (seized by opposite sides at the same time); the snake regurgitated its sibling prey partially digested a few days later (E. Morell, pers. comm. to TMRC, 2008).

*Historical Overview of Bat Predation*

The first reference to possible bat predation by this species was made by W. Palmer in Miller (1904). This author described a cave "a few miles east of Baracoa," eastern Cuba, containing a large bat colony (likely a hot cave according to his description and the bat species collected there: e.g., *P. poeyi*) and pointed out: "On one side of the vertical opening of this cave grew a large tree whose roots descended like a stream into the cavity. The people of the neighborhood assured me that the *majas* (the Cuban boa, *Epicrates angulifer*) coil themselves among these roots and grab at the bats as they fly out. I was told that a snake frequently secures a bat in this manner." Barbour and Ramsden (1919) made reference to a similar situation, described by Palmer and Riley, associated with the bat caves of Guanajay, Artemisa Province, where "the Boas were said by the country folk to take their station at the mouth of the caves and by lunging forward to catch bats from the stream which pours forth just after dusk." The later authors also cited the observations of V. J. Rodriguez ("assistant in the Museo Poey at the University of Havana") in a cave near Maisí (probably also a hot cave according to his description): "A large *majá* was ensconced in a hole at about the level of one's shoulder and with several feet of his body projecting he was making vicious lunges at the passing and repassing bats which we had stirred up." Schwartz and Ogren (1956) found a boa during the day lying extended in an extremely low and narrow tunnel connecting several smaller chambers with the main part of a cave at Guajimico (southern Cienfuegos Province) containing abundant Jamaican fruit-eating bats. During later visits to that same cave several boas were observed on the floor of the same narrow tunnel described by Schwartz and Ogren (1956), but predation was not confirmed (TMRC, pers. obs.).

Hardy (1957) was the first to confirm bat predation by this species, and by any *Chilabothrus*, in a hot cave at Guanayara, Trinidad, Sancti Spíritus Province. He observed a total of forty-one boas either at the cave entrance or in the immediate vicinity of a narrow passage connecting two chambers harboring a large colony of the phyllostomid bat *P. poeyi*. Three snakes were observed during the evening hours constricting or swallowing bats and "the digestive tracts of two others were emptied by forced regurgitation and found to contain, respectively, three and nine bats."

Silva-Taboada and Koopman (1964) and Silva-Taboada (1979) mentioned (based on personal observations and testimonies from local farmers near Omaja, Las Tunas Province) that this species crawls up to the foliage of the "Jata" palm tree (*Copernicia* × *vespertilionum*, a hybrid between *Co. rigida* and *Co. gigas*), which serves as diurnal roosting site for large colonies (up to 3000 individuals; Silva Taboada 1979) of two mollosid bat species (*Mormopterus minutus* and *Nyctinomops laticaudatus*), apparently to feed on them. Additional observations at the plains north of Sancti Spíritus and Ciego de Ávila Provinces corroborate that snakes up to about 1.5 m TL frequent the foliage of these palm trees (including *Co. rigida*) occupied by bat colonies, but bat predation has not yet been confirmed in this context (A. Falcón, pers. comm. to TMRC, 1 April 2020).

Sheplan and Schwartz (1974) commented that bats possibly form a major portion of the diet of cave-associated boas and mentioned a specimen captured at night near the entrance of a cave at "Finca Morales," 12.8 km northwest of Trinidad, central Cuba. This snake contained six bats belonging to at least three different species (two *Mormoops blainvillii*, one *Brachyphylla nana*, two *P. poeyi*, and a small unidentifiable bat) and represents the only record of multiple prey species found in a single boa.

Rodríguez-Cabrera et al. (2015) reported seven very small snakes ranging in size from 505 to 980 mm SVL (some of them with visible umbilical scars) either predating on bats or foraging in two caves in central Cuba; the prey items included three juveniles *Erophylla sezekorni* and three adult *P. poeyi*. The above suggests that the species is capable of taking advantage of these concentrations of food shortly after birth. At the same time, apparently there are some limitations for bat hunting in *C. angulifer* as boas approach 2.0 m TL, which corresponds to small adult snakes (see Reproduction; Rodríguez-Cabrera et al. 2016b; 2020b). Koenig and Schwartz (2003) mentioned that "hunting for bats appears restricted to juveniles" and listed two possible factors that might be responsible: (1) physical constraints of the hunting perch, such as the imposition of gravity on blood circulation when larger snakes hang vertically (see also Chandler and Tolson 1990; Lillywhite and Henderson 1993); and (2) bats represent low- or marginal-energy prey items for adult snakes (see also Cundall and Greene 2002). Bat caves seem to sustain a particular boa cohort only. The forty-one boas collected by Hardy (1957) at Guanayara ranged "from four to eight feet" TL (i.e., 1.22–2.44 m TL). The mean SVL of nineteen individuals measured by Berovides Álvarez and Carbonell Paneque (1998) in a hot cave was 1.56 m, with no significant differences between sexes. The estimated size of nine individuals associated with a sinkhole cave containing a colony of *Artibeus jamaicensis* in Desembarco del Granma National Park ranged from 1.1 to 2.1 m TL (Dinets 2017). Finally, in a sample of nearly 150 boas found associated with bat caves all over the Cuban archipelago from 2002 to 2019, none of them exceeded 2.0 m SVL (mean 1396 ± 287 mm, range 505–1840 mm, $n = 147$), with no significant differences between sexes either (Rodríguez-Cabrera et al. 2015, 2020b). It seems that once the boas exceed an optimal body size that allows them to use bats as a food resource efficiently, they are forced to drop these small and hard-to-capture prey from their diets and abandon bat caves in search of more energetically rewarding prey such as rats, hutias, birds, and domestic animals.

Dinets (2017) described bat hunting by a group of nine boas associated with a sinkhole cave at Desembarco del Granma National Park, confirming thirty-one predation events on *A. jamaicensis*. He reported foraging activity both after sunset and before dawn, which coincides with the exit and return times of the bats (Silva-Taboada 1979). Each boa captured up to a maximum of one bat per hunting session. Since the nine boas were easily recognizable by individual markings, over the eight-day survey five boas were observed consuming four bats each, three boas got three bats each, and one boa got only two bats (V. Dinets in Rodríguez-Cabrera et al. 2020b). He also described that "if multiple boas were present, they positioned themselves in the same part of the passage, forming a barrier," showing a significant preference to aggregate. Successful predation increased significantly when multiple boas co-occurred in the same passage at the same time, leading the author to state that intentional coordinated hunting was the background reason. Nonetheless, owing to the small sample size of hunting sessions involving a single boa ($n = 3$ versus $n = 13$ with multiple boas), additional studies are required to confidently assert that coordinated hunting occurs. Instead of cooperation, Rodríguez-Durán

(1996) and Puente-Rolón and Bird-Picó (2004) repeatedly observed piracy events involving bat predation by Puerto Rican boas (*C. inornatus*).

The way Cuban boas forage for bats in caves or at cave entrances slightly differs from that reported for the closely related Puerto Rican boa (*C. inornatus*) and the Jamaican boa (*C. subflavus*) (see the respective species accounts). These latter two species coil their tails around vines, branches, roots, and rock outcrops and hang down from two-thirds to three-fourths of their bodies (>75%), with the heads slightly lifted (Rodríguez and Reagan 1984; Rodríguez-Durán 1996; Prior and Gibson 1997; Vareschi and Janetzky 1998; Koenig and Schwartz 2003; Dávalos and Eriksson 2004; Puente-Rolón and Bird-Picó 2004). Sometimes they adopt a double S-shape striking posture once the bats commence to emerge. The Puerto Rican boa frequently makes a side-to-side movement while hanging (Rodríguez-Durán 1996; Puente-Rolón and Bird-Picó 2004). Cuban boas position themselves like the two species mentioned above, but they only hang down about one-third of the anterior body, and oscillatory movements have never been observed. Cuban boas also may stretch this one-third of the anterior body either horizontally, obliquely, or vertically with the head directed upward (Rodríguez-Cabrera et al. 2020b). This latter mode is usually adopted when foraging on the ground or from shelving rocks in narrow passages where the cave roof is low enough so the bats are forced to fly close to the foraging boas in order to leave the cave (see Rodríguez-Cabrera et al. 2020b for a review).

Cave-associated *Chilabothrus* that forage for flying bats (i.e., *C. angulifer*, *C. inornatus*, and *C. subflavus*) seem to use this hanging or stretched portion of their bodies as a "tactile antenna." In all the cases observed, it was always necessary for the bats to touch or even to collide with the snakes to trigger the attack, which consisted of a rapid swing with the mouth open toward the origin of the impact. Rodríguez and Reagan (1984) observed the emerging bats frequently colliding with vegetation and foraging Puerto Rican boas and suggested that these collisions may enhance prey capture by the boas. Rodríguez-Durán (1996) and Puente-Rolón and Bird-Picó (2004) also mentioned that Puerto Rican boas try to capture the bats only after they collide with the snakes. Vareschi and Janetzky (1998) observed an increase in the frequency of strikes by Jamaican boas with an increase in the density of bats exiting the cave they studied. Although these authors suggested that the snakes did not strike at individual bats, we suspect that the increase of strikes may be related to an increase in the frequency of touches that comes with a higher density of flying bats. Prior and Gibson (1997) observed a similar behavior in another Jamaican boa and mentioned that only when the bats passed very close or collided with the snake did the snake attempt to capture them. Thermal and chemical prey detection in *Chilabothrus* species seems more important when using an active foraging mode or when employing an ambush strategy not related to hunting flying bats (e.g., forest birds, rodents; see Rodríguez-Cabrera et al. 2020b).

Most of the over two hundred recorded successful predation events by *Chilabothrus* involving bats in caves of the region were phytophagous phyllostomids (Hardy 1957; Sheplan and Schwartz 1974; Rodríguez and Reagan 1984; Rodríguez-Durán 1996; Prior and Gibson 1997; Vareschi and Janetzky 1998; Koenig and Schwartz 2003; Dávalos and Eriksson 2004; Puente-Rolón and Bird-Picó 2004; Mancina 2011; Rodríguez-Cabrera et al. 2015, 2020b; Dinets 2017). Most insectivorous bats forming large colonies in caves of the region (Mormoopidae, Molossidae) have a relatively low body mass (average ca. 9.0 g), large tail membranes, very low wing loading, and low average aspect ratio, which allow highly maneuverable flight. On the contrary, phytophagous phyllostomids are relatively heavy bats (average ca. 30 g in species of the region) with reduced tail membranes, high wing loading, and low aspect ratio, which confer them a limited maneuverability with respect to insectivorous bats (Norberg and Rayner 1987; Mancina 2004; Mancina et al. 2012). All the above suggests that phytophagous phyllostomids are more vulnerable to predation by boas while exiting or entering caves than are insectivorous bats (Rodríguez-Cabrera et al. 2020b).

*Predators and Defensive Behavior*

No predators other than humans have been reported for adult Cuban boas. It is considered a top predator on this archipelago even in modern times (Morgan et al. 1980; Petersen et al. 2007, 2015; Rodríguez-Cabrera et al. 2016a). This snake can even prey on raptors (e.g., *Buteo jamaicensis*, *Caracara cheriway*) and introduced predatory mammals such as cats, dogs, and pigs (Rodríguez-Cabrera et al. 2020b; Table 5.3). The Cuban boa is large and aggressive enough at birth as to be hard-to-capture prey by any potential predator (see Reproduction). Cuban crocodiles (i.e., *Crocodylus acutus* and *C. rhombifer*), some diurnal and nocturnal raptors, feral dogs, feral pigs, and the largest individuals of the alien, invasive Old World catfishes of the genus *Clarias* might be among the few species on the Cuban archipelago capable of predating upon juvenile Cuban boas, but no such records exist. The Broad-winged Hawk (*Buteo platypterus*) is a confirmed predator of the closely related Puerto Rican boa (Puente-Rolón 2012) and has been observed also preying on unidentified snakes in Cuba (N. Navarro, in litt. to TMRC, 11 June 2020). The Red-tailed Hawk (*Buteo jamaicensis*), a relatively common raptor in Cuba, has been widely documented as a snake predator as well, including in the West Indies (e.g., Fitch et al. 1946; Knight and Erickson 1976; Fitch and Bare 1978; Sherrod 1978; Santana and Temple 1988; Gallardo et al. 2019). Wiley and Wiley (1981) reported remains of *Chilabothrus* sp. (referred to as "*Epicrates fordi*," but see the respective account) in the diet of Ridgway's Hawk (*Buteo ridgwayi*) in the Dominican Republic. The Barn Owl (*Tyto alba*) has been previously reported preying upon other boids with a body size equivalent to that of young Cuban boas. Wiley (2010) found remains of both the Hispaniolan boa (*C. striatus*) and the Hispaniolan desert boa (*C. fordii*), among other snakes, in freshly regurgitated pellets of the Barn Owl and the Ashy-faced Owl (*Tyto glaucops*) in the Dominican Republic. Da Costa and Henderson (2013) described an instance of predation by the Barn Owl upon an adult *Corallus hortulana* about 1.0 m TL in Brazil. Arredondo Antúnez and Chirino Flores (2002) found vertebrae of the Cuban racer (*Cubophis cantherigerus*), the second largest snake in Cuba (to 1.3 m SVL; Domínguez Díaz and Moreno García 2006), in seven freshly regurgitated pellets of the Barn Owl in the central part of the island. Gundlach (1883) reported to have found snakes in the stomachs of the Cuban Black Hawk (*Buteogallus gundlachi*), but he did not offer details on the taxonomic identity (see Schnell 1994 for a review). Consistent with that, Gerhardt et al. (1993) repeatedly observed *Boa imperator* (referred to as *B. constrictor*) and other snakes among the prey fed to nestling Great Black Hawks (*Buteogallus urubitinga*) in Guatemala. Scars are frequently observed in free ranging Cuban boas, but whether they are caused by struggling prey or potential predators is unknown.

Endoparasites reported for this species include a sporozoan *Haemogregarina* sp., which may infect up to 1.8% of the erythrocytes (Zajicek and Mauri Méndez 1969), and the cestode *Ophiotaenia nattereri* (Freze and Rysavy 1976; Coy Otero and Lorenzo Hernández 1982); ectoparasites include three species of hard ticks: *Amblyomma quadricavum*, *A. albopictum*, and *Dermacentor nitens* (Pérez Vigueras 1934; Černý 1966, 1969; Coy Otero 1999; Guglielmone et al. 2003). In captivity, the species can be affected by dysecdysis (insufficient shedding), mites, and a kind of fungus that only affects the scales on the head (possibly *Ophidiomyces ophidiicola*) (TMRC, unpubl. data).

The Cuban boa is well known for its bad temper (e.g., Barbour and Ramsden 1919; Lando and Williams 1969; Huff 1976), especially among neonates. Neonates are able to strike repeatedly just a few minutes after birth (Sheplan and Schwartz 1974; Polo Leal and Moreno 2007; Morell Savall 2009). They are prone to flee if given the chance, but once they feel themselves cornered adopt an S-shaped defensive posture with the mouth opened, constantly hissing, and start striking repeatedly toward the potential enemy (Fig. 5.52). Strikes are not a mere warning such as in other *Chilabothrus* species: they actually bite if the potential enemy is within range and may even occasionally hold their bites firmly for several minutes. Like other *Chilabothrus* species, juveniles and even young adults (>1.5 m SVL) of this species may show balling behavior when handled and may hold this posture for several hours if the disturbance persists but may also suddenly shift to aggressive behavior. They can also void considerable amounts of a greenish-yellowish musk from the cloaca with a very strong smell when captured (Gundlach 1880; Garrido and Schwartz 1969; Sheplan and Schwartz 1974; Tolson 1987). For some hyperallergic people, this musk may produce localized skin irritation after contact (A. Hernández Gómez, pers. comm. to TMRC). Every boa has a different temper, and very occasionally tame individuals are found in the wild, but they are unpredictable nonetheless. Wild-caught tame individuals maintained in captivity for some time may suddenly bite and turn aggressive without any apparent reason.

**Figure 5.52.** *Chilabothrus angulifer* is well known for its bad temper; neonates are able to strike repeatedly just a few minutes after birth (*top left*) and striking is also the primary defense of cornered adults (*top right*). However, occasionally both adults (*bottom left*) and juveniles (*bottom right*) will ball up tightly when handled. Photographs by Tomás M. Rodríguez-Cabrera and Aliesky del Río (top right).

*Reproduction*

As demonstrated in most *Chilabothrus* species, female Cuban boas seem to reproduce at least biennially (Huff 1976; Tolson and Teubner 1987; Petersen et al. 2015; Rodríguez-Cabrera et al. 2016b). Viviparous snakes in general invest a large proportion of energy in reproduction, and females in particular allocate large amounts of amino acids and other nutrients to the embryos during development (e.g., Lourdais et al. 2002, 2004, 2005). Female Cuban boas are reported to lose up to 60% of their original body weight during pregnancy (Nowinski 1977; Frynta et al. 2016). In order to accumulate the appropriate amount of lipid reserves and muscular mass, female Cuban boas require a larger size (minimum 1.3 m SVL, 1.7 kg) and probably more time (>5 years) than males (minimum 1.15 m SVL, 0.8 kg; >3 years) to reach sexual maturity in the wild (see growth rates below) (Tolson and Teubner 1987; Rehák 1996; Frynta et al. 2016; Rodríguez-Cabrera et al. 2016b). Tolson and Teubner (1987) suggested that reproductive maturity in female *Chilabothrus* spp. is determined by body size and not age, mentioning two females born in captivity that contained enlarged ovarian follicles (>40 mm diameter) shortly after having reached 3 years of age. Petersen et al. (2015) mentioned that free-ranging male Cuban boas are capable of reproducing every year.

The reproductive activity in species of the genus *Chilabothrus* is seasonal. The first mating pairs or breeding aggregations of Cuban boas are seen at the onset of the rainy season; courtship and mating in the wild have been observed mostly from mid-April to early June (Tolson 1992b; Tolson and Henderson 1993; Morell Savall et al. 1998; Petersen et al. 2015; Rodríguez-Cabrera et al. 2016b). During this period, males may show an intense agonistic behavior (highly ritualized male–male combats) and interference competition for females (Tolson 1983; Tolson and Henderson 1993; Rodríguez-Cabrera et al. 2016b). Breeding aggregations involving multiple males around a single female have been reported since the nineteenth century. Gundlach (1880) commented that these aggregations might have led to a folk legend of a snake with multiple heads. Rodríguez-Cabrera et al. (2016b) reported a breeding aggregation of nine boas (one female 1.52 m SVL, 1.8 kg, and eight males ranging from 1.15 to 1.58 m SVL, 0.8–1.6 kg). Moreover, it is not uncommon to hear people from rural areas of the country talk about this phenomenon, particularly during the early wet season. Breeding aggregations have been repeatedly observed under captivity conditions as well (Tolson 1983; Tolson and Henderson 1993). The courtship process may last for several weeks (Tolson 1992b; Tolson and Henderson 1993; Petersen et al. 2015; Rodríguez-Cabrera et al. 2016b). Tolson and Henderson (1993) described a tactile-chase initial phase in the courtship process in which the male crawls alongside the female and combines tongue flicks with undulations, body looping, chin rubbing, and so on, which is followed by a tactile-alignment phase when the female is stationary; then, tail search starts. *Chilabothrus angulifer* is exceptional because a male merely drapes his tail over the female or encloses her with but a single coil, contrary to other species in which four to five coils have been reported. Vergner (1989) noted that the male "walks" with his pelvic spurs over the back of the female.

There are certain factors described as triggering reproductive behavior in *Chilabothrus* species in general, at least under controlled captive conditions (Teubner 1986; Tolson and Teubner 1987; Tolson 1992b, 1994; Tolson and Henderson 1993). A winter drop in temperature (to around 20 °C) and humidity (to around 40%), including drinking water offered once a week only, followed by a raise in humidity above 80% from May to September (core of the rainy season in Cuba), increases male plasma testosterone levels and stimulates testicular recrudescence. Multiple males in a breeding group stimulate reproductive activity by favoring competition. It has been demonstrated that plasma testosterone levels of alpha males (usually the largest in a group) increase in the presence of a female and other males, especially during male–male ritualized combats; plasma testosterone levels of the smaller males usually decrease. Placing of these managed breeding groups in spacious enclosures (minimum size: 122 × 122 × 244 cm) also increased the chances of reproductive success. Nonetheless, breeding pairs have been successfully bred in smaller enclosures. At the same time, the presence of males, their agonistic interactions, and a prolonged courtship are necessary to stimulate follicular maturation and ovulation in females.

The phenomenon called "false pregnancy" or "pseudo-pregnancy" has been repeatedly reported in this species (Huff 1976; Tolson 1983; Tolson and Teubner 1987; TMRC, pers. obs.). A female develops a large posterior swelling in the abdomen after copulation (a strong indication of pregnancy in *Chilabothrus*), but the abdomen suddenly begins to return to its normal condition showing no further signs of pregnancy. What actually happens is that the female, apparently under the stimulus of the male, develops mature ovarian follicles (the reason for the swelling), but ovulation never take place and follicular atresia (reabsorption) occurs a short time after since fertilization did not occur. Tolson (1983, 1994) suggested that false pregnancy might be due to an insufficient courtship by the male. The presence of a male nearby can stimulate follicular maturation in a female, but more interaction is required to trigger ovulation.

The duration of pregnancy may vary from 152 to 252 days and is highly dependent on the thermoregulatory activities of the female (Huff 1976; Nowinski 1977; Tolson 1983, 1992b, 1994; Vergner 1989; Tolson and Henderson 1993). However, more typical gestation lengths range between 150 and 180 days (Huff 1976; Tolson 1983; Petersen et al. 2015; Rodríguez-Cabrera et al. 2016b), and sperm storage has been suggested for very long gestations (e.g., Nowinski 1977; Tolson 1992b; Tolson and Henderson 1993). Petersen et al. (2015) commented that pregnant females usually seek out sunny spots in the grass or in forest clearings in late afternoon or mid-morning to bask and elevate body temperature. Since both reproductive activity and very young boas have been recorded in or near hot caves (Rodríguez-Cabrera et al. 2015, 2016b), it is possible that pregnant females also make use of the wide thermal gradient in these caves during their thermoregulatory activities.

Considerable evidence suggests a strong synchrony in parturitions, with most births occurring from September to December but especially in October and November (Huff 1976; Nowinski 1977; Murphy et al. 1978; Bloxam and Tonge 1981; Morell Savall et al. 1998; Polo Leal and Moreno 2007; Rodríguez-Cabrera et al. 2016b). Only occasionally has parturition been recorded in January and February and always under captive conditions (Tolson 1983; Polo Leal and Moreno 2007). The appearance of neonates during the late rainy season–early dry season might seem unusual when compared with the hatching period observed in most other West Indian reptiles at the onset or mid-rainy season (see Henderson and Powell 2009 for a review), the time of greatest prey abundance. However, neonates of the Cuban boa are large enough to consume relatively large prey (i.e., bats, rodents, birds) without having any dependence on the peak of hatchling lizards and frogs. This synchrony seems more likely a strategy ensuring the occurrence of gestation during the warmer months since temperature is critical for embryonic development (Rodríguez-Cabrera et al. 2016b).

Litter size varies from one to twenty-eight, depending on the mother's size (Huff 1976; Nowinski 1977; Murphy et al. 1978; Bloxam and Tonge 1981; Tolson 1983, 1992b, 1994; Tolson and Henderson 1993; Morell Savall et al. 1998; Polo Leal and Moreno 2007; Rodríguez-Cabrera et al. 2016b). Young females around 1.4 m SVL and 2.0 kg usually produce two to four offspring; when they reach around 1.8 m SVL they may produce five to seven offspring, and when they exceed 2.5 m SVL they may produce ten to fifteen offspring. Large females up to 3.65 m TL and 20 kg captured on the U.S. Naval Station Guantanamo Bay have produced litters of up to twenty babies (Petersen et al. 2015). It has been predicted that a very large female, such as the one killed on the U.S. Naval Station Guantanamo Bay in 1989 (4.851 m TL; Tolson and Henderson 1993) would have produced more than thirty babies (Petersen et al. 2015). Neonates frequently exceed 600 mm SVL (505–750 mm) and 180 g (88–237 g), although more typical body masses range between 130 and 160 g (Vergner 1978; Bloxam and Tonge 1981; Horlbeck 1988; Tolson 1992b; Tolson and Henderson 1993; Polo Leal and Moreno 2007; Morell Savall 2009; Frynta et al. 2016; Rodríguez-Cabrera et al. 2016b). According to Letsch (1986), juveniles lose about 5% of their weight during the first 10 days. The smallest Cuban boa ever recorded in the wild (with a visible umbilical scar) was a free-ranging juvenile (with visible umbilical scar), which measured 505 mm SVL and weighed 80 g, observed swallowing a freshly captured bat (see Activity and Trophic Ecology) in May 2008 (probably a few months after birth) (Rodríguez-Cabrera et al. 2015). However, Rodríguez-Cabrera et al. (2022) reported even smaller neonate sizes (384 and 402 mm SVL, 42.5 and 54.8 g) in a pair of nonidentical twins in captivity. These twins were encapsulated in the same fetal membranes and were connected to a single yolk sac by separate umbilical cords but showed strong phenotypic discordances. They were also substantially smaller than their normal siblings: that is, between 24% and 30% in length and between 71% and 79% in mass. Relative clutch masses (total mass of the litter divided by the mass of the female) for this species have ranged from 0.145 to 0.415, but no data on relative clutch masses for large females (>3.0 m SVL) have been published (Tolson 1992b; Tolson and Henderson 1993).

Vergner (1989) stated that before parturition the female Cuban boa is nervous. He described that one female gave birth to four neonates in 2 hours and 53 minutes: pushing out the first of them took 33 minutes; in total, there were time intervals of 14–55 minutes to give birth to each of them. Neonates have a considerable amount of yolk that absorbs in a few days (Morell Savall 2009; Petersen et al. 2015; Frynta et al. 2016). The first shed occurs after yolk absorption, usually before the 20th day after birth (9–37 days; Sheplan and Schwartz 1974; Huff 1976; Bloxam and Tonge 1981; Vergner 1978; Polo Leal and Moreno 2007). During this time the babies remain close together, without eating or drinking; only after the first shed they do readily accept food (Bloxam and Tonge 1981; Morell Savall 2009). The same pattern seems to occur in the wild. In November 2006 four neonates were found together under a log in the botanical garden at Universidad Central "Marta Abreu" de Las Villas, central Cuba. All four boas had the eyes and skin dull and whitish, indicative of being close to their neonatal shed. This grouping suggests that they had remained together for several days. Repeated observations of several young snakes (with visible umbilical scars and apparently belonging to the same litter) in a particular locality at the same time (Lando and Williams 1969; Rodríguez-Cabrera et al. 2015) might be the result of an incipient dispersal after they split up from the clutch.

Frynta et al. (2016) observed in a captive colony (42 mothers, 62 litters, and 306 babies) that the sex ratio was highly influenced by the mother's size (large mothers tend to produce females and smaller ones tend to produce males). These authors attributed such variations in the sex ratio to what they called a modification of the local-resource competition hypothesis, a kind of adaptive maternal manipulation. In such a case, any deviation from the primary sex ratio would be explained by selective mortality of the embryos or fetuses. They also found no sex difference in body shape and size at birth. However, later in a characterization of the karyotype of the Cuban boa, Augstenová et al. (2019) found that this species has 2n = 36 chromosomes (16 macro- and 20 microchromosomes) and that sexual chromosomes are absent. This suggests that environmental sex determination (temperature and hydric conditions) might be the cause for the differential sex ratios in the species and may at least partially explain the observations of Frynta et al. (2016). Rovatsos et al. (2015) included this species in an analysis demonstrating conservation of sex chromosomes through Z-specific (pseudo) autosomal genes also present in other caenophidian snakes. Seixas et al. (2020) showed that the Cuban boa is capable of reproducing via facultative parthenogenesis, at least under captive conditions, which might be another reason for the production of male- or female-biased litters (see also Frynta et al. 2016).

Young boas kept in captivity, subjected to a forced sedentary lifestyle, usually show growth rates approaching 25–30 mm per month (Tolson 1992b; Tolson and Henderson 1993; Morell Savall et al. 1998; Polo Leal and Moreno 2007; Morell Savall 2009). Some authors have reported growth rates approaching 10 mm during the first 3–4 months of life but decreasing progressively as the snake ages (Morell Savall et al. 1998). Vergner (1989) mentioned that in the first year the most active juveniles that were born in captivity increased 130 mm in size and weighted 1.2 kg. Observations in captivity suggested that total length shows a linear increase, while the body mass shows an exponential increase, at least during the first year (Polo Leal and Moreno 2007). The differential feeding rate (4–20 times/yr) of seven snakes born in captivity was directly correlated with both the growth rate (7–19 mm/mo) and the frequency of sheds (5–7 sheds/yr; Morell Savall 2009). A captive female in Cuba grew from 1420 mm TL in 2017 to 1830 mm in 2020: 410 mm in 1202 days (about three years), for an average growth rate of about 10 mm/mo (A. Hernández Gómez, pers. comm. to LMD and TMRC). However, marked and recaptured individuals associated with bat caves have shown growth rates considerably lower than those observed in captivity. The average increase in length per month of eight marked and recaptured cave-associated boas ranged from 7 to 20 mm (mean 14.8 ± 4.1 mm; Rodríguez-Cabrera et al. 2016b). Three females ranging in size from 1.25 to 1.39 mm SVL marked in bat caves of central Cuba were recaptured nearly 5 and 7 years later, showing an average growth rate <2 mm/mo (Rodríguez-Cabrera et al. 2020b). However, a marked juvenile male of 665 mm SVL was recaptured ~5 years later (58 months) and had increased 895 mm, for an average growth rate of 15 mm/mo. This is reasonable to expect if we take into account that free-ranging boas must partition their energy budget among various other activities besides growth, such as foraging, competition, defense, reproduction, and interaction with unpredictable climatic variables such as temperature and humidity. However, as occurs in captivity, free-ranging juveniles seem to show a much higher growth rate than do subadults and adults. Once they approach sexual maturation the average growth rate decreases significantly. A genetic component determining the maximum adult size seems to exist also in this species. Some individuals born in captivity show differential growth rates despite being kept under the same conditions and frequency of meals. A female born in captivity at 660 mm TL (Morell Savall 2009) had reached 2.45 m TL and 9.5 kg before her 8th year (Rodríguez-Cabrera et al. 2016b), and she died at 2.6 m TL shortly after her 19th year (E. Morell Savall, pers. comm. to TMRC, 22 April 2020). This female showed a growth rate of 19 mm/mo during her first year (Morell Savall 2009); during the next 7 years (79 months), until she mated and gave birth (Rodríguez-Cabrera et al.

2016b), she maintained approximately that same average growth rate (ca. 20 mm/mo); after that and until her death more than 11 years later (140 months), her average growth rate was barely 1 mm/mo.

The only available longevity data are from boas kept in captivity. Shaw (1957) reported a specimen from the San Diego Zoo still alive at 18 years old. Bowler (1977) reported an age of 22 years and 7 months, whereas Henderson and Powell (2009) stated a lifespan of more than 24 years. Frynta et al. (2016) mentioned individuals of up to 27 years old in a captive-born colony kept in the Czech Republic. Petersen et al. (2015) commented that boas may live more than 30 years in the wild. A female born in captivity in September 2000 (Morell Savall 2009) died apparently of dysecdysis in January 2020 (19 years, 4 months; E. Morell Savall, pers. comm. to TMRC, 22 April 2020). Considering its low growth rate even under maximized food intake in captivity (see Reproduction), Cuban boas exceeding 4.0 m TL in nature must certainly approach 50 years in age at minimum.

*Conservation Status*

Like most boid snakes, *C. angulifer* is seriously impacted by human activities. It was first assessed as Near Threatened on the IUCN Red List (Day and Tolson 1996), but it was recently moved to the category of Least Concern (Fong 2021) based on its large extent of occurrence (~223,381 km$^2$) and on the thoughts that it has "a stable population, it occurs in several protected areas, it can tolerate certain disturbed situations, and the threats do not affect it fast enough to qualify for a more threatened category." However, some of these assumptions have little support and require a more thorough revision. In 2012 it was also listed as Near Threatened in the Libro Rojo de los Vertebrados de Cuba (Polo Leal and Rodríguez Cabrera 2012). The species is listed on Appendix II of CITES. Some Cuban regulations (Ministerio de Justicia 2011) have been written to protect boas and other native species from being poached. According to data from CITES (Outhwaite 2019), *C. angulifer* (along with *Corallus grenadensis*; see account) was the species that accounted for the majority of reptiles in the trade in endemic Caribbean reptiles, with 204 live individuals imported between 2013 and 2017. Of note is that this only represents reported imports.

When not killed, the species can tolerate a high degree of human disturbance, coupling its feeding habits to the available prey. But the latter has consequences. In rural areas the species is persecuted because of the historical human–predator conflict, since it preys on domestic animals (Rodríguez-Cabrera et al. 2020b). The habitat loss and scarcity of hutias in some areas due to overhunting by humans and interference competition with introduced feral dogs (Borroto-Páez 2009) might be pushing the boas, particularly the largest individuals, to farms, small villages, and towns in search of alternative food sources (Rodríguez-Cabrera et al. 2020b). This increases the rate of encounters with humans, which almost always end in the boa's death under a machete blade. In more urbanized areas it is not rare to find boas killed on the roads (Petersen et al. 2015; Rodríguez-Cabrera et al. 2016b; TMRC, pers. obs.). The accelerated development of human settlements and the tourism sector, with the subsequent construction of luxurious hotels and golf courses, are continuously taking big pieces of natural habitats from this and other native species.

In Cayo Santa María, north of Villa Clara Province, the boas and the hutias (*Ca. pilorides*) used to be relatively common. With the development of the tourism sector during the last decades, less and less area covered by native forest remain on the easternmost part of the cay. Also, escaped dogs and cats have become damaging pests in the area as they tend to form feral packs, particularly dogs. Nowadays hutias are very scarce on the cay, as are the boas (A. Arias Barreto, pers. comm. to TMRC, 19 April 2020). The occurrence of some boas deep in the mangrove south of the cay where the hutias persist might be the consequence of a displacement due to the interference competition with introduced mammalian predators (A. Arias Barreto, pers. comm. to TMRC, 19 April 2020).

Hot caves, where the largest concentrations of Cuban boas ever reported occur (see Abundance), are exploited for bat guano to be used as fertilizer. During this process, many of the people entering those caves are afraid of the snakes or simply harbor some resentment toward them because of the loss of domestic animals by predation. In any case, many boas are killed indiscriminately by bat guano harvesters. More recently, boas are being captured to be used in some religions of African origin, in which the snakes must certainly die as part of a ceremony or offering (Rivalta González et al. 2003). Many "boa hunters" also know about the existence of these boa concentrations in hot caves and use them as regular sources of snakes to supply the growing religious market (TMRC, pers. obs.). Other well-known hot caves that were subject of important research in the past are now completely altered and modified as human refuges during war. The latter caves no longer function as hot caves, and their native faunas are gone. Quarrying, particularly for marble and other construction materials, is another cause of cave destruction in Cuba. Hot caves offer year-round favorable and stable conditions (i.e., shelter, food, thermal gradient for pregnant females) to support large numbers of Cuban boas of several age classes, including the most vulnerable ones (i.e., juveniles). Puente-Rolón et al. (2013) found that cave-associated (higher density) populations of the closely related Puerto Rican boa show as much genetic diversity as surface populations (lower density). Owing to the close genetic relationship and similar behavior and foraging modes between these two species, the same phenomenon might be happening with cave-associated populations of the Cuban boa. All the above strongly suggests that hot caves and their associated fauna represent critical sites for the conservation of the species all over the country.

In rural areas the species has traditionally been killed to obtain its fat (which has attributed medicinal properties; at least eight different afflictions are treated with this) and the skin (to be used in leatherwork, and it also has some presumed medicinal properties) (Poey 1866; Linares Rodríguez et al. 2011). Although the species is not especially persecuted for food purposes, it is consumed by some people from the countryside. Berovides Álvarez and Carbonell Paneque (1998) and Linares Rodríguez et al. (2011) calculated the proportional mass of the body parts (meat, skin, fat) in *C. angulifer* and concluded that over 90% of its body is potentially usable by humans.

Alberts et al. (2001) reported the status of the Cuban boa on the U.S. Naval Station Guantanamo Bay. These authors concluded that since 1995, when considerable portions of habitat were cleared on the base to create temporary housing for Haitian and Cuban migrants, a decrease in the number of boas encountered in biannual visits was evident. They also suggested, although without formal population surveys, that it was possible that the decline in the boa population might be related to an abnormally high concentration of hutias (*Ca. pilorides*) in the area (see Activity and Trophic Ecology). Consistent with that, Comas González et al. (1989, 1994) reported mean relative densities of 35.6–73.6 hutias/ha in second-growth forests at Sierra del Chorrillo Managed Resource Protected Area, Camagüey Province. These authors recorded severe damages caused to the native vegetation, with over 80% of the total trees sampled presenting signs of having been attacked by these rodents; several trees were even killed by the hutias. Rodríguez-Cabrera et al. (2016a) commented that these rodents have the potential to explode demographically when associated with highly productive ecosystems where the incidence of natural controllers is low. The Cuban boa has been reported from this same area (Rodríguez Schettino et al. 2013), but apparently the population of this predator here is insufficient to control the hutia population. Comas González et al. (1989) commented that this high density of hutias could be related to the strict protection of the species in this area and the fight against introduced feral dogs, probably the most important competitors with the Cuban boa in natural ecosystems.

*Myths and Legends*

*Chilabothrus angulifer* is probably the reptile species with the most colorful and numerous folk legends in Cuba (Buide 1985; Rivalta González et al. 2003). The most famous of all stories related to this species is that of the "Madre de Aguas" (i.e., Mother of Waters), about a huge magic snake with horns that lives in the rivers and calls like a rooster. According to people from the countryside, this Madre de Aguas is so big that it is capable of swallowing a calf. Another common story is that this snake slips under the sheets to approach a sleeping post-parturient woman holding her baby and, while the snake places the tip of its tail in the baby's mouth to avoid him/her crying, it sucks the milk from the mother's breast. Another story tells that a boa swallowed the entire leg of a man who fell asleep under a tree, and as consequence that leg later atrophied. They also say that if a pigeon captured by this boa opens its wings forming a cross, the snake will be unable to swallow it. According to another folk legend the Cuban boa is capable of paralyzing its prey using hypnosis, a myth attributed to snakes elsewhere world-

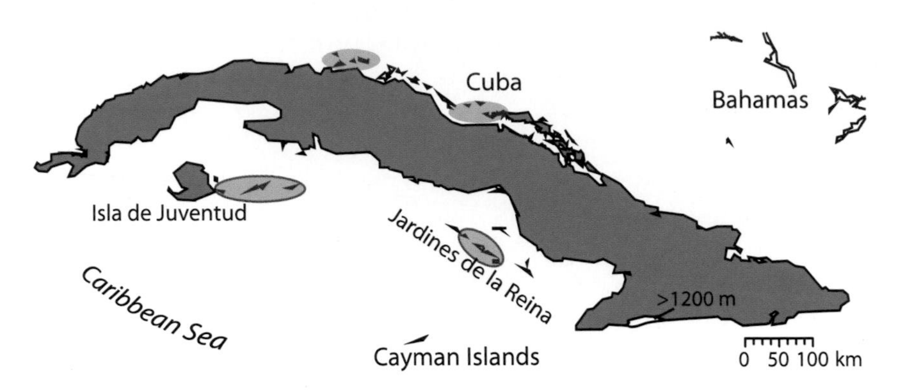


**Figure 5.53.** Range of *Chilabothrus angulifer* on Cuba and satellite islands, up to an elevation of ~1200 m. Note that the highest elevations occur in southeast Cuba. Boas have been reported on some, but not all, of the cays around the main island of Cuba.

wide. Yet another myth (shared with other snakes) says that if its body is cut in several parts, the snake is able to join those parts again and continue living normally.

There are stories repeated in rural areas all over the country related to Cuban boas stalking prehensile-tailed hutias (*Mysateles prehensilis*) in the trees. This hutia species is very arboreal and agile. They say that once a hutia is cornered by a boa toward the tip of a branch, it jumps to the ground and the boa follows it, capturing the hutia in mid-air and reaching the ground already coiling its prey (Gundlach 1880). This folk legend might be largely based on the somewhat frequent behavior whereby a boa will drop to the ground once it captures a relatively large and heavy prey item in the trees, so that it may finish swallowing (Rodríguez-Cabrera et al. 2020b).

Similar to the Madre de Aguas, Amerindians in the region of Guyana consider the green anaconda a sacred spirit of the waters called "Espíritu de las Aguas" (see Robiou Lamarche 2004 for a review). Among the Taíno (who likely originate from Guayana or thereabouts), there was an insular equivalent of the great snake controller of the weather called *Iguanaboína* (possibly also *Atabey* in one of its forms), one of whose names also means Madre de Aguas. There was also a Carib deity called *Obubera*, who was an equivalent of this sacred river giant snake called *Camudi* in Caribbean South America and was certainly inspired by large species of *Chilabothrus* occurring on Cuba, Hispaniola, and Puerto Rico (Robiou Lamarche 2004).

A further story about a snake with multiple heads is probably related to the breeding aggregations frequently observed at the onset of the rainy season (see Reproduction). Nonetheless, the news of a two-headed baby Cuban boa (650 mm TL) captured by a local man at Bolondrón village, Matanzas Province, caused sensation in the region and the newspapers of the country in January 2007, especially because a picture of the animal accompanied the article (de Jesús 2007; Juventud Rebelde 2007). The man found the snake on 20 December 2006 and was still keeping it when the journalists interviewed him. He stated that he fed the snake with little frogs and egg yolk by both heads, and he saw it drinking water with one of the heads. No further news has been published about this neonate, but it represents one of only two cases of bicephalia reported in wild Cuban boas. Gundlach (1880) also made reference to another bicephalous young *C. angulifer* that refused to eat and died. Figure 5.53 shows the distribution of *Chilabothrus angulifer*.

## HISPANIOLA

The second largest of the Greater Antilles, the island of Hispaniola stretches 650 km across the belt of the northern Caribbean. The main island encompasses an area of 76,500 km$^2$ and is administered by two nations, the Republic of Haiti in the west and the Dominican Republic in the east. The highest point is Pico Duarte, rising 3087 m asl and representing the highest peak in the Caribbean, while the lowest point is 46 m below sea level, the site of a series of saline lakes in the Valle de Neiba/Plaine du Cul-de-Sac. The island is further bisected by a series of ridges and valleys, yielding a complex landscape of montane pine forests, lowland tropical wet forest, and tropical dry forest in bands from roughly west–northwest to the east.

There are three larger islands surrounding Hispaniola, including Île de la Gonâve (689 km$^2$, max. 778 m asl) in the Gulf of Gonâve; Île de la Tortue (178 km$^2$, max. 459 m asl) to the northwest; and Isla Saona (105 km$^2$, max. 459 m asl) to the southeast. All three of these islands have boas. Navassa Island, administered by the United States, lies 56 km off the western tip of the Tiburon Peninsula and shares a herpetofaunal affinity with Hispaniola, although no boas are recorded there. A number of smaller satellite islands surround Hispaniola, many with boa populations. A series of submerged banks, the Silver, Navidad, and Mouchoir, are geologically part of the Lucayan Archipelago (although administered by the Dominican Republic and disputed by the Turks and Caicos) and stretch across >250 km of sea between the Samaná Peninsula and the Turks Bank. These banks, when emergent during periods of lower sea level, would have brought the Lucayan Archipelago to within only 60 km of Hispaniola.

The present-day island of Hispaniola is the result of a complex geologic history, probably involving at least four Eocene paleo islands, culminating in the collision of two large paleo islands. The south island, what is now the Tiburon Peninsula, and the north island collided in the Miocene, shaping much of the topographical complexity during the Pliocene and Pleistocene. The juncture between these two islands lies in the vicinity of the Sierra de Neiba and the Enriquillo Trough (Valle de Neiba/Plaine du Cul-de-Sac). Humans have been present for at least 800 years, having built a large civilization of possibly up to a million Taíno people prior to its collapse following the introduction of European disease. Further oppressive and violent persecution led to the loss of the once-vibrant Taíno culture.

Biogeographically, the island of Hispaniola is especially rich and well positioned. Its proximity to Puerto Rico to the east, the Lucayan Archipelago to the north, and Cuba to the west has likely facilitated a number of important biogeographic dispersals to and from other Greater Antillean islands. This, combined with a long paleo history of emergence since the Miocene, topographical complexity, and a huge variety of habitats has generated a tremendous amount of diversity on the island. The herpetofaunal diversity has long been appreciated, particularly after chronicling by Cochran (1941). Four extant boa species are found here, all of the genus *Chilabothrus*, and three have enough intraspecific morphological or genetic divergence and diversity to warrant recognition of subspecies.

**Figure 5.54.** Coastal habitat on the Samaná Peninsula, Dominican Republic, home to both *Chilabothrus striatus* and *C. gracilis*. Photo by R. Graham Reynolds.

## HISPANIOLAN VINEBOA

*Chilabothrus ampelophis* Landestoy T., Reynolds, and Henderson 2021

### *Taxonomy*

This is the most recently discovered West Indian boa and the first to be described from Hispaniola in 133 years; the first and only specimens were discovered in situ in 2020. Its closest relatives are the Hispaniolan endemics *C. fordii* and *C. gracilis*. Based on molecular phylogenetic analysis, the species diverged from an ancestor shared with *C. fordii* ~3.76 MY during the Pliocene (Landestoy T. et al. 2021b).

### *Etymology*

The epithet is from ancient Greek *ampelos*, meaning vine, in allusion to the slender body and head shape and given the abundance of vines in the habitat at the type locality. The suffix *-ophis* refers to a snake, and the species name is translated as "vinesnake."

### *Description*

This species is among the smallest of the West Indian boas and is possibly the smallest member of the family Boidae. It has very small body size (SVL to 683 mm); a slender habitus (body laterally compressed); a head distinctive from neck and flat in profile view; and a snout that is attenuated and rather long (Figs 5.55, 5.56). The eyes are slightly protuberant to the level of the frontal plane (in profile view) and directed anterolaterally. The species has a moderate to high ventral scale count (263–272) and moderate to high number of subcaudals (86–93). The dorsal scale rows at midbody number 38–40, supralabials 15, infralabials 15–16, loreals 4–5, circumorbitals 13–16, preoculars 1–2, and supraoculars 4/4–8/8 (Landestoy T. et al. 2021b).

The dorsal ground color is gray with scales densely stippled with dark brown or black. The primary dorsal elements form a series of dark-brown to black diagonal blotches that are usually fused, thus creating a zigzag pattern, each edged with very pale scales. A lateral dark-brown to black stripe that originates on the head

**Figure 5.55.** *Chilabothrus ampelophis* from southwestern Dominican Republic. Photo by Miguel Landestoy.

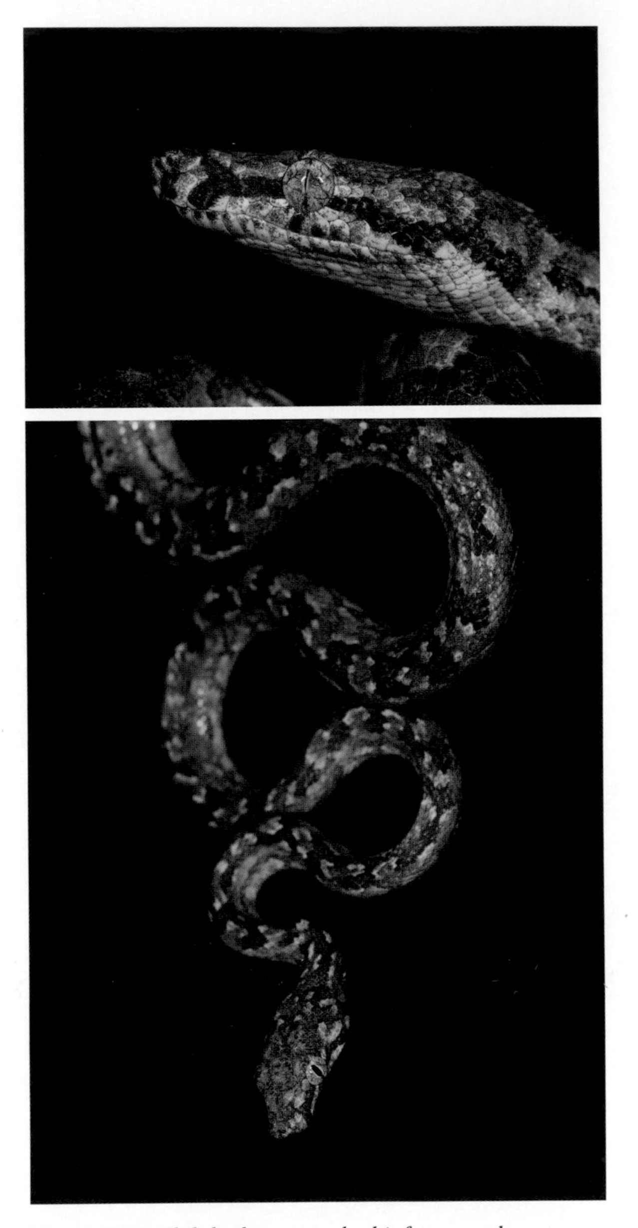

**Figure 5.56.** *Chilabothrus ampelophis* from southwestern Dominican Republic. *Top*: Head portrait showing the lateral scale patterns. *Bottom*: Note the similarity in body position to other highly arboreal boids, such as *Corallus cookii*. Photos by Miguel Landestoy.

repeatedly and irregularly branch toward dorsolateral and lateroventral areas, forming reticulated markings laterally. The dorsal surface of the head is dark brown to blackish overall; the central parietal area has a faint, pale rectangle (at times bisected posteriorly) followed by diffuse pale spots arranged transversely in the occipital region; pale lobes enter the parietal areas from the temporal region. Laterally, there are two horizontal Y-shaped stripes, one of which is a postocular pale-cream stripe and the other a preocular dark-brown to black stripe. The venter is whitish, with dark-brown to blackish stippling overall and peripheral smudges on the ventral scales (Landestoy T. et al., 2021b).

*Distribution*

This species is known from only five specimens, all of which were found a few kilometers north of Pedernales (80–105 m elevation) in Pedernales Province, Dominican Republic. The species occurs very close to the border between the Dominican Republic and Haiti. The extent of this species' distribution is unknown but likely conforms to the karstic foothill dry forests of the southern Sierra del Bahoruco. Extensive herpetological work has been conducted at Parque Nacional Jaragua as well as other locations along the Barahona Peninsula, which suggests that the species is highly restricted to the vicinity of the type locality.

*Habitat*

The type locality occurs in a karst uplift that lies along the low plains associated with the Río Pedernales, which forms the main landmark for the border between the Dominican Republic and Haiti. The vegetation in the lowest areas is more mesic, but this changes with the rise of the limestone bedrock where vegetation rapidly becomes xerophytic (acacias, cacti), although relatively large trees, vines, and undergrowth also occur there (Fig. 5.57).

One boa (357 mm SVL) was found stretched out horizontally at 1.5 m on a thin branch of the legume *Senna atomaria* beside a narrow logging trail on the slope of a hill. It is unknown if the snake was actively foraging or in a sit-and-wait posture to ambush prey. A boa (629 mm SVL) was encountered at 2340 hours stretched out at 3.0 m in a *Phyllostylon rhamnoides* tree. One individual (486 mm SVL) was found at 2240 hours, coiled and apparently inactive at 1.25 m

**Figure 5.57.**
*Top*: *Chilabothrus ampelophis* are very small and slender and can be quite hard to locate in their natural habitat (note the individual in the center of the photo). *Bottom*: Typical habitat of *C. ampelophis* in the southwestern foothills of the Sierra del Bahoruco. Photos by Miguel Landestoy.

in a young *Acacia skleroxyla* tree. The tree was surrounded and overlapped by vines, under cover of ~70% of the main canopy of an 8-m tall *P. rhamnoides*. At 2000 hours, another boa (697 mm SVL) was spotted stretched out and moving slowly at 3.5 m in a ~6.0 m tall *Bursera simaruba* tree. Another *C. ampelophis* (564 mm SVL) was found at 1930 hours stretched out at 2.0 m and moving slowly from a *P. rhamnoides* to a cactus (Landestoy T. et al., 2021b).

*Abundance*

We have no data on abundance, but all five individuals were encountered within an 800-m stretch of habitat from one another.

*Activity and Trophic Ecology*

The diurnal lizards *Anolis alumina*, *A. brevirostris*, *A. cybotes*, and *A. strahmi* are commonly encountered roosting at night in small bushes and vines and represent potential prey. Additional potential prey includes

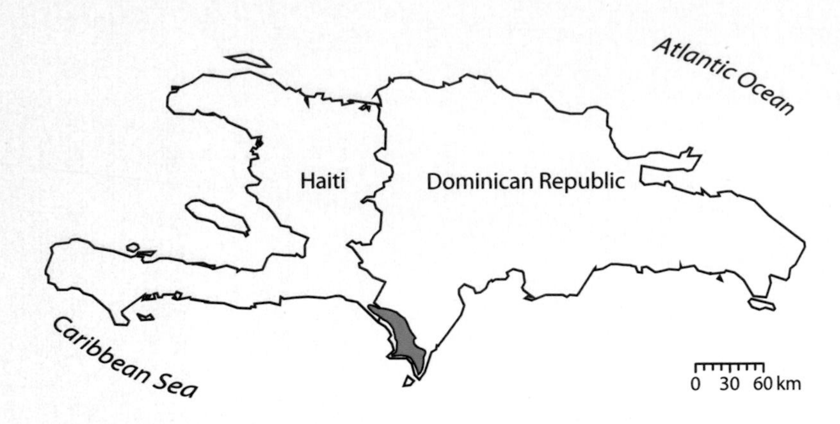


**Figure 5.58.** Range of *Chilabothrus ampelophis* in the southwestern Dominican Republic. The species possibly occurs just across the border into extreme southeastern Haiti, although the habitat there is degraded. This species is largely restricted to limestone dry forest.

the frogs *Osteopilus dominicensis* and *Eleutherodactylus alcoae*, as well as the gecko *Aristelliger expectatus*. While in captivity, two *C. ampelophis* consumed *A. brevirostris* offered to them.

All five individuals of *C. ampelophis* were encountered at night, and four of the five were active (one was coiled and inactive). Based on these limited data, it is likely that *C. ampelophis* is an active forager that hunts for quiescent prey, primarily *Anolis* lizards, at night (Landestoy T. et al. 2021b). Lizards in general, and anoles in particular, are abundant and ubiquitous and are important components in the diets of many West Indian snakes, including boas (Henderson 2015). The foraging behavior and diet of *C. ampelophis* likely parallels that of the slender and highly arboreal *C. gracilis*.

*Predators and Defensive Behavior*

Although no predation has been observed, we suspect that predators of *Chilabothrus ampelophis* are likely the same as those that might prey on *C. fordii* and *C. gracilis* (e.g., raptors, rats, cats).

While being photographed, the tails of two of the boas became rigidly straight while the posterior portion of the snakes' bodies were hanging off branches. We suggest this behavior may have the purpose of either balancing or crypsis (i.e., simulating a smaller branch from the one supporting most of the boa's body). One individual whipped and undulated the tail when grabbed anteriorly, possibly in an attempt to distract attention away from the head. Also, when handled, most individuals coiled into a tight ball with the head hidden. Voiding of the cloaca and musking were also employed as defense strategies. Only one of the boas attempted to bite, and it was in a pre-shed condition (Landestoy T. et al. 2021b).

*Reproduction*

Nothing is known about reproduction in this species, but we assume it is viviparous.

*Conservation Status*

This species has not yet been assessed on the IUCN Red List, but Landestoy T. et al. (2021b) suggested that this species fits the criteria for Critically Endangered. Like all boas, it should be considered a CITES Appendix II species pending further research. The type locality lies along the Dominico-Haitian border (Fig. 5.58), and there is intensive agriculture and cutting of trees for charcoal production at the type locality of *C. ampelophis*. Locally, there is habitat alteration by the removal of small trees and bushes of *Amyris* spp., the essence of which is used in the perfume industry. The final destination of all of these forest products is Haiti. Also, free-roaming cattle may negatively affect the habitat. Other endangered species are found here (*Anolis strahmi*, an undescribed *Tropidophis*, and *Peltophryne armata*). We anticipate that the habitat of this new species is under threat from these resource exploitation activities, and we urge additional work to further characterize the conservation status of the species (Landestoy T. et al. 2021b). Figure 5.58 shows the distribution of *Chilabothrus ampelophis*.

## HISPANIOLAN DESERT BOA

*Chilabothrus fordii* (Günther 1861)
*Chilabothrus fordii fordii* (Günther 1861)
*Chilabothrus fordii agametus* (Sheplan and Schwartz 1974)
*Chilabothrus fordii manototus* (Schwartz 1979)

### *Taxonomy*

The original name for this species was *Pelophilus fordii*. Taxonomic changes included *Chilabothrus maculatus* Fischer 1888 and *Epicrates fordi* by Boulenger 1893 (subsequently referred to as *Epicrates fordii*); Stull (1935) considered it to be a subspecies of *Ep. inornatus*. Sheplan and Schwartz (1974) were the first to recognize subspecies of *Ep. fordii*. Sheplan and Schwartz (1974) noted that the correct specific name is *fordii* but that *fordi* should have been the proper name as the species is named for the individual G. H. Ford. According to Article 32.5 of the International Code for Zoological Nomenclature (ICZN), the original spelling does not constitute an "inadvertent error"; thus, the original specific epithet *fordii* stands. Subsequent authors have used both spellings, occasionally using the spellings interchangeably in separate treatments. Earlier works frequently used *fordi* (e.g., Schwartz 1979; Henderson et al. 1987; Tolson 1992b; Tzika et al. 2008; Tolson and Henderson 2011), while a mixture of older and most recent works recognize ICZN authority and use the spelling *fordii* (e.g., Tolson 1987; Kluge 1989; Reynolds et al. 2013a, 2014a, 2015, 2016a).

### *Etymology*

The species name *fordii* is a patronym honoring G. H. Ford, the illustrator of the plate accompanying the original description (Wetherbee 1989). The subspecies name *agametus* is Greek for bachelor and refers to the unique male holotype (Sheplan and Schwartz 1974). The subspecies name *manototus* is Greek for "most rare" and is in reference to the apparent scarcity of these boas on Ile à Cabrit (Schwartz 1979). There are no distinct common names for the subspecies (Hedges et al. 2019). There is no specific vernacular name for *C. fordii* in the Dominican Republic as it is often believed to be a juvenile *C. striatus* (M. A. Landestoy T., in litt. to RWH, 21 August 2018). In Haiti it is likely known as "koulèv" or "sèpan."

### *Description*

A small boa with known maximum SVL of 860 mm in males and 730 mm in females, although it is presumed that the species can reach slightly longer sizes (<1.0 m SVL). Dorsal scale rows are 31–39 at midbody, ventrals 231–261 in males and 236–259 in females; subcaudals are 69–85 in males and 70–89 in females; supralabials 11–15 (usually 13); and infralabials 12–16 (usually 14).

*Chilabothrus fordii fordii*: Dorsal scale rows at midbody 31–39; ventrals 231–259 in males, 231–259 in females. The dorsal ground color is gray with 58–88 milk-chocolate brown dorsal body blotches outlined in black and paler gray (Fig. 5.59). The blotches are usually discrete, but in many boas they are variously fused to give a chain-like effect. The color of the dorsal blotches is more intense in young snakes (solid color and somewhat reddish brown). Laterally a series of irregular blotches or other markings are of the same color as the dorsal blotches (Figs. 5.60, 5.61). There are 13–27 blotches on the tail. The head pattern consists of a loreal–postocular stripe that ends abruptly on the temporal region and a "scroll-like" or "lyre-shaped" dark-brown figure on top of the head. The dorsal head pattern ends abruptly posteriorly and is the first of the series of dorsal blotches (Fig. 5.60). The venter is dark gray with brown spots, flecks, and streaks (Sheplan and Schwartz 1974).

*Chilabothrus fordii agametus*: Dorsal scale rows at midbody 38, ventrals 261, subcaudals 82, dorsal body blotches 92, and tail blotches 17. In preservative the dorsum is dull grayish tan, the dorsal dark blotches medium brown and outlined in black; the blotches are sometimes fused across the back to form saddles; lower sides with a longitudinal series of irregular dark markings. The venter is dull gray with diffuse and irregularly spaced darker spots posteriorly. The dorsal and ventral surfaces of the tail have the same patterns and colors as the body. The head has a dark-brown loreal–postocular stripe; upper surface of the head is darker gray with vague pale gray vermiculations entering the top of the head coloring posteriorly. The chin and throat are nearly patternless except for dark gray smudging on the mental and infralabials (Sheplan and Schwartz 1974).

*Chilabothrus fordii manototus*: Dorsal scale rows at midbody 33–36; ventrals 257–263 in females; 72–74 dorsal body blotches. The dorsal ground color is dark

**Figure 5.59.** *Chilabothrus fordii* from La Descubierta near Lago Enriquillo, Independencia Province, Dominican Republic. Photo by Miguel A. Landestoy.

**Figure 5.60.** *Top*: *Chilabothrus fordii* from La Descubierta near Lago Enriquillo, Independencia Province, Dominican Republic. Photo by Miguel A. Landestoy. *Bottom*: *Chilabothrus fordii* from 10.4 km northwest Ca Soleil, Départment l'Artibonite, Haiti (130 m asl). Photo by S. Blair Hedges.

gray to pale brown; dorsal blotches brown to reddish brown with brown to dark-brown edges. Each blotch is outlined by pale gray. There is an incomplete lateral or ventrolateral series of smaller blotches irregularly present. The dorsal surface of the head is reddish brown with a prominent postocular stripe to the angle of the jaws on each side. The venter is dirty white to pale gray with scattered dark-gray, square to rectangular blotches on most scales and restricted individual scales (usually near their lateral margins) (Schwartz 1979). The iris is grayish with brown flecks (Sheplan and Schwartz 1974).

*Distribution*

*Chilabothrus fordii fordii*: Found in more xeric and low-lying regions across Hispaniola, excluding the Tiburon Peninsula, Île à Cabrit, and west of Cap-Haitien. It is also found on a number of Hispaniolan satellite islands (Île de la Gonâve, Isla Catalina, and Isla Saona (Fig. 5.62); Powell et al. 1999; Tolson and Henderson 1993; Henderson and Powell 2004).

*Chilabothrus fordii agametus*: Occurs in low-lying areas near the city of Môle Saint-Nicholas in the Pointe de Nord-Ouest, Haiti. It might also occur east toward Cap-Haitien; little is known about this subspecies.

*Chilabothrus fordii manototus*: Endemic to Île à Cabrit, Département de l'Ouest, Haiti (Schwartz 1979; Tolson and Henderson 1993). Île à Cabrit is a small island <0.5 km off the coast of Haiti in Port-au-Prince Bay near the town of Aubry. This subspecies is the least well characterized.

*Habitat*

As its English common name implies, *Chilabothrus fordii* is frequently found in relatively dry habitats, but it does occur in both xeric and mesic situations (Schwartz and Henderson 1991). It has been encountered at 1.5 m in a fire-scarred *Sabal palmetto*, dead on a road in a *Musa* grove, dead on a road in an area irrigated for rice culture, and beneath rocks on grassy hillsides (Sheplan and Schwartz 1974); in the stem of an *Agave* (Schwartz and Henderson 1991); on Isla Saona, at 1.2–1.5 m at mangrove edge (Henderson and Sajdak 1983); on Île à Cabrit at the base of a grass tussock in acacia–cactus scrub and another under a large dead cactus trunk on an exposed hillside (Schwartz 1979); beneath limestone rocks partially embedded in soil; and from the interior of an *Aloe* clump bordering a garden in a populated area near Manneville, Haiti (Tolson and Henderson 1993).

**Figure 5.61.** *Chilabothrus fordii* from Rabo de Gato, Puerto Escondido, Independencia Province Dominican Republic. Photos by Pedro Genaro Rodríguez.

*Abundance*

Barbour (1935) stated that *C. fordii* was "certainly very rare." Barbour's assessments of rarity must be taken with a healthy dose of skepticism; he often had little or no firsthand experience with a given species in the field. According to Sheplan and Schwartz (1974), "Under suitable ecological situations (primarily xeric),

[*C. fordii*] is an abundant snake." More recently, M. A. Landestoy T. (in litt. 21 August 2018), like Sheplan and Schwartz, considered the species to be locally common. On a single night he encountered three *C. fordii* active on fences and each within ~100 m of each other near Lago Enriquillo. In the last few years, he encountered three individuals during one night at Boca Canasta (south of Baní, Peravia Province). These largely anecdotal assessments by trained herpetologists must be extrapolated cautiously; even rare (threatened) species can occur in small enclaves in which they are locally common.

**Figure 5.62.** *Chilabothrus fordii* from the outskirts of Mano Juan, Isla Saona, Dominican Republic. Note the darker coloration; it is not clear if this is a regional characteristic of this population. Photo by R. Sajdak.

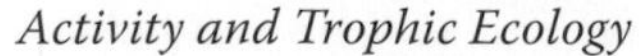

*Activity and Trophic Ecology*

*Chilabothrus fordii* is nocturnal. It forages on the ground as well as in low (<2.0 m) vegetation. On Isla Saona (~2.0 km off the southeastern tip of the Dominican Republic), three *C. f. fordii* were observed moving slowly through low (1.5–2.0 m) vegetation at mangrove edge; presumably they were foraging for sleeping anoles (Henderson and Sajdak 1983). We know little about the diet of *C. fordii* in nature, but analysis of stomach contents yielded two *Mus musculus* and four *Anolis* (*A. cybotes* and *Anolis* sp.; Henderson et al. 1987). In the field, *C. fordii* accepted a gecko (*Aristelliger expectatus*) that was offered by hand (M. A. Landestoy T., in litt. to RWH, 21 August 2018).

*Predators and Defensive Behavior*

Humans, of course, are likely the most significant predator of *C. fordii*. In the Dominican Republic, Wiley (2010) recovered evidence of *C. fordii* in the diets of the Ashy-faced Owl (*Tyto glaucops*) and the Barn Owl (*T. alba*); remains of three *C. fordii* were recovered compared with twenty-five records for *C. striatus*. Although Wiley and Wiley (1981) reported remains of *C. fordii* in the diet of Ridgway's Hawk (*Buteo ridgwayi*), the record is outside the known range of *C. fordii*; likely it was either *C. gracilis* or a juvenile *C. striatus*.

*Reproduction*

A female from Frères (Haiti) produced seven young on 25 June; female from near Pétionville (Haiti) was gravid on 16 June (Sheplan and Schwartz 1974); five females from near Manneville (Haiti) gave birth 24–28 May

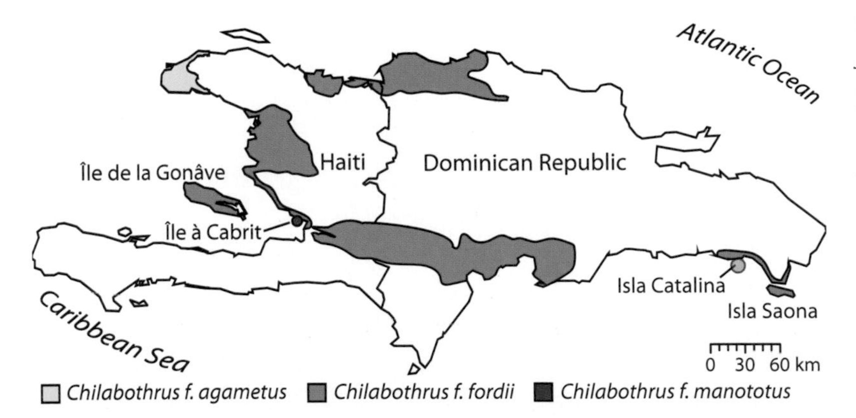


**Figure 5.63**. Range of *Chilabothrus fordii* on Hispaniola and satellite islands. The range is approximated by an isocline of less than 1000 mm maximum annual precipitation, which corresponds well to the known range of the species. Note that two subspecies are known only from the type localities. *Chilabothrus f. manototus* is restricted to Île à Cabrit near Port-au-Prince, Haiti.

with litter sizes of three to four; neonate mass was 3.6–5.3 g, and relative clutch mass was 0.416–0.493. A female from near Pétionville gave birth to four young on 24 August; neonate mass was 3.8–4.3 g, and relative clutch mass was 0.263 (Tolson and Teubner 1987; Tolson and Henderson 1993). A captive male and female >15 years old were still reproductively active (Tolson 1991b; Tolson and Henderson 1993). A wild-caught pair (from near Port-au-Prince, Haiti) gave birth to five offspring 20 July 1973, measuring 231–246 mm SVL and 4.2–4.4 g. The same pair gave birth again to four offspring 29 August 1975, measuring 196–207 mm SVL and 4.0–5.1 g (Murphy and Guese 1977; Murphy et al. 1978).

### *Conservation Status*

The species has been evaluated for listing on the IUCN Red List as Near Threatened (Landestoy T. et al. 2021a) and as Endangered on the National Red List of the Dominican Republic (Ministerio de Medio Ambiente y Recursos Naturales de la República Domincana 2011). The subspecies *C. f. manototus* is of significant conservation concern if indeed it is restricted to a single small island. As for many snake species in the West Indies, the long-term future (and the conservation status) of *C. fordii* is in flux. Habitat loss or fragmentation are likely the major threats to *C. fordii*; a portion of its range does occur in protected areas, which should confer some security (S. J. Incháustegui, in litt. to RWH, 19 May 2020). Figure 5.63 shows the distribution of *Chilabothrus fordii*.

## HISPANIOLAN GRACILE BOA

*Chilabothrus gracilis* Fischer 1888
*Chilabothrus gracilis gracilis* Fischer 1888
*Chilabothrus gracilis hapalus* Sheplan and Schwartz 1974

### *Taxonomy*

Originally described as *Chilabothrus gracilis*; Boulenger (1893) listed it as *Epicrates gracilis*. The first use of the trinomial was by Stull (1935) when she considered the taxon *monensis* to be a subspecies of *Ep. gracilis*; Sheplan and Schwartz (1974) provided a convincing argument that *monensis* was not a subspecies of *Ep. gracilis*. Sheplan and Schwartz (1974) reviewed the species and partitioned it into two subspecies (*Ep. g. gracilis* and *Ep. g. hapalus*).

### *Etymology*

The species' name *gracilis* is from the Latin for slender or thin, certainly in reference to its slender morphology. The subspecies name *hapalus* is from the Greek for gentle or delicate in reference to the demeanor of these snakes (Sheplan and Schwartz 1974). There is no vernacular name for this species in the Dominican Republic, as it is assumed to be a juvenile of *Chilabothrus striatus* (known locally as "culebra jabá," or regionally as "culebra colorá"); *C. gracilis* could just be referred to as "culebra" (M. A. Landestoy T., in litt. to RWH, 17 August 2018). In Haitian Creole it is likely known as "koulèv."

### *Description*

*Chilabothrus gracilis gracilis* (Fig. 5.64, 5.65): The body is elongated and laterally compressed. Dorsal scale rows at midbody 33–43; ventrals 271–286 in males, 271–286 in females; subcaudals 91–97 in males, 92–102 in females; males to 700 mm SVL and females to 718 mm SVL. The dorsal ground color is tan with 62–100 dorsal blotches and 21–35 dark brown tail blotches; the body blotches are subcircular to ovate. Laterally, there is a row of dark markings that are variable in shape. The head has a dark stripe from between the eyes and extending posteriorly onto the occiput, a dark transverse line across the base of the snout, a pair of dark lines behind the eye with the dorsal-most extending onto the temporal and the more ventral

**Figure 5.64.** *Chilabothrus g. gracilis* from the Río Gurabo, Santiago Province, Dominican Republic. Note the patterning, iridescence, and large eyes. Photo by Luis M. Díaz.

forming a diagonal line to the angle of the jaws. The venter is yellow tan to brownish and nearly immaculate except for brownish suffusions. The iris is pale buff (Fig. 5.64). Juvenile *C. g. gracilis* are more brightly colored than adults; the dorsum may be orange tan to tannish orange and the venter orange (Sheplan and Schwartz 1974; Schwartz and Henderson 1991). The tongue is grayish pink with white tips.

*Chilabothrus gracilis hapalus*: The body is elongated and laterally compressed. Dorsal scale rows at midbody 34–47; ventrals 278–304 in males, 279–296 in females; subcaudals 96–111 in males, 90–100 in females; males to 870 mm SVL and females to 905 mm SVL. Dorsal ground color is brown with darker brown subcircular to ovate blotches; lateral surfaces with small and diffuse widely scattered dark markings. The top of the head has a dark median spot anterior to eyes,

**Figure 5.65.** *Chilabothrus g. gracilis* from near Caño Hondo, Sababan de la Mar, Hato Mayor Province, Dominican Republic. Photo by Pedro Genaro Rodríguez.

an irregular pair of lines from posterior to the eyes to the occiput, and a dark line on each side from the posterior margin of the eye to the upper temporal region. The venter is brown and heavily clouded with dark gray except on anterior-most ventrals, which are tan to brownish (Sheplan and Schwartz 1974; Schwartz and Henderson 1991).

*Chilabothrus gracilis* from the east coast of the Barahona Peninsula may represent intergrades between the two subspecies (Schwartz and Henderson 1991).

*Distribution*

This species is apparently restricted to the main island of Hispaniola and is not found on satellite islands.

*Chilabothrus gracilis gracilis* is known from areas north of the Cul-de-Sac/Valle de Neiba plain on Hispaniola. Most localities from which it is known are coastal or near coastal on the northern portion of the island; max. elev. is ~175 m (Sheplan and Schwartz 1974). The type locality is from Cap-Haïtien, Haiti.

*Chilabothrus gracilis hapalus* is known from the Tiburon (southwest) Peninsula in Haiti (north coast) and the Barahona Peninsula in the Dominican Republic. Although primarily known from coastal localities, the type locality occurs at ~300 m on the southern slopes of the Massif de la Hotte (Sheplan and Schwartz 1974).

*Habitat*

This species appears to favor lowland wooded habitats, often associated with rivers and streams. It can be found sympatrically with *C. striatus*. It has been encountered in bushes, trees, and vine tangles at heights ranging from 0.5 m to 4.6 m (Fig. 5.66). One was taken from the trunk of a large *Ficus* adjacent to a road in a grassy pasture with a few scattered bushes. By day *C. gracilis* was found sequestered in the knothole of a fence post, in a rock pile, dead on a road, and in a solution hole in limestone rock (Sheplan and Schwartz 1974; Schwartz and Henderson 1991; Tolson and Henderson 1993; RGR, pers. obs.) (Fig. 5.67). Also, it has been found at 5.0 m in vine tangles hanging from trees and utility wires and at 3.0 m in acacia scrub (Henderson and Powell 2009). In early March 2014, three adult boas of undetermined sex (>600 mm SVL) were found together near midday (a possible mating aggregation).

**Figure 5.66.** *Chilabothrus gracilis* are proportioned similarly to highly arboreal colubrid snakes, with long and skinny bodies, large eyes, and the ability to bridge large gaps between branches. Individual from near Caño Hondo, Sababan de la Mar, Hato Mayor Province, Dominican Republic. Photo by Pedro Genaro Rodríguez.

**Figure 5.67.** *Chilabothrus g. gracilis* from El Bretón, María Trinidad Sánchez, Dominican Republic; the individual was found under the bark of a fence post during the day in coastal deciduous forest. Photos by Raimundo López-Silvero.

**Figure 5.68.** Habitat of *Chilabothrus gracilis* in the northern Dominican Republic. *Top*: El Bretón; *bottom*: Parque Los Haitises. Photos by Rolando Teruel.

They were in a dry palm leaf sheath partially detached about 30 cm above the ground, at the edge of a patch of subcoastal lowland rainforest on volcanic soil (Fig. 5.68) ~3 km west of Sabana de la Mar, Hato Mayor, Dominican Republic (R. Teruel, pers. comm. to TMRC, 17 July 2020). In mid-September 2016, an adult male (>600 mm SVL) was found at midday under the bark of a fence pole 60 cm above the ground at the edge of a coastal broadleaf semideciduous forest on limestone karst (Fig. 5.68) along a road near El Bretón, María Trinidad Sánchez, Dominican Republic (R. Teruel, pers. comm. to TMRC, 17 July 2020). In October 2013, two individuals were found in the vegetation surrounding an abandoned apartment village adjacent to the beach and a mangrove swamp near Miches, Dominican Republic (RGR, pers. obs.).

*Abundance*

In 1914, Barbour described *C. gracilis* as "evidently a rare species." Twenty-seven years later Cochran (1941) considered *C. gracilis* to be "exceedingly rare," but 33 years post-Cochran, Sheplan and Schwartz (1974) considered it to be "locally abundant." In dense thickets in the hills of Los Patos (Barahona Province), two *C. gracilis* were captured per night on consecutive nights (M. A. Landestoy T., in litt. to RWH, 21 August 2018).

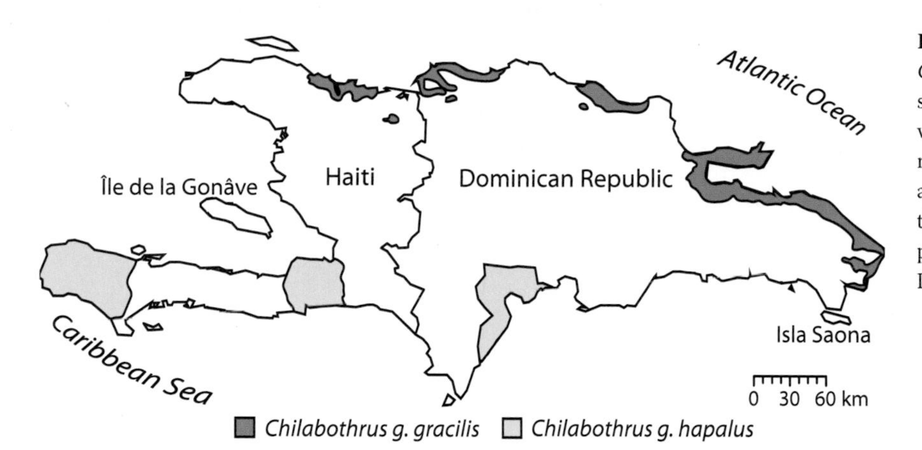


**Figure 5.69.** Range of *Chilabothrus gracilis* subspecies on Hispaniola, which is frequently found in more mesic situations and along waterways. We note that the taxonomic status of populations from Barahona Peninsula are unknown.

Based on personal (RWH) experience, *C. gracilis* is moderately common; that is, in appropriate habitat at night you could expect to find one individual in several hours of searching. At a site 8 km northwest of Paraíso (Barahona Province), RWH established a snake market, and over 8 days he was brought sixty *C. striatus* and four *C. gracilis*. Presumably, all the boas were captured during the day, so the preponderance of the much larger (and more conspicuous) *C. striatus* is perhaps not unexpected.

*Activity and Trophic Ecology*

*Chilabothrus gracilis* is nocturnal, although Sheplan and Schwartz (1974) report one crawling on a barbed-wire fence by day. It is highly arboreal and has been observed actively foraging at heights from 0.5 to 4.6 m in trees, bushes, and vine tangles. Little is known about diet, but six of six prey items recovered from preserved *C. gracilis* were *Anolis* lizards (*A. cybotes* and *A. distichus*; Henderson et al. 1987). It would not be surprising to learn that it also takes small frogs (e.g., *Eleutherodactylus* spp.; small *Osteopilus dominicensis*).

*Predators and Defensive Behavior*

We are unaware of any natural predators, but its size likely makes *C. gracilis* meal size for a variety of birds and small mammals. When encountered by humans, it is usually killed (M. A. Landestoy T., in litt. to RWH, 21 August 2018). *Chilabothrus gracilis* is likely among the most innocuous species of boa in the West Indies. Its primary defensive behavior is curling up into a ball with its head toward the center of the ball. It may include voiding its cloaca, but we have not experienced that behavior.

*Reproduction*

A female (730 mm SVL) that was pregnant when collected at Limbé, Haiti, gave birth to five young on 22 September; SVL of the neonates ranged from 246 to 254 mm and mass was 3.8–4.1 g (Murphy et al. 1978). Each of two females collected near Limbé, Haiti, gave birth to three living young and two infertile egg masses on 21 and 24 October; relative clutch mass ($n$ = 2) was 0.192 and 0.224, and neonate mass ranged from 2.0 to 2.3 g (Tolson and Henderson 1993).

*Conservation Status*

This species was assessed as Near Threatened on the IUCN Red List (Henderson et al. 2021b) and assessed as Endangered in the National Red List of the Dominican Republic (Ministerio de Medio Ambiente y Recursos Naturales de la República Domincana 2011). This species' proclivity for dense undergrowth and thickets may render it vulnerable to habitat alteration (M. A. Landestoy T., in litt. to RWH, 21 August 2018). Habitat loss or fragmentation are likely the major threats to *C. gracilis*; a portion of its range does occur in protected areas, which should confer it some protection (S. J. Incháustegui, in litt. to RWH, 19 May 2020). Figure 5.69 shows the distribution of *Chilabothrus gracilis*.

## HISPANIOLAN BOA

*Chilabothrus striatus* (Fischer 1856)
Tiburon boa (*Chilabothrus striatus exagistus*) Sheplan and Schwartz 1974
Hispaniolan boa (*Chilabothrus striatus striatus*) Fischer 1856
Haitian boa (*Chilabothrus striatus warreni*) Sheplan and Schwartz 1974

### *Taxonomy*

This species was originally described as *Homalochilus striatus* and subsequently was placed into the genus *Epicrates* by Boulenger (1893). The species was recognized from the island of Hispaniola as well as the Great Bahamas Bank. Sheplan and Schwartz (1974) partitioned *C. striatus* sensu lato into eight subspecies; three from Hispaniola and five from the Bahamas Archipelago, based largely on color pattern differences and some squamation characters. Reynolds et al. (2013a) restricted *C. striatus* to Hispaniola, recognizing the Bahamas populations as *C. strigilatus*.

### *Etymology*

The specific epithet *striatus* is Latin for striped or grooved and refers to the dorsal coloration of the holotype. The subspecific epithet *warreni* is a Latinized patronym honoring C. R. Warren, a naturalist who collected specimens for Albert Schwartz on Île de la Tortue, including the type. The subspecific epithet *exagistus* is from the Greek for "mystical" and is an apparent reference to local vodoun (= voodoo) culture (Sheplan and Schwartz 1974). In the pet trade this species is referred to as the "Haitian boa," "Dominican boa," and "Dominican red mountain boa." Local names for this species include "culebrón" and "culebra" (Dominican Republic) and "koulèv" or "sèpan" (Haiti).

### *Description*

This is the second largest member of the genus *Chilabothrus* behind *C. angulifer* and possibly tied with *C. strigilatus*, attaining a verifiable length of up to 2000 mm SVL, although a compelling record of 2489 mm SVL exists (Ottenwalder 1985; Reynolds et al. 2016a). Fernández de Oviedo (1851b) reported finding a very large (>6.0 m TL) freshly killed snake in 1515 around the mouth of the Neiba River, in the foothills of the "Pedernales" range (= Sierra de Bahoruco). No species other than *C. striatus* on Hispaniola could possibly fit Oviedo's description. Although Oviedo's accounts must be taken cautiously, this boa must certainly have attained much larger sizes in the past than the maximum record reported in modern times (see Rodríguez-Cabrera et al. 2016a for a review). It is not known if sexual size dimorphism exists in length, but females likely grow more massive than males. Neonates are 370–508 mm SVL. Few data sets of sizes exist in the published literature, but RGR compiled data from *C. s. striatus* and *C. s. exagistus* from wild specimens (*n* = 13) and museum specimens from the MCZ (*n* = 33) for the accounts below.

*Chilabothrus striatus exagistus* (Fig 5.70): Body size in this subspecies is up to 1711 mm SVL, with 271–286 ventrals, 10–11 circumorbital scales, and usually 1 loreal scale. The color pattern resembles other members of the species, although it is frequently considerably faded or reduced. The dorsal ground color is tan to brown, with very faint blotches except in juveniles. Museum specimens averaged 1429 mm SVL for adults (*n* = 5, range 1295–1626 mm).

*Chilabothrus striatus striatus* (Fig 5.71): Body size may exceed 2000 mm SVL, with a wide head and robust body. Wild and museum specimens averaged 1460 mm SVL for adults (*n* = 29, range 914–1905) and 592 mm SVL for juveniles (*n* = 15, range 425–749). Coloration is extremely variable, although generally consists of a lighter dorsal background color and a darker blotching pattern on the dorsum, with blotches frequently outlined by dark or black scales and occasional striping near the anterior. Dorsal ground color ranges from tan to orange to red, often with a mix of scales of different colors. The venter is usually light gray and white, with occasional mottling, although in red and orange individuals the venter matches the dorsal color. Red or orange individuals are often thought to be from the mountains, hence a common name "red mountain boa," although in reality, red boas can be found throughout the range of this subspecies (RGR, pers. obs.). Most boa populations consist of a mix of colors, with most individuals being tan to brown and some individuals being red to orange. Others can have a mix of scales with all of these colors. Calico individ-

**Figure 5.70.** *Chilabothrus striatus exagistus* from Île-à-Vache, Haiti. Few individuals have been photographed from this island. Photos by René Durocher.

uals are produced in the pet trade. The tongue is dark gray with white tips. There are 11 circumorbital scales and frequently 2 loreals.

*Chilabothrus striatus warreni* (Fig 5.72): Few individuals have been documented, but *C. s. warreni* likely reaches a smaller size than other subspecies—the largest individual documented was a female 1411 mm in SVL. There are usually 11 circumorbitals and 2 loreals. Like *C. s. exagistus*, the color pattern is considerably faded, with faint blotches or anterior stripes present that minimally contrast with the tan to light brown dorsal ground color.

*Distribution*

Hispaniolan boas are widely distributed across the island of Hispaniola and some of its satellite islands. The three subspecies are separated geographically, with two on the main island of Hispaniola and one restricted to a large satellite island.

*Chilabothrus striatus exagistus* was described as being restricted to the western end of the Tiburon Peninsula, which was part of a paleo island that joined with Hispaniola in the Miocene near the present-day Lago Enriquillo on the border of Haiti and the Dominican Republic. This subspecies is also found on Île à

**Figure 5.71.** *Chilabothrus s. striatus* showing the range of colors present in this subspecies. *Top left*: Some *C. s. striatus* have lots of red or orange coloration, such as this individual from Parque del Este. Photo by R. Graham Reynolds. *Bottom left*: Adult female *C. s. striatus* from Miches, Dominican Republic, showing a medium level of red scalation. Photo by R. Graham Reynolds. *Top right*: Some individuals can be quite dark, such as this one from Parque del Este. Photo by R. Graham Reynolds. *Bottom right*: Lighter individuals can also be found intermixed with other color morphs, such as this individual from near Santo Domingo. Photo by Luis M. Díaz.

**Figure 5.72.** *Chilabothrus striatus warreni* from Île de la Tortue, Haiti. This is one of the only photographs of this subspecies. Photo by Rhea Warren.

**Figure 5.73.** South paleo island *Chilabothrus s. striatus* from Oviedo, Pedernales Province, Dominican Republic. While it is yet unclear which subspecies is represented in this region, fewer red individuals are seen in that area. Photos by Pedro Genaro Rodríguez.

Vache, located 10 km off the southwest coast of Haiti. However, molecular genetic data suggest that this subspecies might inhabit the entirety of the Tiburon Peninsula (RGR and S. Pasachnik, unpubl. data).

*Chilabothrus striatus striatus* is thought to be restricted to the Hispaniolan paleo island (Fig. 5.73) and hence is now considered to be found roughly north and east of an imaginary diagonal line passing south of Île de la Gonâve, through Lago Enriquillo, and through Barahona, Dominican Republic. It is also found on the satellite islands of Île de la Gonâve (Haiti; 692 $km^2$) and Isla Saona (Dominican Republic; 105 $km^2$). One individual has been found on Vieques Island, Puerto Rico—an introduction from the pet trade (see Chapter 3).

*Chilabothrus striatus warreni* is endemic to Île de la Tortue (Tortuga), a 188 $km^2$ island politically part of Haiti and located off the northwest coast of Hispaniola.

*Habitat*

Hispaniolan boas (Fig. 5.74) are habitat generalists and can be found from the densest rainforest to dry scrub to sugarcane fields and banana groves to urban parks; they occur up to an elevation of 1220 m. In these environments boas will use nearly all available habitats: arboreal habitats from the tops of large trees to smaller shrubs, vines, and branches; terrestrial habitats such as rock walls and debris piles; and even subterranean habitats such as caves. Boas can use relatively small

**Figure 5.74.** *Top*: Individual *Chilabothrus s. striatus* can have a tremendous variety of scale colors on the dorsum, as shown in this individual from Barahona, while the venter tends to be plain-colored. Photo by Jon Suh. *Bottom*: Some individuals are stunningly red/orange against darker dorsal patterns. Photo by Matthijs Kuijpers.

patches of habitat, including single-tree-wide strips of vegetation lining extensive sugarcane fields. In areas such as Hato Mayor Province in the Dominican Republic, large monoculture plantations of sugarcane are lined by creeks and irrigation ditches that often have a bit of native vegetation, usually only one or two trees wide, that will support boa populations, albeit at lower density than less-disturbed habitats. Boas are relatively easily found in mature forests in protected areas, including Parque del Este and Los Haitises, as well as in forested areas near Pedernales, Barahona, and Las Galeras in the Dominican Republic. Boas will occupy small patches of forest surrounded by urbanization—for example being readily found on the grounds of the Santo Domingo Zoo. They are also readily found in secondary growth and open areas near settlements in Haiti (Tolson and Henderson 2011).

*Abundance*

This species is more abundant and more frequently encountered in lowland moist tropical forest habitats, although they can be found with relatively lower frequency in most other habitat types in their elevational range. Nighttime surveys in forested or edge habitats in the Dominican Republic frequently yield at least one individual, and many nocturnal researchers, such as those studying *Anolis* lizards, encounter boas regularly. Despite the relative ease with which they are encountered in the Dominican Republic and certain areas in Haiti, there are few published abundance estimates for this species.

During 8 days in March 1985, RWH established a snake market near Paraiso on the Barahona Peninsula, Dominican Republic; it resulted in sixty *C. striatus* of all size classes except very young individuals. In May 2013, RGR conducted 16 person-hours (p-h) of nocturnal survey in tropical moist forest at Las Galeras on the Samana Peninsula that yielded six boas, including four juveniles. Of note, this survey also produced two observations of wild Hispaniolan hutia (*Plagiodontia aedium*). Eighteen p-h of surveying in Parque del Este in May 2013, also by RGR, yielded a single adult female, while 4 hours of surveying by four people near Miches yielded two boas. Among other species of West Indian boas, this species is relatively easy to find under survey effort in appropriate habitat in the Dominican Republic. Almost any day of driving in the summertime on rural roads in the lowlands of the Dominican Republic will yield at least one road-killed boa, which reinforces how widespread and abundant the species is even in seemingly marginal habitats. Less is known regarding Haitian populations, although presumably the populations are less dense and more localized owing to extensive deforestation there. Tolson and Henderson (2011) note that the species exhibits unusually high density around the villages of Limbe and Thomazeau in Haiti.

*Activity and Trophic Ecology*

Like all West Indian boas, this species is largely nocturnal, with occasional crepuscular activity. By day, *C. striatus* is known to coil in limestone crevices, in bird nests, in tree hollows (Fig. 5.75), in clumps of vegetation, or in the shade on broad horizontal branches from 5 to 20 m above ground (Henderson et al. 1987). Boas are occasionally found coiled under roof over-

**Figure 5.75.** *Left*: *Chilabothrus s. striatus* is an excellent climber and actively hunts in vegetation at night, such as this large female from Parque del Este. Photo by R. Graham Reynolds. *Right*: *Chilabothrus s. striatus* sharing a mango (*Mangifera indica*) tree hole with a tarantula (probably *Phormictopus cancerides*), south of Bani, Dominican Republic. Photo by Miguel Landestoy.

hangs during the day. Unfortunately, the most common diurnal observation of boas is dead on roads. At night boas are much more readily found and will use all aspects of appropriate habitat, from the ground (or even under ground cover or in caves) to the tops of trees. Nocturnal surveys frequently located foraging boas from ground level to about 3 m up in trees, bushes, and shrubs (Fig. 5.75). Boas use higher parts of the canopy to forage as well, although they are less frequently observed higher up because of the challenge of surveying tree canopies. Nevertheless, much of the prey that boas use occurs lower in the forest. Many species of *Anolis* lizards sleep along thin branches and on leaves lower in the canopy, and rats frequently move from ground to trees and could be easily intercepted in the lower parts of the forest.

This species is both an active and ambush forager, using different strategies depending on the boa's age and the prey being sought. For example, sleeping *Anolis* lizards or birds are hunted using active foraging strategies, where the boa investigates branches of trees very slowly using tongue flicks to locate the prey. Sleeping birds and *Anolis* are sensitive to even slight movements of the branch on which they are sleeping, and a jostle will cause them to awake and flee. For *Anolis* lizards this usually means jumping off the perch rather than running. Rats and other mammals are often captured using an ambush technique, by which the boa locates a potential active runway used by the animal and waits with the head resting flat on the branch (RGR, pers. obs.).

The diet of *C. striatus* has been relatively well studied, at least in the context of other members of the genus. A caveat is that this dietary information is known only for *C. s. striatus*, and the diet of the other subspecies is not well characterized at all. This taxon exhibits ontogenetic shifts in diet, moving to different preferred prey items as they age. Interestingly, although the species is considered a generalist, in that the adults will take a wide variety of prey, the juveniles are specialists (Henderson et al. 1987; Reynolds et al. 2016a). Young boas, those smaller than 600 mm SVL, feed almost exclusively on *Anolis* lizards, an abundant and diverse prey group on Hispaniola. Older boas, those over 800 mm SVL, switch to a more generalist diet, preying on birds, introduced *Mus* and *Rattus*, and native mammals such as hutia and possibly solenodon (*Solenodon paradoxus*, although predation on this species has never been confirmed). In a free-flight aviary at the National Zoo in the Dominican Republic, free-living *C. striatus* were responsible for the predation on at least ten birds over an 8-month span. They consumed psittacids, columbids, phasianids, and cracids, demonstrating opportunistic feeding (Ottenwalder 1980; Henderson et al. 1987). Adults will also consume bats and will use an ambush technique at the mouths of bat caves to intercept bats leaving roosts (Ottenwalder in Henderson and Powell 2009). The native prey base on Hispaniola has changed dramatically in the last few centuries, such that the typical adult diet has shifted away from many of the native species of vertebrates (aside from birds) toward introduced species such as rats (see Chapter 2). For example, the large iguanid *Cyclura* was previously much more abundant and widely distributed on Hispaniola, with *Cy. ricordi* found in more xeric scrub areas and *Cy. cornuta* along the coasts and in tropical dry forests. Other West Indian boas are known to consume *Cyclura* (e.g., *Chilabothrus angulifer, C. chrysogaster, C. subflavus*), and rock iguanas were probably an important and abundant dietary item historically for many of the larger West Indian boas in the Greater Antilles. However, populations of *Cyclura* have plummeted over the last century, and many populations are so restricted or at such low abundance that they likely constitute a minimal component of the diet of West Indian boas at present (with a few exceptions in the Bahamas and Turks and Caicos; Knapp and Owens 2004; Reynolds 2011d).

*Predators and Defensive Behavior*

Hispaniolan boas will void the contents of their cloaca when picked up, although they rarely strike. Native predators are poorly known, but birds of prey (hawks, owls) are confirmed predators as well as presumably other predatory birds such as herons. Wiley (2010) repeatedly found *C. striatus* in the diets of both the Barn Owl (*Tyto alba*) and the Ashy-faced Owl (*T. glaucops*) in equivalent proportions ($n$ = 13 and $n$ = 12, respectively) in the Dominican Republic. Females in captivity are known to consume offspring (Hanlon 1964), although it is not known whether this occurs in the wild. Juvenile boas are likely preyed upon by rats, and cats threaten all but the largest boas (Tolson and Henderson 2011)

*Reproduction*

Reproductive seasonality among most members of the genus *Chilabothrus* is roughly similar, with females likely reproducing every other year with mating occurring around March (Tolson 1992b). Males are known to engage in sometimes violent combat for access to females, and once courtship is initiated it can last for days to weeks (Tolson 1980; Teubner and Tolson 1984). Male reproductive behavior and recrudescence is likely encouraged by temperature and humidity drops in the winter. Teubner and Tolson (1984) found that a drop in average temperature of 5–10 °C, followed by an increase in ambient humidity in the spring, was required to stimulate increase in plasma testosterone. This increase in testosterone is also associated with dominance in ritual combat, with larger males having higher testosterone levels than younger males; younger males exhibit increases in testosterone only after larger males have already reproduced (Teubner 1986).

Female *C. striatus* become gravid in the summer and give birth in the fall. Gestation lasts between 192 and 224 days. Peak parturition is likely around October, and females give birth to five to thirty-one offspring, although they are likely capable of larger litter sizes. Neonates are 12–19 g in mass. Lifespan is at least 22 years.

*Conservation*

This species is listed on CITES Appendix II. It is listed as Least Concern on the IUCN Red List, owing to presumably robust population sizes and widespread distribution (Landestoy T. et al. 2018). The single largest local threat to the species appears to be habitat destruction. Boas are also routinely killed by humans, and the majority of human encounters result in the death of the snake. Being the largest boa on Hispaniola and the most widely distributed, it is the one most likely to be

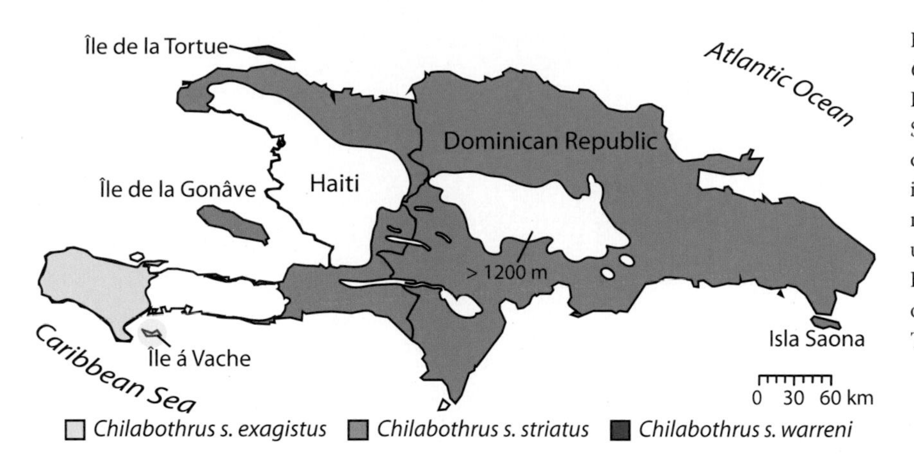


**Figure 5.76.** Range of *Chilabothrus striatus* on Hispaniola and satellite islands. Subspecies are shown in different colors. *Chilabothrus s. striatus* is widespread, occurring across most of the Dominican Republic up to an elevation of ~1200 m. Boas are not recorded from parts of western Haiti and the central Tiburon Peninsula.

in conflict with humans, especially as it is frequently found in urbanized areas. According to S. J. Inchaústegui (in litt. 19 May 2020), *C. striatus* in the Dominican Republic is killed for its fat, which is rendered into oil, ostensibly for relief of joint pain. It is sometimes killed because it feeds on domestic fowl but also for human consumption. Snakes are frequently found dead on roads; most drives in the countryside of the Dominican Republic result in the observation of one or more roadkill carcasses (RGR, pers. obs.). The species is popular in the pet trade, often traded as "Dominican red mountain boas," which refers to individuals with lots of red coloration, which can come from many regions in the Dominican Republic (not necessarily from the mountains; see Description above). Brown or pale specimens are often traded as "Haitian boas." While many boas in captivity in the United States and Europe are captive bred, specimens are still being regularly taken from the wild for the pet trade (Landestoy T. et al. 2018). Further, in the Dominican Republic many people keep snakes captured on their property as pets or to display to tourists (RGR, pers. obs.).

While the species itself is likely secure in some areas of Hispaniola, most information about this species comes from studies and observations of the subspecies *C. s. striatus* in the Dominican Republic. Populations of *C. s. exagistus* are poorly studied, and next to nothing is known of *C. s. warreni* on Île de la Tortue. These two subspecies are likely being much more dramatically impacted by disturbances such as forest destruction than *C. s. striatus*. *Chilabothrus s. warreni* is known only from the highly impoverished Haitian island of Île de la Tortue. Although no population data exist for the species, nor any published population surveys, we expect that the subspecies is likely at serious risk of extirpation. Fifty years ago, between January 1968 and September 1970, C. Rhea Warren made four visits to Île de la Tortue. Over 6 days of collecting, he encountered five *C. striatus* (in litt. to RWH, October 2001). *Chilabothrus s. exagistus* is also poorly studied and is probably exclusively found in southern Haiti, where extensive deforestation has led to the loss of most forest cover. Figure 5.76 shows the distribution of *Chilabothrus striatus*.

## PUERTO RICO, MONA, AND THE VIRGIN ISLANDS

From a zoogeographic perspective, the Puerto Rico region consists of the Puerto Rico Bank and the island of Mona to the west. In this region are three political units. Puerto Rico (an unincorporated territory of the United States) administers Isla de Mona, the main island of Puerto Rico, and the Spanish Virgin Islands (Culebra, Vieques, and satellite islands). The U.S. Virgin Islands (USVI; also unincorporated territory of the United States) administers the next group of islands to the east, including St. Thomas and St. John as well as many smaller satellites. Farthest to the east are islands from Jost van Dyke and Tortola stretching northeastward to Anegada, administered as the British Virgin Islands (BVI), a British Overseas Territory. Isla de Mona and Monito are separate island banks of 56 km$^2$ (max. elev. 80 m) and 0.16 km$^2$ (max. elev. 65 m), respectively, that lie in the Mona Passage between Puerto Rico 69 km to the east and Hispaniola 65 km to the west. These islands are characterized by tropical dry forest and tropical dry scrub. The main island of Puerto Rico is the smallest of the four Greater Antillean islands (9104 km$^2$, max. elev. 1338 m) and consists of a wide variety of habitats—from montane rainforest to xeric scrubland. To the east lies the Puerto Rico Bank, a shallow-water platform supporting the Spanish, U.S. (exclusive of St. Croix, which lies on its own bank), and British Virgin Islands. This region stretches ~173 km and is composed of dozens of islands and islets supporting a dense growth of tropical dry forest as well as tropical dry scrub on Anegada. The largest island on the Puerto Rico Bank is St. Thomas, with an area of 83 km$^2$ and max. elev. of 523 m; St. Croix is larger at 214 km$^2$, but it lies on its own bank and has no native boa species (but see Chapter 3). This region supports three species of *Chilabothrus* boas.

**Figure 5.77.** Several species of West Indian boas, including the Puerto Rican boa (*Chilabothrus inornatus*) have adapted a hunting behavior to capture bats from midair near caves. Photo © Days Edge Productions.

## VIRGIN ISLANDS BOA

*Chilabothrus granti* (Stull 1933)

### *Taxonomy*

The Virgin Islands boa was originally described as *Epicrates inornatus granti* (Stull 1933) based on a specimen collected in 1932 from Tortola, British Virgin Islands. The collector (Grant 1932a) remarked that it was a "great surprise to find a boa so far east of the eastern record of this family, *Epicrates inornatus* (Reinhardt) of Puerto Rico." As a footnote, Stull (1933) noted that C. Grant told her that *Ep. i. granti* was gray, whereas *Ep. inornatus* was brown. Sheplan and Schwartz (1974) revisited the Puerto Rican boas in their revision of the West Indian boas and determined that *Ep. monensis* from Mona Island was a distinct species from *Ep. inornatus* from Puerto Rico, and further that *granti* was best described as the subspecies *Ep. monensis granti*. The name *Chilabothrus granti* was occasionally used to refer to the populations of *C. monensis* occurring in the Virgin Islands without formal justification (i.e., anecdotally) in the literature. Reynolds et al. (2015) showed that populations of *C. m. monensis* and *C. m. granti* have significant genetic differentiation and suggested they be recognized as separated species. In a concurrent publication, Rodríguez-Robles et al. (2015) further suggested that *C. granti* be recognized as a separate species based on molecular genetic species delimitation and phenotypic differences but did not elaborate on the taxonomic history from Stull (1933) nor offer a formal description. Reynolds and Henderson (2018) provide additional reference for recognizing the species based on collective genetic and phenotypic information.

### *Etymology*

The epithet *granti* is a patronym honoring the individual Chapman Grant, who obtained a specimen from Tortola that serves as the holotype (Grant 1932a; Stull 1933). This species is colloquially known as the "Virgin Islands treeboa" and the "VI boa" in the USVI, and as "night snake" in the BVI (Grant 1932a).

### *Description*

This is a relatively small member of the genus *Chilabothrus* with a long, slender, and delicate body shape relative to other species. The maximum reported sizes for this species come from translocated populations in the USVI (Tolson 2005), which is either a result of the species growing larger (or surviving longer) in these populations or more individuals having been measured in these populations. The maximum male size is 1112 mm SVL, and female size is 1066 mm SVL—at these sizes the animals appear decidedly more robust than the more typically sized adult animals 600–800 mm SVL. Animals from the BVI are anecdotally thought to be larger than those typically seen in Puerto Rico and the USVI. Anecdotal reports of animals up to 2 m from Tortola remain unconfirmed, but specimens up to 910 mm SVL have been recorded from Tortola (Lazell 2005). A photograph of a recently killed animal there suggests a size well over 1 m SVL (RGR, unpubl. data).

The dorsal ground color is usually gray or grayish brown, although some individuals can be quite dark brown (Sheplan and Schwartz 1974). Relative lightness and darkness can also change depending on time of day, with individuals often appearing lighter in color at night and darker during the day (D. Smith, in litt. to RGR, 10 October 2020). This circadian change in color phase has not been reported in any other species of *Chilabothrus* but is common in snakes of the genus *Tropidophis* in the Caribbean (Rehák 1987; Hedges et al. 1989). The dorsal color pattern consists of highly contrasting blotches arranged on the flanks and dorsum outlined with black scales (Fig. 5.78). These blotches are usually very dark brown. The dorsal color pattern is indistinguishable from *C. monensis*, although the color pattern on the top of the head is sometimes diagnostic (Sheplan and Schwartz 1974). The venter ranges from unstippled to heavily stippled with gray or brown. Neonates do not undergo the dramatic ontogenetic color change seen in other members of the genus. The juveniles do, however, have much more contrasting dorsal color patterns than do adults, with a light-gray dorsal ground color and highly contrasting dark dorsal blotches (Fig. 5.79). The tail is more darkly pigmented, tapers to a very fine point, and is prehensile and capable of being tightly coiled. The pupil is black and the iris is mottled gray and streaked

**Figure 5.78.** Adult female *Chilabothrus granti* from Puerto Rico. Photo by R. Graham Reynolds.

with brown in both adults and juveniles; like most species in the genus the eyes are noticeably large relative to the head size in juveniles and appear to "bulge" out of the head. The tongue is dark gray to black with white tips starting at the fork.

The species is diagnosed by the following characters: 5 intersupraocular scales, 41–47 midbody dorsal scale rows, modally 2 supralabials contacting the eye (Sheplan and Schwartz 1974). Sheplan and Schwartz (1974) suggest that the number of blotches is variable between *C. granti* and *C. monensis*, although we find that blotching is sufficiently variable as to preclude its use as a diagnostic character.

*Distribution*

This species is endemic to the Puerto Rico Bank (Rivero 1998; Mayer 2012). It has been definitively recorded from the following islands (also see Mayer 2012): Puerto Rico (Tolson 1991c), Cayo Diablo, Islas Cordilleras (Tolson 1985; Tolson and Piñero 1985), Culebra (Tolson 1991c), St. Thomas (Sheplan and Schwartz 1974; Nellis et al. 1983), and Tortola (Grant 1932a). The population on Culebra was only discovered in 1991, and the population on Puerto Rico was found in 1988 (Tolson 1991c). A boa was reported from Outer Brass Island (north of St. Thomas) but Tolson (1991c) was unable to locate animals over the course of five surveys there. Compelling anecdotal reports without specimens exist for Water Island in the USVI (C. Pacheco, in litt.), Great St. James in the USVI, and for the following islands in the BVI: Guana (Grant 1932a; Mayer and Lazell 1988; Lazell 2005), Necker (Lazell 1983), Great Camanoe (Lazell 1983; Mayer and Lazell 1988; B. Barker in Henderson and Powell 2009), Jost van Dyke (Mayer and Lazell 1988; Island Resources Foundation 2009), and Virgin Gorda (S. Lazell in Nellis et al. 1983). Translocated populations were established on Cayo Ratones in the Islas Cordilleras in 1993, and a small unnamed cay in the USVI in 2002 (Tolson et al. 2008). The Cayo Ratones population is likely

**Figure 5.79.** *Top*: Juvenile *Chilabothrus granti* from near St. Thomas, U.S. Virgin Islands (USVI). Photo by Dustin Smith. *Bottom*: Close-up of a juvenile *C. granti* from Culebra Island. Photo by Alberto R. Puente-Rolón.

extinct following the passage of Hurricane Maria in 2017 and the reestablishment of rats on that island (J. P. Zegarra, pers. comm. to RGR). High densities of rats likely prevent boa persistence on other smaller islands (Tolson 1986).

On larger islands the species is restricted to one or several localities, while on small islands such as Cayo Diablo the species is found across the entire island. On Puerto Rico, the species is restricted to a very small patch of forest near Río Grande, and no documented records exist anywhere else on the island (but see Chapter 3). Some anecdotal reports suggest that boas might occur near Humacao and Fajardo, although this has not yet been confirmed and seems unlikely. Nevertheless, in 1998 a young *C. granti* was brought by a citizen to the Humacao Wildlife Refuge, but there was no information on the origin of the individual, which died soon thereafter (identification confirmed by ARPR). It is plausible that the species was much more widespread in the lowlands of Puerto Rico historically (Nellis et al. 1983), prior to the near-complete deforestation of the island in the early twentieth century (Koenig 1953). Alternatively, Tolson and Piñero (1985) suggested that it might have gone extinct during the Pleistocene along with other xeric-adapted species. On Culebra the species has primarily, although not exclusively (Ríos-Franceschi et al. 2016), been found near Monte Resaca. On St. Thomas boas are restricted to the southeastern end of the island in the vicinity of Red Hook (Harvey and Platenberg 2009; Platenberg and Harvey 2010).

In the BVI fewer recent records are available. On Guana Island, no individual has been officially recorded although anecdotal accounts exist from the 1930s (Grant 1932b), and a snake fitting the description of a boa was apparently killed by groundskeepers there in the 1980s (Lazell 1980; Mayer and Lazell 1988). No boas have been seen on Guana since, although a focused survey of this island has not occurred recently (J. D. Lazell, in litt. to RGR). Similarly, it is unclear whether the species still occurs on Necker, Great Camanoe, and Virgin Gorda. Records from Necker and Virgin Gorda are anecdotal observations by a botanist (John G. H. Smith in Mayer and Lazell 1988). According to Mayer and Lazell (1988), the Great Camanoe record is based on a specimen collected there and kept as a pet. But Henderson and Powell (2009) relate a note from B. Barker who observed a boa hanging from a rock wall on the island; no further documentation is available however. A recent publication (Island Resources Foundation 2009) asserts matter-of-factly that Virgin Islands boas are found on Jost van Dyke, although no photos or other evidence are offered. Mayer and Lazell (1988) also suggest that a boa was captured on Jost van Dyke; no official records nor photos of these boas exist, hence these observations are considered anecdotal. A potentially widespread population occurs on Tortola (Lazell 2005), and the species has been found there recently (Rodríguez-Robles et al. 2015; D. Barber in litt. to RGR, 2019).

*Habitat*

This species is found exclusively in xeric habitat (Nellis et al. 1983), preferring subtropical dry forest and coastal forest over subtropical moist forest. Virgin Islands boas are arboreal and hence require habitat with dense, connected canopy—indeed the forest structure is more important than the tree species present. The canopy does not need to be tall, nor does the species prefer primary forest. Most observations are along edge habitat (i.e., road cuts). The only population present on the island of Puerto Rico is found along a road in subtropical moist forest, near a hill that is surrounded by urbanization and tourism development. Habitat modeling on St. Thomas demonstrated that boas use only a portion of available habitat on that island, preferring woody vegetation developed over nonstony soils (Harvey and Platenberg 2009; Platenberg and Harvey 2010). Boas are also more frequently found on gentle slopes and elevations below 150 m, although it should be noted that "flat" areas on St. Thomas are usually the first to be heavily developed. These authors (Harvey and Platenberg 2009; Platenberg and Harvey 2010) also found that boas had about 800 ha of available habitat on St. Thomas but that nearly three-quarters of that has been developed.

On Culebra the species has been sighted in secondary forest near pastureland (Tolson 1992a) and along road cuts. On Tortola the species occurs in all habitats from mangrove forest to subtropical dry forest on Mt. Sage and might particularly favor forested dry washes (locally called "ghuts" or "guts") on the sides of Mt. Sage (USFWS 2009). On small islands, the species will use all available habitat types, including vegetation in the wrack line along the shore, although it seems to prefer *Coccoloba uvifera* (Tolson 1991a). The species is also found in *Cassine xylocarpa*, *Caesalpinea*, *Cassythia*, *Suriana*, *Opuntia*, *Avicennia germinans*, and *Conocarpus erectus* (Tolson 1991c; RGR, pers. obs.). Boas are occasionally encountered on rocks on the ground on small islands as well (D. Smith, in litt. to RGR). The species takes diurnal refuge in termite nests, in palm axils, under debris (Tolson and Piñero 1985), under rocks, in bromeliad axils, in tree holes, and in vine tangles (Fig. 5.80). Of almost singular importance to boa presence or absence, beyond appropriate vegetation and climate, seem to be (1) an absence or low number of rats (Tolson 1988) and (2) a high density of *Anolis cristatellus* (Tolson 1991c). Where rat density is high on smaller islands, boas are not able to persist, despite adult boas occasionally taking rats as prey (see Activity and Trophic Ecology).

**Figure 5.80.** *Top*: Close up of patterning on *Chilabothrus granti* from near St. Thomas, USVI. Photo by Dustin Smith. *Bottom*: Typical resting posture coiled within a bromeliad near St. Thomas. Photo by R. Graham Reynolds.

*Abundance*

This species is quite cryptic, and its preferred arboreal habitat, particularly in vines and tangled vegetation, can make it very difficult to find in nocturnal settings and nearly impossible to detect in diurnal settings. When active, boas often have only parts of their bodies exposed on top of the vegetation on which they are climbing; it is difficult to spot the animal by looking for a snake body shape against the background of vegetation. In contrast, on small cays the species is very easy to find at night, likely owing to the relatively higher population densities and the shorter vegetation. Virgin Islands boas were suggested to be "exceedingly rare" by MacLean (1982), who knew of eleven specimens of this species, and Nellis et al. (1983), who recorded twelve specimens. Subsequent review by the USFWS (1986b) found a total of only seventy-one specimen records.

Virgin Islands boas are considered to be common on Tortola (Lazell 2005), although this assertion is not corroborated by encounter rates (that the authors are aware of). Nevertheless, individuals can be found with some effort (D. Barber, pers. comm. to RGR).

On St. Thomas the species has never been found under survey effort (R. Platenberg, pers. comm. to RGR). It is, however, found several times per year either by property owners or dead on the roads near Red Hook. Between 1980 and 2006, 114 confirmed boa sightings were reported to the USVI Division of Fish and Wildlife on St. Thomas, many of which were dead (Platenberg and Harvey 2010).

On Culebra, García (1992) estimated an encounter rate of 0.72 boas/p-h, while a survey in 2001 estimated an encounter rate of 0.01 boas/h (Puente-Rolón 2001). Surveys of Monte Resaca, Culebra, occurring twice per month for one year did not yield any sightings (Ríos-Franceschi et al. 2016).

Three nights of surveys of the Puerto Rico population by the authors (RGR and ARPR) in 2013, constituting 12 p-h, yielded only two sightings (0.17 boas/h). Meanwhile a survey by Island Conservation (2018) found an encounter rate of 0.13 boas/h. Most surveys at this site by trained herpetologists yield between 0 and 2 individuals per night.

The population of boas on Cayo Diablo (Islas Cordilleras, Puerto Rico) has been estimated at 105–125 boas per hectare (Tolson 1991a, 2004), which would also make this the highest density of any natural population of any species of West Indian boa. However, a survey conducted by Island Conservation in 2018 found encounter rates of only 0.25 boas/h on Cayo Diablo (Island Conservation 2018). An introduced population on Cayo Ratones (Islas Cordilleras, Puerto Rico) was previously considered to have a population of around five hundred boas (Tolson et al. 2008) 15 years after forty-one snakes were translocated to the cay. Nevertheless, no boas were found there in 2018 (Island Conservation 2018). Black rats were found to have recolonized the island, suggesting that this translocation has failed in the long term (~25 years post-release).

A third reintroduced boa population on an unnamed island in the USVI, founded by forty-two translocated individuals (see Conservation below), is considered to have the densest population of Virgin Islands boas, with an estimated population density of 202 boas/ha in 2004 (Tolson 2005). The density has since declined to an estimated 45–110 boas/ha in 2018 (D. Smith et al., unpubl. data), still a very high density. In 2018, surveys of this island found an encounter rate of 0.57 boas/h (twenty boas in 35 hours, nine males, nine females, and two juveniles; none exceeding 750 mm SVL). Some of this decline might be due to shifts in the prey base on the island as well as impacts from hurricanes Irma and Maria in 2017 (D. Smith, pers. comm. to RGR).

The 2009 5-year review (USFWS 2009) concluded that only 1300–1500 Virgin Islands boas (Fig. 5.81) existed on Puerto Rico and the USVI; estimates of population size in the BVI are hampered by lack of survey data. Most researchers familiar with the species consider the Puerto Rico and USVI populations to be far smaller than the 2009 estimates as of now (2021), particularly following the apparent loss of the Cayo Ratones population (five hundred animals) and the apparent steep decline in the population on Cayo Diablo, coupled with ongoing mortality and shrinking habitat on St. Thomas.

*Activity and Trophic Ecology*

Like all West Indian boas, this species is largely nocturnal, although occasional observations suggest that it might rarely move or feed during the day. Virgin Islands boas mostly feed on *Anolis* lizards, particularly the abundant *A. cristatellus* (USFWS 1986b). They will also take the lizards *Pholidoscelis exsul* and hatchling *Iguana iguana*. Indeed, *P. exsul* and *I. iguana* used to occur on the unnamed Virgin Island cay where boas were translocated, but those species no longer occur there (RGR pers. obs.) and could have been extirpated by the boas. Virgin Islands boas will also take introduced mammalian prey including *Mus musculus* (Sheplan and Schwartz 1974; MacLean 1982) as well as *Rattus rattus* (C. Walton in litt., photographic evidence from Tortola). Tolson and Henderson (1993) reported a Virgin Islands boa containing a Yellow Warbler (*Setophaga petechia*), which constitutes the only case of birds in the diet of this species.

Boas employ two strategies to find prey at night—active hunting and ambush predation (Figs. 5.82,

**Figure 5.81.** *Chilabothrus granti* have some variation in the lightness or darkness of their dorsal coloration, although this can also vary between day and night within an individual. These two boas from Puerto Rico show these extremes. Top photo by José R. Almodovar; bottom photo by J. P. Zegarra.

5.83). Boas actively hunt by moving slowly along thin branches, tongue flicking to find sleeping *Anolis cristatellus* (Fig. 5.83). Boas are also occasionally observed immobile at the base of a tree or branch, with the head placed flat against the branch, usually at an intersection of branches, with minimal tongue flicking and no movement (Fig. 5.82). This is most likely an attempt to intercept prey, possibly nocturnal geckos, moving along the branch; this strategy would not work for capturing *A. cristatellus,* and no prey captures have been observed by boas using this strategy. Juvenile boas are frequently observed foraging lower in the vegetation, while adults are frequently >2 m high (Tolson 1991c).

*Predators and Defensive Behavior*

Potential natural predators of Virgin Islands boas include Yellow-crowned Night-Herons (*Nyctanassa violacea*) and probably Puerto Rican Screech Owls (*Otus nudipes*); the possibility exists that Puerto Rican racers (*Borikenophis portoricensis*) might also consume boas (Tolson 1991a). Introduced predators almost certainly cause most boa mortality owing to predation. Cats (*Felis catus*) are known predators of boas, especially on St. Thomas where numerous dead boas taken to wildlife officials appear to have been killed by them. Rats (*Rattus rattus*) are known to eat boas, although it has been suggested that rats more commonly impact Virgin Islands boas by consuming their prey (*Anolis*). Nevertheless, the extinction of the translocated population of boas on Cayo Ratones coincided with the reintroduction of rats, while the anole population appears unchanged (J. L. Herrera, pers. comm. to RGR). Cats and rats might represent a dynamic balance that allows boas to persist under some conditions on islands such as Puerto Rico, Culebra, and St. Thomas, if cats largely predate rats and thus keep the rat population low (Tolson 1991c). The introduced small Indian mongoose (*Urva auropunctata*) likely does not prey on boas, as it is a terrestrial, diurnal predator and would be unlikely to encounter boas (Tolson 1988).

*Chilabothrus granti* will ball up when handled but will never strike. When first handled they will void the contents of the cloaca but exhibit no other defensive behaviors.

*Reproduction*

This species reproduces biennially in the wild and can achieve a longevity over 30 years, remaining reproductively competent the entire time (Tolson 1996a). Captive individuals can reach 40 years in age (D. Smith, in litt. to RGR). Females can reach maturity in three years (USFWS 1986b), or potentially even faster (Tolson 1986), and are reproductively competent at sizes as small as 521 mm SVL (Tolson 1986), although smaller females produce smaller litters. Courtship occurs in the spring, from February to May, and individuals kept in captivity away from others will not cycle. Gravid females on Cayo Diablo prefer to sequester themselves in termite mounds (Tolson 1986) and will bask in the morning and afternoon (Tolson 1989) in August and September. Young are born alive in the late summer

**Figure 5.82.** *Chilabothrus granti* are highly arboreal; this individual is in a typical ambush position, near St. Thomas, USVI. Photo by R. Graham Reynolds.

**Figure 5.83.** *Chilabothrus granti* foraging along the coast, with the lights of St. Thomas, USVI, in the background. Photo by R. Graham Reynolds.

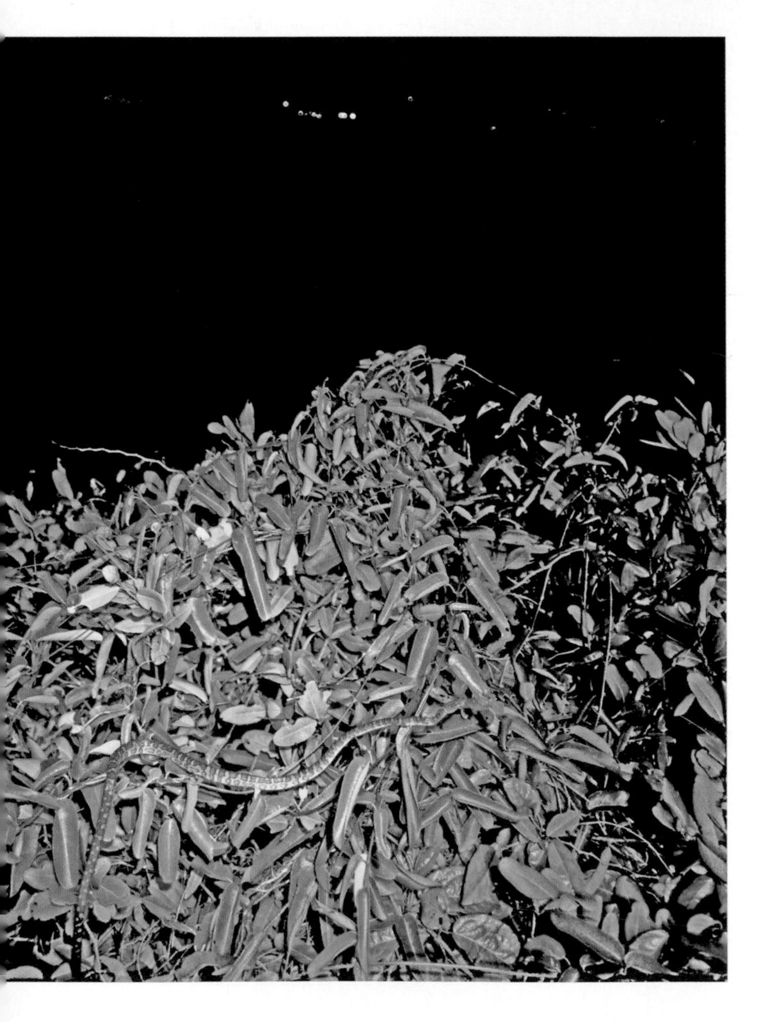

and early fall following a gestation of about 132 days (Tolson 1989), with relatively small litter sizes of two to ten offspring. Neonates are long (200–300 mm SVL) and thin (2–7 g mass). Like other members of the genus, they shed soon after parturition and wait several weeks before feeding (Tolson 1989). Although females likely breed every other year, under ideal conditions they are capable of breeding annually (Tolson and Piñero 1985). There is quite a bit of information on captive propagation of this species, which was bred in a small colony at the Toledo Zoo for several decades (Tolson 1989). Neonates can be induced to achieve 100 g mass in 1 year with heavy feeding, and captive boas can become sexually mature in 2 years.

*Conservation*

*Chilabothrus granti* is listed on CITES Appendix I and is listed as Endangered on the IUCN Red List (Tolson 1996a; Platenberg 2021). It was listed on the U.S. Endangered Species List (USFWS 1980) in 1970, and again in 1979, based on a disjunct distribution, habitat loss, and the impacts of invasive predators. Virgin Island boas are protected in the USVI by the Virgin Islands Endangered and Indigenous Species Act, and in Puerto Rico it is protected as a critically endangered species. The single largest threat to the species appears to be habitat destruction, with impacts by rats a close second. They are also routinely killed by humans, and many human encounters result in the death of the snake. Many records of this species outside of translocation sites are road-killed individuals or boas that have been killed by property owners. Nellis et al. (1983) noted that collection for the pet trade is a potential threat, although the species is not known to be actively (or at least openly) in the pet trade presently outside Puerto Rico (see Chapter 3). Hurricanes could potentially threaten small island populations, either through direct mortality or habitat loss. Hurricane Hugo (1989) caused the loss of 2000 $m^2$ of habitat on Cayo Diablo as well as the loss of every mature *Coccoloba* tree (Tolson 1991c). Hurricane Maria might have contributed to the loss of boas on Cayo Ratones.

On St. Thomas and Culebra, much of the habitat used by boas is privately owned, making surveying more difficult and putting additional pressure on the species owing to habitat conversion to commercial, housing, or tourism development. The apparent loss of one translocated population (on Cayo Ratones, see below) is disturbing, as it suggests that these small island populations might be especially vulnerable to extinction if rats recolonize the islands following eradication. The loss of that population also represented the loss of nearly one-third of the 2009 estimated population size of the species. Rat eradication has met with mixed success in the USVI, and rats have been shown occasionally to recolonize islands following removal (Savidge et al. 2012), with between a 10% and 20% eradication failure on the Puerto Rican Bank (Island Conservation 2018).

This species has been the focus of significant conservation attention (e.g., Tolson 1996c). Recovery plans have been published by the U.S. Fish and Wildlife Service in 1980 (USFWS 1980) and 1986 (USFWS 1986b). Five-year reviews have been completed in 1991 (USFWS 1991) and 2009 (USFWS 2009). A comprehensive USFWS Species Status Assessment was published in 2018 (USFWS 2018) with detailed conservation concerns and actions for *C. granti*. However, in 2020 the species was proposed for downlisting based on the findings in the 2018 Species Status Assessment. The USFWS received public comments related to this proposal and were subsequently sued so that a public hearing could be held. Comments received on this proposal during both periods robustly illustrated that the information in the 2018 report showed that the species is actually doing much worse, not better, than it was in 2009. An outcome of this process is still pending as of December 2021. In 2019 partners from several organizations launched a series of workshops to produce a Conservation Action Plan for the species, which is expected to be formalized in 2022 and will include recommendations for conservation and management in the USVI.

Captive reproduction and translocation programs have been in place since the 1990s (see below). These efforts have undoubtedly benefitted the species, although populations on St. Thomas, Tortola, and Culebra remain imperiled. The population on Puerto Rico is likely extremely small and has not expanded its range in at least three decades of observations (ARPR, pers. obs.). On small islands, the species has persisted on Cayo Diablo, although recent surveys (2018) suggest that the population size is much smaller now than it was a decade ago. Cayo Ratones, one of two translocated populations in the USVI, has likely gone extinct 25 years after the release of forty-one boas (Island Conservation 2018), while the other (unnamed island in the USVI) remains robust, although apparently without any large adults (RGR, pers. obs.). The largest boa found there over three recent surveys (2017–2018) was 750 mm SVL—two-thirds the size of the animals that previously occurred there in 2004 (Tolson 2005).

The current status of the species is almost certainly worse than in 2009 when the last 5-year review was completed and recommended downlisting the species from endangered to threatened (USFWS 2009). Importantly, at least one significant population has apparently gone extinct on Cayo Ratones since that time, possibly representing as much as one-third of the entire boa population in Puerto Rico and the USVI. Reynolds et al. (2015) offered recommendations of islands that could potentially serve as reintroduction sites with some invasive species remediation, such as the removal of rats. The species will benefit from additional applied conservation, such as invasive species mitigation and reintroduction to some of these areas.

*Captive Breeding and Reintroduction*

Uniquely among West Indian boas, *C. granti* has been the focus of captive breeding, island restoration, and reintroduction/translocation since 1985 (USFWS 2018). This joint effort between USFWS, Puerto Rican

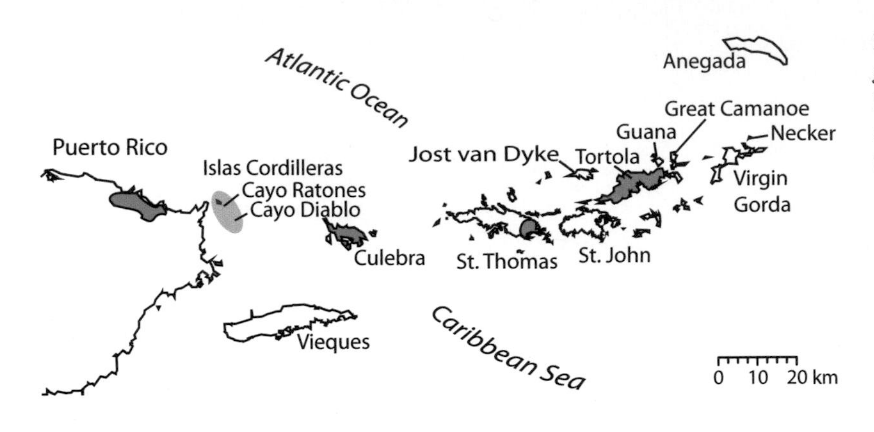


**Figure 5.84**. Range of *Chilabothrus granti*, located entirely on the eastern Puerto Rico Bank. Note that the species is found only in a small area of St. Thomas.

Departmento de Recursos Naturales y Ambientales, Virgin Islands Department of Natural Resources, the Toledo Zoological Gardens, and others involved propagating the species at the Toledo Zoo (Tolson 1989) and introducing captive-reared offspring to islands in Puerto Rico and the USVI. Starting with founder individuals from St. Thomas and Cayo Diablo, managed as separate populations, appropriate reproductive conditions in captivity were worked out with the goal of establishing new populations on rat-free islands. Neonate offspring from Cayo Diablo boas were used to establish a population on nearby Cayo Ratones, Islas Cordilleras, with an initial release of twenty-eight individuals and a subsequent release of thirteen boas. This population was estimated to contain five hundred boas by 2004, showing that reintroduction might meet with success (but see below). Founder individuals from St. Thomas were used to produce eleven neonates that were reintroduced to the nearby unnamed island in the USVI. Additional individuals brought to Department of Natural Resources on St. Thomas ($n$ = 31) were also released onto the island between 2002 and 2003. This population was estimated to grow to 168 boas as quickly as 2004. These programs, including regular surveying and reintroductions, wound down around 2004.

In 2017, a consortium of conservation scientists and wildlife officials began a new captive propagation, monitoring, and restoration program for the species. Focusing on renewed surveys in Puerto Rico, Islas Cordilleras, and the USVI, they have generated new, and frankly alarming, data on the status of the species on some of these islands. For example, Zegarra et al. (USFWS 2018) found that rats were able to recolonize Cayo Ratones in the Islas Cordilleras sometime between 2007 and 2017 and that following the passages of hurricanes Maria and Irma in 2017 no boas were found on Cayo Ratones during a series of three separate surveys (six total survey nights) over a period of 2 years. This strongly suggests that the reintroduced population has been extirpated, presumably by rats (although rats are not conclusively demonstrated as the agent of this lost population). Smith, Zegarra, N. Angeli of the USVI Department of Planning and Natural Resources, and others have been working on the unnamed island in the USVI (unnamed to protect its location and identity and thus to protect the boas) since 2017, reviving the last studies of that population, which ended in 2004. They have found that, although the population has declined in density since 2004, it is still among the densest population of boas in the Caribbean (although the total population size is less than one hundred boas on an island that is a mere 0.83 ha).

A new captive breeding and reintroduction campaign began in 2017 with boas captured from the unnamed island being reared (for eventual propagation) in off-exhibit dedicated endangered species propagation facilities at the North Carolina Zoo (Asheboro), the St. Louis Zoo (Missouri), and the Ft. Worth Zoo (Texas). Offspring from these animals will be used for reintroductions and translocations following additional planned rat eradications in Puerto Rico and the USVI. Reintroduction and translocation efforts are being supported by expertise from Island Conservation to identify suitable invasive species control measures and reintroduction sites (Island Conservation 2018). Such applied conservation actions are crucial to the continued survival of *C. granti*. Figure 5.84 shows the distribution of *Chilabothrus granti*.

## PUERTO RICAN BOA

*Chilabothrus inornatus* (Reinhardt 1843)

### *Taxonomy*

The Puerto Rican boa (Fig. 5.85) was described by Reinhardt (1843) as *Boa inornata*. In 1844, Duméril and Bibron reclassified the species as a member of the genus *Chilabothrus*. In 1893, Boulenger reclassified the species in the genus *Epicrates*. The species was considered to be conspecific with *Ep. subflavus* from Jamaica until Stejneger (1901) separated them and redescribed the Puerto Rican boa as *Ep. inornatus*.

### *Etymology*

The epithet *inornatus* refers to being unadorned or undecorated, likely in reference to the idea that individuals lack patterns or colors (possibly the type specimen appeared drab). Nevertheless, this is an inaccurate description, as the species exhibits a variety of different color morphs and blotching patterns (see below). The species is commonly referred to as "boa puertorriqueña" or "culebrón."

### *Description*

This species is the largest snake on the Puerto Rican Bank. The maximum total length on record is 2200 mm and the largest mass is around 6.0 kg (ARPR, pers. obs.). There are anecdotal, although unsubstantiated, reports of animals reaching sizes of 3.0 m, although such sizes are unlikely. The average SVL for the species is 1.4 m (Wiley 2003; Wunderle et. al. 2004; Puente-Rolón 2012; Mulero Oliveras 2019). Huff (1980) observed that females were longer than males, but recent data have not found any significant sexual size dimorphism in SVL for adult boas (Puente-Rolón 2012; Mulero Oliveras 2019). Wiley (2003) found that boas from various sites across Puerto Rico (most from the Luquillo Mountains) were on average 1370 mm SVL ($n$ = 49) and 952 g in mass. The longest individual was a male, but the females were significantly heavier than males. Among thirty-four adult boas (fifteen females, nineteen males) measured in the northern karst forest from Hatillo to Cupey, RGR and ARPR (unpubl. data) found an average SVL of 1377 mm and an average mass of 1426 g. Puente-Rolón (2012) found the average SVL of males and females was nearly identical at about 1433 mm ($n$ = 48); although male mass (1461 g; range 681–2500 g; $n$ = 25) was less than average female mass (1764 g; range 585–3500 g; $n$ = 23), he found no sexual size differences of statistical significance. It is unknown if regional variation in sizes exists, although it seems unlikely based on these studies from different areas of Puerto Rico. Nevertheless, it appears that cave-associated boas in the northern karst region are, on average, smaller than free-ranging boas in the same

**Figure 5.85.** *Top*: Adult female *Chilabothrus inornatus* from the northern karst region of Puerto Rico. *Bottom*: Juvenile *C. inornatus* from semi-urban dry forest near Río Grande, Puerto Rico. Photos by R. Graham Reynolds.

**Figure 5.86.** *Chilabothrus inornatus* can exhibit a range of color morphs. *Top*: Highly unusual hypomelanistic *Chilabothrus inornatus* from the northern karst region, feeding on a bat. Note the few dark scales and very light dorsal scales. *Bottom*: Adult female *C. inornatus* displaying an unusual reddish coloration. Photos by Alberto R. Puente-Rolón.

region at 1376 mm SVL versus 1437 mm SVL, respectively (Puente-Rolón 2012).

Neonates of the species have an average SVL of 340 mm and a tail length of 70 mm. Neonates average 13.6 g in mass and are not sexually dimorphic at that stage (Puente-Rolón 2012).

The color of this species is highly variable, although most individuals have some combination of tan to very dark brown dorsal color, often with several of these colors in a mottle. Occasional individuals are quite reddish, and one hypomelanistic individual has been found (Fig. 5.86). Some individuals can appear very dark and patternless, while others may show seventy to eighty irregular or diffuse markings outlined by a dark brown border across the dorsum (Fig. 5.85). These patterns can appear as larger blotches or as thinner lines, almost like stripes. In lighter individuals there is a dark postocular stripe indistinctly connected with a mediolateral longitudinal line on the neck that eventually dissipates into other pattern elements. Neonates are bright orange and typically have visible, although faint, contrasting color patterning on the dorsum. Juveniles are a dull orange that will fade with age and frequently exhibit distinct patterning on the dorsum and lateral aspects, including a distinct postocular stripe (Figs. 5.85, 5.106). Boas with these various color and pattern elements are not restricted to certain geographic areas, as all color and pattern combinations can potentially be found throughout the range of the species. While all members of the genus *Chilabothrus* can have an iridescent sheen when exposed to sunlight, darker-colored *C. inornatus* appear particularly striking in this regard as the rainbow sheen contrasts with the dark body color.

Scale rows around the midbody range from 38 to 42 (Rivero 2006). Ventral scales range between 261 to 271 and can be light or dark brown, even nearly black (Rivero 2006). Caudal scales range from 67 to 73. Supralabials fade into pale brownish gray at the commissure. The iris is silvery gray clouded with dusky speckling, and the tongue in adults is dark gray to blue gray with white tips.

### *Distribution*

The species has a wide distribution across the island of Puerto Rico, from the coast up to a documented elevation of ~1050 m asl (Wiley 2003; Henderson and Powell 2009), although elevations above 400 m support relatively few boas or much lower densities than lower elevations (Wiley 2003; RGR, pers. obs.). Boas have been recorded from nearly all the seventy-eight municipalities of Puerto Rico, including the larger islands of Vieques and Culebra. In the last century, the islands of Puerto Rico and Vieques were significantly deforested for agriculture (Birdsey and Weaver 1987), including sugarcane and coffee plantations, which likely severely impacted the range of this species. Boas were presumably extirpated from Culebra and Vieques, as well as large areas across the lowlands of Puerto Rico. Happily, many forests have regrown, and boas can now be found across most of their historical range, although often in fragmented populations. There are recent records of Puerto Rican boas from the islands of Vieques ($n$ = 2) and Culebra ($n$ = 1), although these are likely waifs and not representative of reestablished populations (ARPR, pers. obs.; J. P. Zegarra in litt. to RGR, 8 February 2021). Puerto Rican boas do not occur on other satellite islands on the Puerto Rico Bank, such as Caja de Muertos or the Cordillera islands east of Fajardo. However, Wiley (2003) reported finding a boa on a mangrove cay 0.5 km off coast in the southwest of the island; although the exact location is unknown, we presume it is located in Bahía Montalva. In their range boas are restricted to areas with some forest or vegetation, although populations can persist in heavily urbanized areas, including in and around San Juan and Mayaguez. Pockets of boa populations persist even when completely surrounded by urbanization, such as within parks or on karst hilltops like San Patricio west of downtown San Juan (Tyler et al. 2016; Tavárez and Elbakidze 2019). In the Luquillo Mountains, the average elevation where boas were found by Wiley (2003) was 250 m, and boa density decreases at elevations above this in the Sierra Luquillo (RGR, pers. obs.), dropping out entirely in the higher forests (Wiley 2003).

*Habitat*

The Puerto Rican boa is a habitat generalist and can be observed in wet, moist, and dry mature and secondary forest (Fig. 5.87). Pasturelands, shrub areas, agricultural lands, and wetlands (mangrove and *Pterocarpus* forests) also have been reported as habit used by the species (Wiley 2003; Gould et al. 2008; Henderson and Powell 2009). It has been estimated that the available habitat to support boa populations on the island is 46% of the geographic area, although only 9% of this is within protected lands (Gould et al. 2008). Highly urban areas with fragments of forest cover have been successfully supporting populations of this species (Mulero Oliveras 2019). Structurally, the Puerto Rican boa has been reported to prefer trees with larger diameter at breast height, higher vine coverage (unattached and attached to trunk), higher density of surrounding vegetation, and more crown contact with adjacent trees (Wunderle et al. 2004; Mulero Oliveras 2019). Boas will also use rock walls, tree holes, and cavities in the ground to rest more frequently than they use other cover, such as leaf litter (Puente-Rolón and Bird-Picó 2004). In particular, rock walls are thought to be a valuable habitat feature (Puente-Rolón and Bird-Picó 2004).

While reported nearly island-wide, the species is more frequently seen in the northern karst region on the island (Wiley 2003; Gould et al. 2008; USFWS 2011), where forests develop over limestone "haystack hills" or "mogotes" that provide a great deal of subterranean habitat, including caves. Indeed, boas are easier to observe in this region and become more infrequent toward the central mountain areas and least frequent in the southern dry areas. However, some "hot spots" can be found in the areas with denser populations, frequently associated with limestones outcrops or caves.

Caves in the north and other limestone forests on the island are a favored habitat used by this species and provide conditions to observe aggregations of boas. There are at least twenty-one caves in the northern karst region with documented records of boas (Rodríguez-Durán 1996; Puente-Rolón and Bird-Picó 2004; Puente-Rolón 2012). Boas use both exterior and interior parts of caves—they are frequently seen basking or resting around the mouth of caves or in the forest near caves. Boas will also occupy the interior of hot caves and can be found resting on the floor and walls of chambers in caves. At night, some boas use cave mouths to feed, capturing bats in midflight (Rodríguez and Reagan 1984; Rodríguez-Duran 1996; Puente-Rolón and Bird-Picó 2004).

*Abundance*

In the late-nineteenth century, Garman (1887) wrote that Puerto Rican boas were "very common along the streams on the branches of the trees" in Bayamon, Puerto Rico. But abundances seem to have rapidly changed, and for much of the twentieth century this species was thought to be rare. The islands of Puerto Rico and Vieques were almost completely deforested at the turn of the twentieth century, retaining an estimated 1% of the original forest cover (Wadsworth

**Figure 5.87.** Northern karst forest of Puerto Rico near Rio Abajo, showing the "haystack hill" mogotes composed of limestone and covered with dense forest. This is excellent habitat for *Chilabothrus inornatus*. Photo by R. Graham Reynolds.

1949; Birdsey and Weaver 1987). Further, boas during this period were hunted extensively for skin and fat and pursued by introduced predators (Joglar et al. 2011). In 1973, the USFWS stated that there were less than two hundred individuals on the Island (USFWS 1986a). However, no methodology was presented to support that number.

More recent data give a more accurate estimate of boa abundance. Surveys conducted by Bird-Picó (1994) reported eighty-six boas captured over 5000 p-h of searching, but eighty-two of the eighty-six observations were around caves in the northern karst region. A focal survey was conducted by P. J. Tolson in the U.S. Naval Security Group Activity at Sabana Seca (located just west of San Juan in the northern karst region) from 1996 to 1997. The surveys were conducted in three transects located in karst forest habitat, resulting in the capture and marking of a total of twenty-two snakes during four sampling periods. No marked snakes were recaptured during this period, and Tolson (1997) estimated between 5.23 and 44 boas/ha (the difference coming from different capture–mark–recapture models; Tolson 1997). Ríos-López and Aide (2007) established transects in three distinct types of habitat (reforested valley, old growth valley, and karst hilltop) in Sabana Seca and estimated a mean density of 5.6 boas/ha. Wunderle et al. (2004) marked seventy boas in the lowlands of the Luquillo Experimental Forest during the period of October 1996 to July 2001, with many captures coming from boas found crossing the roads. Despite marking so many animals, only one was recaptured, and the authors recommended performing searches over larger areas (10 to 100 ha) to increase recaptures. This is presumably because the authors supposed the animals had larger home ranges than what their surveys were encompassing. More re-

**Figure 5.88.** *Top*: Boas will often use vines for accessing the mouth of bat caves. Vines serve both as resting places and as holdfasts from which to hunt. Photo by Alberto R. Puente-Rolón. *Bottom*: Young adult *Chilabothrus inornatus* feeding on a bat at a cave-mouth entrance near Arecibo, Puerto Rico. Note the color of the ventral scales. Photo by R. Graham Reynolds.

cently, studies from 2013 to 2017 by Mulero Oliveras (2019) focused on boa populations in fragmented habitats at Fort Buchanan military base adjacent to greater San Juan. In this study, fifty snakes were captured, and thirty-eight of these snakes required an estimated total search effort of 900 p-h. For these fragmented areas the population density was estimated to be 1.24 boas/ha.

Forests of the karst region (Fig. 5.87) are considered crucial to Puerto Rican boas, as the densest populations of boas are reported from this region (Bird-Picó 1994; Rivero 1998). Caves have by far the highest abundance, with Puente-Rolón et al. (2013) having counted between seven and twenty boas at a single cave in a single night (Fig. 5.88). Nevertheless, cave populations seem to be decreasing—caves with human visitation quickly lose robust boa populations, and other caves are impacted by additional negative interactions such as invasive predators. Only the most remote and inaccessible caves still harbor more than twelve boas at any given time (ARPR, pers. obs.)

### *Activity and Trophic Ecology*

Like all West Indian boas, the Puerto Rican boa is mostly nocturnally active. Individuals can be observed basking in the sun during the day, and a favored place would be in patches of sunlight reaching the forest floor near a log or treefall. Remote hot caves are used for diurnal resting, particularly by females (ARPR, pers. obs.). In a recent study on fragmented habitats,

Mulero Oliveras (2019) found differences between juvenile and adult boa encounters. She reported that juveniles were more easily spotted when walking transects during the night, while adults were more easily spotted on the transects during the day. However, it is important to take in account that these diurnal sightings consisted mainly of snakes basking coiled or stretched at urban areas or forest edges. In forested areas, adult snakes are much more easily detected during the night than the day (Wunderle et al. 2004; Puente-Rolón 2012), and juveniles are readily found within a few meters off the ground at night along forested trails in the northern karst region and the eastern dry forest (RGR, pers obs.).

Home range and activity patterns for Puerto Rican boa populations have been studied at four localities using radio telemetry (Puente-Rolón 1999; Puente-Rolón and Bird-Picó 2004; Wunderle et al. 2004). Three of the studies were focused on intact forested habitat, and one was focused on a fragmented habitat. All four studies showed that Puerto Rican boas are more active during the months of April and May, which is considered the peak of the rainy season on the island. This weather and associated increase in activity is likely related to reproduction (mate searching in males and gestation in females) and prey availability (Puente-Rolón and Bird-Picó 2004; Wunderle et al. 2004). Cave-associated populations of boas tend to have smaller home ranges than "surface" populations (Puente-Rolón and Bird-Picó 2004; Wunderle et al. 2004). In a study focused on a cave-associated population near Arecibo, Puente-Rolón and Bird-Picó (2004) found that male and female home range sizes did not differ significantly, although females did have a slightly larger mean home range size of 7890 $m^2$ versus male mean home range size of 5000 $m^2$. Males and females also traveled a similar distance during each move (47.9 ± 18 m for males, 51.5 ± 24 m for females) and showed similar movement patterns during the reproductive period of March to October, moving on average 83 ± 51 m per day (males) and 99 ± 88 m per day (females). Females did move much more than males during the nonreproductive period of November to February. Male snakes spent between 5 and 53 consecutive days without moving across seasons (mean 37 days) while females spent 2–96 days without moving across seasons (mean 47 days).

This species can be considered one of the top predators on the island and seems to employ both active and ambush foraging strategies (Puente-Rolón and Bird-Picó 2004; Wunderle et al. 2004; Henderson and Powell 2009). In captivity, neonates start to eat after their second shed between 4 and 5 weeks after birth (Bloxam and Tonge 1981). Neonates and juveniles feed on *Anolis* lizards (*A. cuvieri*, *A. evermanni*, *A. gundlachi*, and *A. cristatellus*; probably other species as well), ground lizards (*Pholidoscelis* spp.), and treefrogs of the genus *Eleutherodactylus* (Reagan 1984; Wiley 2003; Rivero 2006; Henderson and Powell 2009).

Adult snakes appear to feed mostly on *Rattus*, or at least this makes up the bulk of the diet of individuals that have been examined (Wiley 2003; Puente-Rolón et al. 2016), although *Mus* are also readily consumed. Wiley (2003) found that ~62% of the twenty-six food-containing boas he dissected contained *Rattus*. Puente-Rolón (2012) used stable isotope analysis to determine that both cave-associated and non-cave-associated boas obtained >80% of their energy from *Rattus* and *Mus*.

Boas have also been reported feeding on small- to medium-size birds, including domestic fowl (*Gallus*), Common Ground Doves (*Columbina passerina*), and Cattle Egrets (*Bubulcus ibis*; Wiley 2003; Henderson and Powell 2009; ARPR, pers. obs.). Boas have been observed attempting to prey on nestling Puerto Rican Parrots (*Amazona vittata*) at Bosque de Río Abajo near Arecibo (R. Valentín, in litt. to TMRC, 17 July 2020). Despite not being a large part of the energy intake for boas (Puente-Rolón 2012), bats are a regular food item, and boas actively hunt them at the mouths of caves (Fig. 5.88; Rodríguez and Reagan 1984; Rodríguez-Durán 1996; Puente-Rolón and Bird-Picó 2004). *Monophyllus redmani*, *Erophylla sezekorni*, *Mormoops blainivillii*, and *Pteronotus quadridens* are the most commonly captured bats at a cave in the northern karst forest, although boas occasionally captured *Brachyphylla cavernarum*. Bat hunting is a specific behavioral syndrome that consists of selecting an ambush site and attempting to capture bats as they brush past the boa during emergence in the evening (described in detail in Puente-Rolón and Bird-Picó 2004). Starting at 1730 hours, boas will extend up to two-thirds to three-fourths of their body length from a holdfast on the cave ceiling or a hanging root at the

**Figure 5.89.** Typical hanging behavior by *Chilabothrus inornatus* hunting bats at a cave mouth. Boas will extend nearly two-thirds of their body into midair to await contact with a passing bat. Note that there are other boas in the background, on the left and above the center-frame bat. Photo by Alberto R. Puente-Rolón.

cave mouth (Fig. 5.89). Their bodies will generally be straight, and they will occasionally wave their bodies slowly from side to side until a bat brushes against them, eliciting a fast upward strike and attempt to coil. Most strikes are unsuccessful (ARPR and RGR, pers. obs.), but boas will quickly resume hunting following an unsuccessful strike. Puente-Rolón and Bird-Picó (2004) found that boas required an average of 12.5 minutes to subdue and consume a single bat (Fig. 5.90), and that boas would attempt to capture multiple bats during an evening. They observed a maximum of four bats captured and consumed by a single boa during one evening. Boas will also attempt to take captured bats from nearby boas, and often the pirate boa is successful in obtaining the prey. Boas will repeatedly return to a cave entrance to feed throughout the year, spending the rest of the time in the surrounding forest (Puente-Rolón and Bird-Picó 2004).

Boas also feed on other squamates, including introduced *Iguana iguana* (ARPR, pers. obs.), *Anolis* lizards such as *A. cuvieri*, *A. evermanni*, and *A. gundlachi* (Wiley 2003), and conspecifics (ophiophagy; Acevedo-Torres et al. 2005). Some invertebrates such as land crabs (*Cardisoma guanhumi*) and fireflies (family Lampyridae) have also been documented being eaten by these boas (Reagan 1984; Wiley 2003), although the latter could have been an accidental ingestion. The introduction of potentially competitive *Boa constrictor* to Puerto Rico at present has unknown effects on the trophic ecology or food availability of Puerto Rican boas (Reynolds et al. 2013b), although it is known that introduced *B. constrictor*, like Puerto Rican boas, feed heavily on *Rattus* (Quick et al. 2005; Vega-Ross 2018).

*Predators and Defensive Behavior*

Predation on the Puerto Rican boa can be divided in two stages—juveniles and adults. It is believed that juveniles suffer more predation than do adults. However, data are not available to support this fully, and most of the information is anecdotal or speculative. Native potential predatory birds of juvenile boas suggested by Reagan and Zucca (1982) include the Puerto Rican Lizard-Cuckoo (*Coccyzus vieilloti*), Red-legged Thrush (*Turdus plumbeus*), Yellow-crowned Night-Hheron (*Nyctanassa violacea*), Red-tailed Hawk (*Buteo jamaicensis jamaicensis*), Pearly-eyed Thrasher (*Margarops fuscatus*), and Puerto Rican Screech Owl (*Megascops nudipes*). Only one of these species, the Puerto Rican Lizard-Cuckoo, has been directly observed preying on a neonate (ARPR, pers. obs.). Another potential predator of juveniles might be the Puerto Rican racer (*Borikenophis portoricensis*), as that species has been reported preying on other snake species of similar size (Rodríguez-Robles and Leal 1993).

Introduced predators that may be preying on juveniles as well as on adults of the Puerto Rican boa are cats (*Felis catus*), rats (*Rattus rattus*) and small Indian mongoose (*Urva auropunctata*). A dead boa from the Luquillo Mountains appeared to have been preyed upon by a mongoose (Wiley 2003). Despite the general

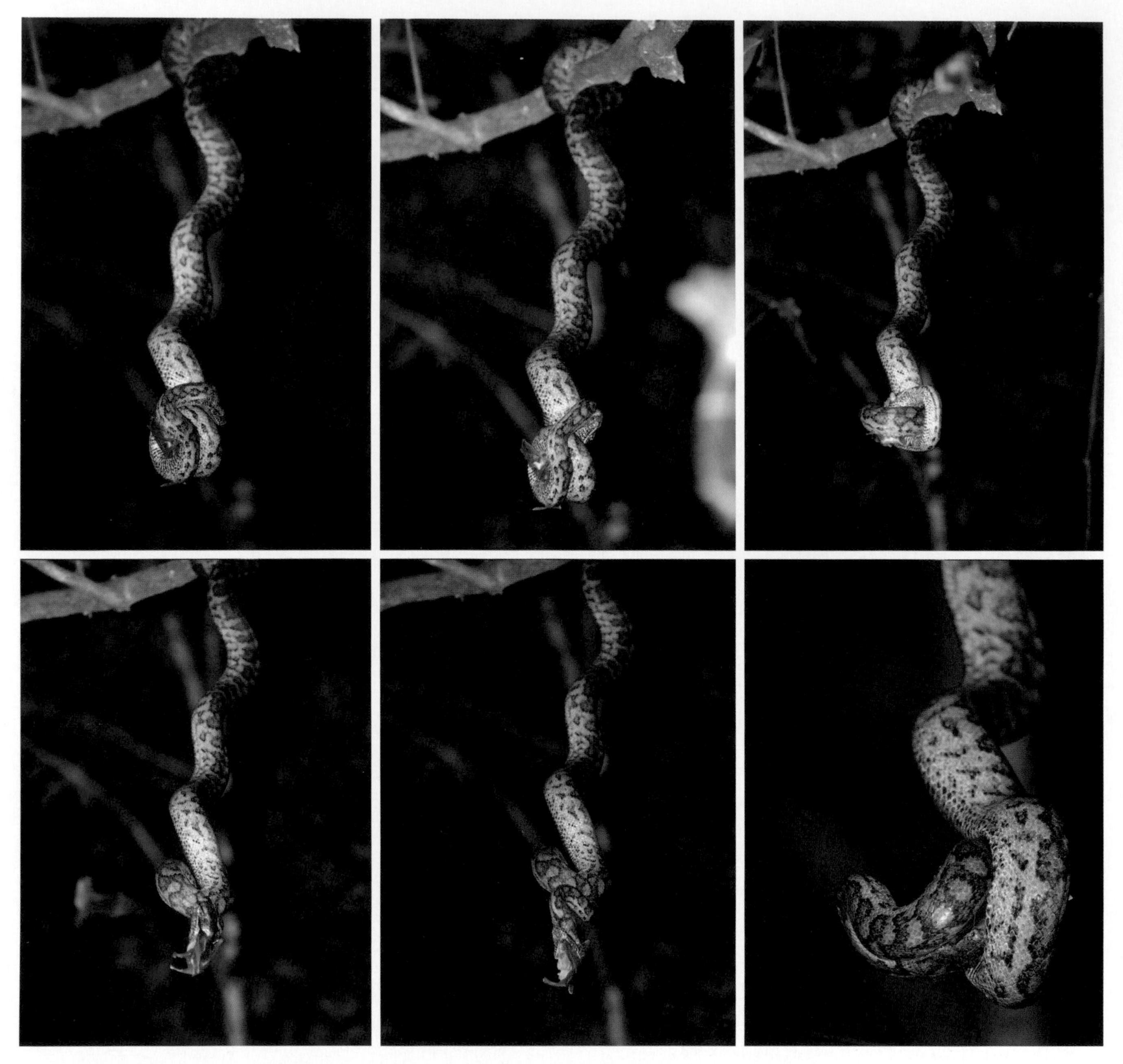

**Figure 5.90.** Sequence of six photos showing a young adult *Chilabothrus inornatus* manipulating and consuming a bat at a cave in the northern karst region (*clockwise from top left*). Following capture with a quick upward strike, the boa will loop two coils around the bat until it is subdued. The boa will then return to a hanging head-down position to adjust the grip on the bat followed by using the coils again to push the bat into its mouth. This entire process lasts on average about 12.5 minutes. Photos by Alberto R. Puente-Rolón.

consensus among wildlife biologists in Puerto Rico that these three introduced species prey heavily on snake populations, there are no robust studies to confirm it. Therefore, it is important to start to study the diets of invasive predators and competitors to determine if they are a threat to boa populations.

Adults face predation as well, including from Red-tailed Hawks (*Buteo jamaicensis jamaicensis*; Gallardo et al. 2019). In 2012, Puente-Rolón was conducting a radio-telemetric study on the species in the northern karst area when one adult male with a SVL of 1470 mm was preyed upon by a Broad-winged Hawk (*Buteo platypterus brunnescens*) that was nesting in the area. Humans are also frequent predators for adults of the species. Grant (1933) reported the hunting of snakes for fat extraction, and later this widespread practice was reconfirmed by Rivero (1998). Some people claim that snake fat has medicinal properties; however, there is no scientific evidence to support this common belief among people in Puerto Rico. Bird-Picó (1994), while conducting a population survey in the southwest part of Puerto Rico, reported the use of Puerto Rican boa meat for the preparation of "empanadillas"—a local food that looks like a fritter (spiced meat wrapped in flour and deep-fried). This is the only direct evidence of snake meat use in Puerto Rico, although anecdotes abound.

Birds will occasionally interfere with boas by "mobbing" them, or continually vocalizing and diving on them in an attempt to drive them off. Several species of passerines as well as Puerto Rican Woodpeckers (*Melanerpes portoricensis*) have been documented to engage in this behavior (Mercado et al. 2002), although it is unknown what impact this might have on the boas.

It is impossible to generalize in terms of the defensive behavior of this species, as some snakes will tolerate handling without any response, while others will try to bite as soon as you make eye contact with the snake. The first defensive response of the Puerto Rican boa generally is to escape. If cornered, adult snakes will assume an S-shaped posture, will make a short hiss, and will strike with their mouth open, but they usually do not complete the bite. Some individuals, like many members of the genus, will ball themselves up when threatened. Almost all individuals (both juveniles and adults; Fig. 5.91) will release an unpleasant musk from their cloaca. The pungency is similar to other large species of *Chilabothrus*, although the musk in *C. inornatus* has been observed to cause nausea in some people. However, the species' major defense is their cryptic behavior combined with camouflage, which serve to reduce interactions with potential predators and with humans.

*Reproduction*

Captive breeding data from the Reptile Breeding Foundation in Canada identified the species as one with a biennial reproductive cycle (Huff 1978, 1979). However, there are numerous records of captive snakes breeding annually (Bloxam and Tonge 1981), presumably owing to idealized conditions and supplemental

**Figure 5.91.** *Top*: Typical adult *Chilabothrus inornatus.* Note the mottled scales and the presence of black, brown, and red scales as well as the postocular stripe. Photo by José R. Almodovar. *Bottom*: Juvenile *C. inornatus* that has not yet transitioned to adult coloration. Note the tongue colors. Photo by Alberto R. Puente-Rolón.

**Figure 5.92.** Wild litter of neonate *Chilabothrus inornatus* from Ciales, Puerto Rico. Photo by Alberto R. Puente-Rolón.

feeding. Courtship in the wild has not been described in detail. Some individuals have been observed mating on tree branches (Bloxam and Tonge 1981; Mulero Oliveras 2019) and on the ground (ARPR, pers. obs.). Two breeding-ball events have been observed, whereby one female was surrounded by four males in each instance (Puente-Rolón 2012). In captivity, courtship lasts for several hours and involves the male rubbing his spurs on the flank of the female while engaging in frequent tongue flicking (Rivero 1998). Like all members of the genus, the species is viviparous, and courtship and mating take place mostly at the beginning of March to late May (Puente-Rolón 2012). Sexually mature females are generally well over 1000 mm SVL and at least 1.0 kg in mass, although Wiley (2003) reported a sexually mature female of 1080 mm SVL and 685 g.

Timing of reproduction and parturition seems to be variable on the island, as neonates for this species have been observed in early March and in late May (Mulero Oliveras 2019). Wiley (2003) reported a gravid female in August, indicating that parturition might run all the way through summer months. Gravid females reduce their movements and spend more time in exposed areas that allow them to thermoregulate (Puente-Rolón and Bird-Picó 2004; Wunderle et al. 2004). The gestation period ranges from 152 to 193 days (Rivero 2006), and the species seems to have one of the larger litter sizes in the genus *Chilabothrus* (Huff 1979; Puente-Rolón 2012). The average litter size is eighteen and can range from twelve to thirty-two (Fig. 5.92; Tolson 1992b; Puente-Rolón 2012), with a maximum of thirty-seven reported (Wiley 2003). Captive litters seem to be smaller than wild litters, averaging twelve to fourteen young (Huff 1978). Neonates have a mean SVL of 342 ± 18 mm, tail length of 70 ± 5.3 mm, and a body mass of 13.61 ± 1.6 g (Puente-Rolón 2012). Studies on the reproduction of the species did not find a relationship between the relative investment per offspring and maternal body size. However, females with better prepartum body condition (relative mass) had larger litter sizes. This means that snakes with more fat reserves are expected to have a larger surplus of energy to invest in reproduction, which translates to more offspring rather than fewer better-provisioned offspring. The lifespan for the species in captivity ranges from 20 to 30 years (Rivero 2006).

*Conservation*

Spanish notes from the 1700s state that boas were common around houses and that people usually allowed the snakes to remain to control rat populations near settlements (Abbad y Lasierra 1788). Decline of this species probably started in late 1800s, when Puerto Rico suffered a period of intense deforestation culminating in nearly complete deforestation by 1910 or so (USFWS 2011). Low encounter rates during the

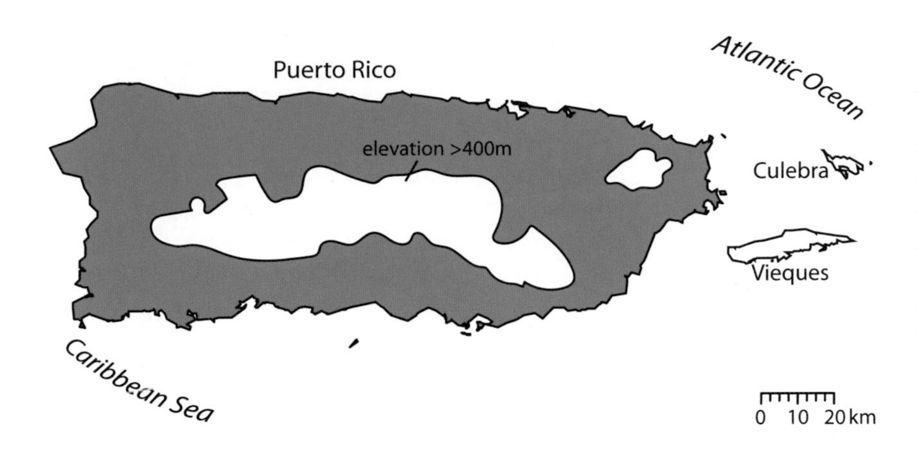


**Figure 5.93**. Range of *Chilabothrus inornatus*, endemic to Puerto Rico up to an elevation of 400–480 m. Boas can be found throughout this range, although they are more common in the north. Despite one report of a boa from Vieques and from Culebra and likely historical occurrence, we do not consider them presently established on these islands.

mid-1900s and shockingly low population size estimates awoke the attention of federal agencies, and the species was listed under the Endangered Species Act (ESA) in 1973 (USFWS 1986a). The Commonwealth of Puerto Rico also protected the species, designating it as threatened by the Law Number 241 of 1999 (Wildlife Law of the Commonwealth of Puerto Rico) and its Regulation Number 6766 (to Govern Threatened and Endangered Species of the Commonwealth of Puerto Rico; Departamento de Recursos Naturales y Ambientales de Puerto Rico 2004). Internationally, the Puerto Rican boa is included on Appendix I of CITES, which lists species that are threatened with extinction. However, the species has been assessed using IUCN criteria as Least Concern on the IUCN Red List (Rodríguez et al. 2018). This presents a very interesting situation in terms of conservation because the species is considered endangered, threatened, and of least concern at the same time. That the species was protected since the 1970s has benefited boa populations, and today there is a general impression that the species is more common than previously thought and is widespread throughout Puerto Rico. Further, populations appear to have rebounded in a number of areas following reforestation over the last century.

Genetic data have recently been brought to bear on Puerto Rican boa conservation. It has been found, for example, that cave-associated populations are important reservoirs of genetic diversity and thus need to be protected (Puente-Rolón et. al. 2013). The same study also found that the common management practice of removing snakes from human populated areas and relocating them to forested areas has meant that little genetic structure is found in these regions. Further, the likely loss of significant amounts of habitat followed by reforestation is reflected in genetic signatures of widespread haplotypes across Puerto Rico (Puente-Rolón et al. 2013).

Aungst et al. (2020), studying ex situ populations of the Puerto Rican boa, found similar genetic diversity in U.S. captive populations to what is observed in wild populations. These findings will help to develop captive breeding strategies to maintain that diversity over the long term. If wild populations decrease again, this information will allow managers to plan and establish a conservation breeding program to add individuals to populations given their genetic background.

Despite a general mitigation of the threats leading to an apparent near extinction of the species a century ago, numerous threats for the species still exist. Habitat loss persists, most impactfully in the form of limestone quarrying (whereby entire mogotes are razed) and landscape fragmentation for agriculture and urbanization. Predation by introduced species such as the mongoose (*Urva auropunctata*) and feral cats (*Felis catus*) remains a significant concern for long-term survival of the species (USFWS 2011). Also, poaching for the pet trade, intentional killing, and harvesting for fat extraction might contribute to local population declines (Wiley 2003; Puente-Rolón and Bird-Picó 2004; Wunderle et al. 2004). Crucially, there are no data or studies available to determine the impact of those actions on boa populations. The first study to try to determine how boas used urban fragmented habitat was conducted by Mulero Oliveras (2019). She studied movements and habitat use and found that snake abundance was lower and that urban snakes move less compared with snakes in continuous forest. In addition, she documented that boas in fragmented urban habitat selected areas with bigger vegetation clumps, good canopy cover, leaf litter, and woody material—all habitat features that are limited in urbanized landscapes. Roads are certainly a threat—over a 13-year study period in the Luquillo Mountains, Wiley (2003) noted that nearly half of the forty-nine boas he found were killed by cars (a further significant fraction were killed by people). Tucker et al. (2020) used demographic simulation to show that increasing urbanization, as little as 8% per decade, could lead to population declines over a time period of 30 years.

Recently, a new threat that might impact boa populations over the near and long term is the introduction and spread of nonnative potential competitors such as *Boa constrictor*, reticulated pythons (*Malayopython reticulatus*), and ball pythons (*Python regius*); all of which have been documented in the wild in Puerto Rico. These species might compete for resources like prey and retreat sites at different life stages. In addition, the emerging infectious disease known as snake fungal disease (*Ophidiomyces ophiodiicola*) adds another variable to consider when developing new recovery plans for the species. This pathogen has already caused the decline and extirpation of some snake populations in North America (Thompson et al. 2018), and the fungus has recently been documented in Puerto Rico. Figure 5.93 shows the distribution of *Chilabothrus inornatus*.

## MONA BOA

*Chilabothrus monensis* (Zenneck 1898)

### *Taxonomy*

The Mona Island boa was originally described by Zenneck (1898) from five specimens collected on Mona Island by Charles Bock between 1891 and 1894 and deposited in the Hamburg Museum. Meerwarth (1901), reclassified the species as *Epicrates fordii* var. *monensis*. Stejneger (1904) recognized the species as *Epicrates monensis*. Schmidt (1926), after reviewing six specimens, suggested *Ep. monensis* to be related to *Ep. fordii*, but two years later (1928) he refers to the species as *Ep. monensis*. In 1932, Grant (Grant 1932b) collected three more specimens and recognized them as *Ep. monensis*. Later in 1933, Stull considered it a subspecies of *Ep. gracilis*, but Mertens (1939) recommended including *Ep. monensis* as a subspecies of *Ep. inornatus*. During a revision of the Puerto Rican boas, Sheplan and Schwartz (1974) reaffirmed *Epicrates monensis* as a valid species and included the Virgin Island boa (*Ep. inornatus granti* at the time) as a subspecies of *Ep. monensis*. Since 1974, taxonomists recognized the Mona Island boa (*Epicrates monensis monensis*) and the Virgin Islands boa (*Epicrates monensis granti*) as subspecies. Even though they were considered subspecies, they were identified separately as *Ep. monensis* and *Ep. granti* (e.g., Mayer 2012) anecdotally and without any formal support. In 2013, Reynolds et al. (2015) showed that both subspecies possessed enough genetic differentiation to be recognized as separate species. Concurrently, Rodríguez-Robles et al. (2015) suggested that both subspecies of *C. monensis* be recognized as separate species based on molecular genetic species delimitation and phenotypic differences. We can state that after 117 years, the taxonomy of this species is finally clear to the scientific community.

### *Etymology*

The epithet *monensis* refers to Mona Island (= Isla de Mona), the site of the type locality. This species is known as the "Mona boa," "Mona Island boa," "boa de Mona," and "boita de Mona."

*Description*

This species is among the small-sized and slender-bodied species in the genus *Chilabothrus*, although these smaller boas evolved largely independently from presumably larger ancestors (Reynolds et al. 2016a). The average adult male size is 903.9 mm SVL (ranges from 700 to 1027 mm SVL) and have an average mass of 174.9 g (ranges from 65 to 314 g). Adult females average 963.8 mm SVL (range 722–1255 mm) and have an average mass of 230.8 g (range 62–426 g) (Tolson et al. 2007). Neonate average size is 314.6 mm SVL (range 283–335 mm) and average mass is 7.7 g (range 5.5–9.0 g). Data on sizes from yearlings to three-year-old snakes range from 420 to 646 mm SVL and from 15 to 63 g in mass (Tolson et al. 2007).

Body coloration in adults is cream or light brown dorsally with 47 to 56 dark brown body blotches and 10 to 14 tail blotches (Fig. 5.94; Sheplan and Schwartz 1974; Tolson et al. 2007). Blotches are saddle-like and can be bridged to the middorsal line but sometimes can be staggered, thereby giving a chain-like effect of reticulation (Sheplan and Schwartz 1974). Some of the caudal blotches form almost complete rings. There is no head pattern, and the first body blotch has an inverted U-shape that breaks at the dorsal midline (Sheplan and Schwartz 1974). The ventral scales are cream or beige colored and sometimes some scattered darker spots can be present. Neonates and juveniles have a very light yellowish-brown body coloration with very dark and distinct brown-black blotches on the dorsum and creamy underside (Fig. 5.95; Rivero et al. 1982). This gives them a more contrasting coloration than adults. Eyes have a black pupil, and the iris has a sandy color streaked with dark brown. The tongue is light pinkish to gray with white tips. Scalation is as follows: 3 post-intersupraoculars, 39 to 42 midbody dorsal scale rows, and 0 to 1 supralabials entering the eye (Tolson and Henderson 1993).

**Figure 5.94.** Adult female *Chilabothrus monensis* from Isla de Mona. Note the cloacal exudate on the dorsum and the color of the tongue. Photo by J. P. Zegarra.

**Figure 5.95.** Juvenile *Chilabothrus monensis*, note the distinct contrasting pattern of the juveniles. Photo by J. P. Zegarra.

*Distribution*

This species is endemic to Mona Island, located in the Mona passage between Puerto Rico and Hispaniola. The island is a raised platform with high cliffs to the north and south. Its western, southwestern, and southeastern areas exhibit less-steep cliffs that descend to coastal lowlands (Aaron 1974). It has an area of 5519 ha and is a Puerto Rico Commonwealth Natural Reserve since 1986. Most of the vegetation is semideciduous scrub consisting of a mixture of trees and shrubs with some large cacti between them. Boas have been observed mainly on the xeric plateau above Sardinera beach, Uvero beach, and Pájaros beach (Tolson 1991c).

*Habitat*

This species uses xeric coastal forest and plateau forest (Campbell and Thompson 1978; Rivero et al. 1982). In this habitat, snakes have been observed at heights over 6 m and also crawling at ground level on limestone boulders (Tolson 2000). Tolson et al. (2007) observed that the species used trees and shrubs (thirty-four species), vines (three species), and one species of bromeliad. Some of the commonly used plant species are *Eugenia axillaris, Antirhea acutata, Bursera simaruba, Caesalpinea monensis, Coccoloba uvifera, Clusia rosea, Ficus citrifolia,* and *Tillandsia utriculata*. However, 57% of snakes captured in the Tolson et al. (2007) study were found on four species of plants (*Antirhea acutata, Clusia rosea, Caesalpinea monensis, Ficus citrifolia*) that were not in high densities at the study area. The diversity of plant species used by *C. monensis* at all age classes highlights the importance of the structural aspects of the habitat for the species (Fig. 5.96). For example, species such as *Ficus citrifolia* and *Clusia rosea* have a combination of aerial roots, compound trunks, and spreading crowns that increase the available space for foraging or movement in the habitat (Tolson et al. 2007).

Perch height for the species ranges from ground level to 8 m, and no difference between perch heights among age classes has been found (Tolson et al. 2007). However, adult snakes seem to be found either on the ground or high in the open canopy. Perch diameter seems to be the habitat variable that reduces competition for food resources among age classes. Smaller snakes <400 mm SVL used smaller perches than young-of-the-year snakes (401–500 mm SVL), subadults (501–700 mm SVL), and adults (>700 mm SVL). Tolson et al. (2007) hypothesized that *C. monensis* will use bromeliads more often during dry periods and fruiting trees during fructification periods, suggesting some seasonality to perch use.

There is no information on the diurnal refuges used by the species. However, it is probable that as in *C. granti*, the species uses termite nests, debris, tree holes, palm axils, and bromeliads. The abundance of caves and cavities in the limestone outcrops present on Mona Island represent excellent diurnal refugia for the species. Campers at Pájaro beach have observed the species resting during the day in caves near the camping sites (ARPR, pers. obs.), and only two individuals have been observed active during the day (Tolson 1991c). Boas appear to avoid refuges that are also used by land crabs and hermit crabs (Tolson 1991c).

*Abundance*

This species is highly cryptic, and when active its detection often occurs because of its contrast with the limestone outcrops or light reflecting off of its exposed body parts, particularly the venter, when climbing on vegetation. Many unsuccessful collection visits to Mona Island from 1932 to 1973 produced the impression that the species was extinct (Rivero 1978). The abundance of optimal habitat on the island represents a challenge to produce a reliable population estimate, so current population numbers are unknown. The only population density estimate was 120 snakes/ha in suitable habitats (Tolson 2000), a surprisingly high estimate. The species does seem to be more abundant than previously thought, as ninety-six individuals were captured in a 5-month period (Tolson et al. 2007).

*Activity and Trophic Ecology*

*Chilabothrus monensis* is nocturnal, and it is more easily detected during the waxing quarter and new moon phases. Additional studies relating moon phases and snake activity are needed because if there is a relationship, efforts to survey for the species can be more effective. Diurnal observations of the species moving inside caves or feeding on the limestone outcrops occur occasionally, although it is unknown how important this habitat is.

Schmidt (1928) detected the tail of the lizard *Anolis monensis* in the stomach of *C. monensis*. Anoles constitutes an important prey item for subadults of the species, and Tolson (1988) showed that areas with high densities of anoles have higher densities of boas. Another prey for subadults is the Mona Island coqui (*Eleutherodactylus monensis*). This frog perches at night on trees, limestone outcrops, and bromeliads. Adult *C. monensis* are robust snakes that can grow large enough to feed on black rats (*Rattus rattus*) and small birds, but these prey types have not yet been confirmed in the diet of this boa (Tolson et al. 2007).

During the night, active hunting and ambush are the two strategies used by boas (Fig. 5.97). Snakes move on tree branches and bromeliads with active tongue flicking looking for sleeping anoles or frogs. Adult snakes on the other hand have been observed in ambush postures on branches of trees with mature fruit that were commonly visited by black rats (Tolson et al. 2007).

*Predators and Defensive Behavior*

Predators for the species can be divided into birds, reptiles, invertebrates, and mammals. Nocturnal birds that may prey on *C. monensis* are the Yellow-crowned Night-Heron (*Nyctanassa violacea*) and Puerto Rican

**Figure 5.96.** *Chilabothrus monensis* from Isla de Mona. Note the tail loss, which is somewhat common in all species of *Chilabothrus*. Photo by J. P. Zegarra.

Screech Owl (*Megascops nudipes*) (Tolson 1988). Potential diurnal bird predators are the Red-tailed Hawk (*Buteo jamaicensis jamaicensis*) and Pearly-eyed Thrasher (*Margarops fuscatus*). The latter species is abundant on Mona Island. The only potential reptile predator is the Mona Island racer (*Borikenophis variegatus*). Nevertheless, there are no studies on the diet habits of this racer, although the other species of racer present on Puerto Rico (*Borikenophis portoricensis*) is known to prey on other snake species. Tolson (1988) identified the land crab (*Gecarcinus* sp.) and hermit crab (*Coenobita clypeatus*) as potential predators. It is important to note that there are no direct observations or solid evidence of any of these species preying on *C. monensis*.

Mona Island has no extant species of native mammalian predators; therefore, all mammal species have been brought by humans. Feral house cats (*Felis catus*) have been reported on Mona Island since 1898 (Hübener 1898). This species hunts during the day and night on Isla Mona, and there is evidence of the species preying on boas. Wiewandt (1977) studied eight cat stomachs and found remains of the Mona Island racer (*B. variegatus*), the anole (*Anolis monensis*), the skink (*Spondilurus monae*), and the ground lizard (*Pholidoscelis alboguttatus*). The first evidence of cat impact on the species was suggested by Tolson (1996b) when

**Figure 5.97.** *Top*: Adult *Chilabothrus monensis* from Isla de Mona. Note the ventral and lateral coloration. Photo by Alberto R. Puente-Rolón. *Bottom*: Foraging adult female *Chilabothrus monensis* from Isla de Mona. Photo by J. P. Zegarra.

he reported that in March 1993 a hunter surprised a cat on the El Faro Road and the cat dropped a 40 mm *C. monensis* head. Scars and injuries probably from cat attacks were reported for 70% of boas captured in the early 1990s (Tolson 1996b). Later, from 1996 to 1999, the Puerto Rico Depeartment of Natural and Environmental Resources studied the stomach contents of 107 cats and did not find *C. monensis* remains (García et al. 2001). Another potential boa predator is the black rat (*Rattus rattus*), but for Mona Island there is no direct evidence of this. Black rats may prey on smaller snakes and be a food resource for adult snakes. Rat capture rate for three nights of trapping was 0.126 rats/trap/h (Tolson et al. 2007). Feral pigs (*Sus scrofa*) could potentially be considered another predator on *C. monensis* because when foraging they destroy important vegetation such as bromeliads, and they might also eat the snakes themselves when encountered (Tolson 1988).

*Chilabothrus monensis* is docile when handled and its first defensive response generally is to escape. On very rare occasions they will strike, although with decided lackluster. Most of the time when disturbed boas will ball up and release an unpleasant musk from their cloaca.

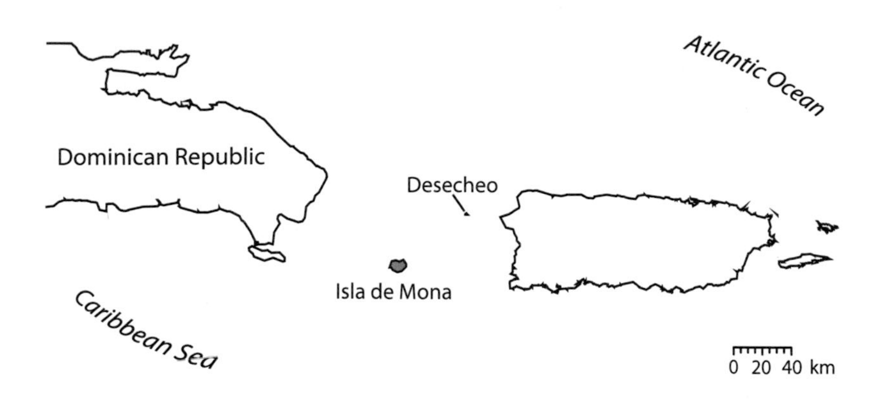


**Figure 5.98.** Range of *Chilabothrus monensis*, endemic to Isla Mona between Puerto Rico and Hispaniola.

*Reproduction*

Information on reproduction for this species is scarce. Rivero et al. (1982) reported a 918 mm SVL female captured on 27 May 1979 at Pájaros beach. The snake was taken to San Juan and during the transport to Mayagüez gave birth to three dead snakes. A few days later while handling the same boa for photos it gave birth to a fourth snake. The average size of those neonates was 249 mm SVL. Ten years later Tolson (1992a) published data on reproduction of the species, but at that time *C. granti* was a subspecies of *C. monensis*, and those data relate more to *C granti*.

Mona Island boas reach sexual maturity close to 800 mm SVL, or about 5 years of age (Tolson et al. 2007). Neonate sizes range from 283 to 335 mm SVL and body mass ranges from 5.5 g to 9.0 g. Gravid female sizes for the species ranges from 905 to 1255 mm SVL with a mass of 164–537 g. Adult males range from 700 to 1027 mm SVL with a mass of 65–314 g. Gravid females have been reported in August (Tolson et al. 2007). There are no data on reproductive phenology for the species, but observations from Rivero and Tolson suggest that reproduction might occur at different times of the year. It is possible that the species has a biennial reproductive cycle and produces litters of two to ten neonates with gestation periods close to 134 days, as in *C. granti*.

*Conservation*

This species was listed as Threatened by the U.S. Fish and Wildlife Service in 1978 (USFWS 1978), and a recovery plan was published in 1984 (USFWS 1984). Its listing was predicated on the presumed impact of the different species of feral mammals (rats, cats, pigs, and goats) and their associated habitat modification on Isla de Mona. The most recent 5-year review was completed in 2014 and still considered the species to be threatened. However, USFWS considers that *C. monensis* has a high recovery potential if management actions are implemented, and since its habitat is protected, it experiences moderate levels of threat relative to other species (USFWS 2014). Nevertheless, the high density of cats on Mona is of concern, as is the continued impact of vegetation damage done by introduced grazers (Tolson 1991c).

The Department of Natural and Environmental Resources of the Commonwealth of Puerto Rico has recognized the species as vulnerable since 1985. After a revision and the approval of Law Number 241 of 1999 (Wildlife Law of the Commonwealth of Puerto Rico) and its Regulation Number 6766 (to Govern Threatened and Endangered Species of the Commonwealth of Puerto Rico; Departamento de Recursos Naturales y Ambientales de Puerto Rico 2004), the species was reclassified as endangered.

Internationally, the Mona Island boa is included on Appendix I of CITES, which identifies species that are endangered and threatened with extinction. The species was considered Endangered on the 1996 IUCN Red List (Tolson 1996b), but is now listed as Near Threatened owing to a lack of evidence for declining habitat quality and some conservation measures in place (Rodríguez et al. 2021). Figure 5.98 shows the distribution of *Chilabothrus monensis*.

## JAMAICA

The island of Jamaica is the third largest of the Greater Antillean Islands and is located to the south and west of the other large islands (Cuba, Hispaniola, and Puerto Rico). The island is 234 km by 80 km, with an area of ~10,900 km$^2$, and is mostly a contiguous island with a few small satellite islands. It is completely surrounded by deep water. During periods of lower sea level, Jamaica would be smaller than the exposed Puerto Rico Bank. Jamaica has a relatively complex geological history, consisting of tens of millions of years as a submerged island bank during which limestone accreted followed by gradual emergence starting in the Miocene with the uplift of the Caribbean Plate. Jamaica probably emerged as a series of several islands, with the eastern end emerging first and the western end emerging last. The island became unified as recently as 8 MY.

Rainfall is heaviest during the summer, although rain occurs throughout the year and hurricanes can bring large amounts of precipitation in the autumn. Precipitation decreases from northeast to southwest. Elevations range from slightly below sea level to 2256 m in the Blue Mountain range, although nearly half the island area lies above 300 m. The highest elevations occur in the Blue Mountains and John Crow Mountains in the east, with elevated plateaus in the central and western parts of the island characterized by complex limestone hills. Two-thirds of Jamaica is composed of limestone substrate, a porous rock that erodes in characteristic patterns leaving lots of topographical complexity in the form of "haystack hills," mountains, valleys, caves, and subterranean rivers. The topographical complexity of the island has led to the development of a tremendous variety of habitat types on the island, from large rivers and marshes to lowland evergreen forest, mangroves, xeric scrub, tropical dry forest, and upland forest. This island is home to a large number of endemic species, including at least sixty-five species of reptiles and amphibians. The Jamaican boa (*Chilabothrus subflavus*) is the only boid occurring on this island and it is an uncommon species that most typically occurs in karst areas covered by dense forests and with an abundance of caves.

## JAMAICAN BOA

*Chilabothrus subflavus* (Stejneger 1901)

### *Taxonomy*

This species was originally thought to be conspecific with *C. inornatus* from Puerto Rico, although Stejneger (1901) examined specimens of both and described *C. subflavus* as a distinct species from Jamaica. No subspecies are recognized. *Chilabothrus subflavus* has no direct close relatives, having diverged from the Hispaniolan/Lucayan clade approximately 17 MY (probably an upper-bound estimate; Reynolds et al. 2016a).

### *Etymology*

The species epithet references the yellow coloration often found on the venter of adults (*flavus* = yellow in Latin), although the yellow coloration can be lateral or even dorsal on the anterior of some individuals. Indeed, the species is locally known as "yellow snake" and "yellow boa." The historical common name was "nanka" (Grant 1940). Neonates are sometimes called "red switchy-tail" (Grant 1940; Newman et al. 2020).

### *Description*

This is among the larger members of the genus, reaching a maximum SVL of about 2050 mm (Reynolds et al. 2016a), with a stout and robust appearance. Reports of larger individuals, closer to 3000 mm, are anecdotal and unverified (e.g., Grant 1940). Miersma (2010) recorded mean male sizes of 1520 mm SVL ($n$ = 4) and 1725 g ($n$ = 4) and mean female sizes of 1550 mm SVL and 1952 g ($n$ = 8). Newman et al. (2019) added seven female measurements, averaging 1490 mm SVL and 1823 g in mass. Neonates are 360–530 mm SVL and, like many other species in the genus, appear very thin and with noticeably large eyes. The number of dorsal scale rows at midbody is 41–47, ventrals 277–283, subcaudals 78–79, circumorbital scales usually 9, and there is a single loreal scale. Sexual dimorphism exists in the size of the spurs, with male spurs being larger than those of females.

This species has a very distinct coloration, with many adults possessing a contrasting color of yellow and black (Fig. 5.99). On the anterior end most dorsal and lateral scales are yellowish to olive colored (rarely orangish, which is more common in the pet trade) in-

**Figure 5.99.** Adult *Chilabothrus subflavus*, showing the distinctive yellowish-orange coloration that transitions to a black tail. Photo by Jeff Lemm.

terspersed with occasionally black scales giving a "dirty yellow" appearance. The head has large iridescent olive-yellow scales outlined with black, and there is a black postorbital stripe. Toward the posterior, the dorsal coloration becomes banded with black cross bands reaching almost to the flanks, which contrast with the yellow in between. Toward the tail these bands become larger and more well defined with less and less yellow visible as thin yellow bands, which eventually dissipate near the vent, with tail becoming completely black. The area where transition to black occurs varies by individual, and some individuals are mostly black with little yellow coloration. The venter largely mirrors the dorsal coloration, being mostly yellow or orange to olive anteriorly and mostly black posteriorly. The iris is yellowish, and the pupils are black. The tongue can be pink or gray with whitish tips. Neonates are born orange or yellowish tan with faint patterning consisting of lighter and darker orange scales (Fig. 5.100). They transition to the adult coloration in 2 years.

*Distribution*

*Chilabothrus subflavus* is endemic to Jamaica, where it historically occurred in forested areas up to ~400 m. There are records of the species from the Goat Islands, a duo of low-lying islands off the southern coast of Jamaica near Hellshire Hills (Barbour 1910), although the species has likely been extirpated there (Gibson et al. 2021). While previously widespread on Jamaica (Gosse 1851), the species is now found in only a few areas of the island (Newman et al. 2016). Newman et al. (2016) determined the current distribution of

**Figure 5.100.** *Top*: Juvenile *Chilabothrus subflavus*; note the bright-orange coloration. *Bottom*: Portrait of a young adult *C. subflavus*. Photos by Matthijs Kuijpers.

*C. subflavus* comprises 1000.6 km$^2$, or about 9% of Jamaica's area. Furthermore, many of the species' "populations are now localized and severely disjointed due to habitat fragmentation and anthropogenic landscape changes" (Newman et al. 2020). Since 1995 boas have been recorded in all parishes except St. Mary and St. Andrew, although the most frequent sightings come from Cockpit Country, the Blue Mountains, and the John Crow Mountains, where more habitat remains (Gibson 1996b; Miersma 2010). Newman et al. (2016) determined that, cumulatively, 93% of known areas of occurrence were in Cockpit Country, Blue Mountains, Yallah Mountains, Hellshire Hills, and Portland Bight and "may constitute the remaining strongholds for *C. subflavus*."

*Habitat*

Of the twenty-one forest habitat types recognized on Jamaica, boas are known to use only five (Newman et al. 2016). This species is most frequently associated with mesic forests developed over limestone substrate (Fig. 5.101). This habitat provides a structurally and topographically complex matrix of forests and hills that allow boas to move between different forest types (upland and bottomland) and different hill slopes and slope aspects. Boas frequently move between caves and treetops and will use arboreal, terrestrial, and subterranean habitats. Boas are rarely found in agricultural areas, and a radio-tracking study (Miersma 2010; Koenig 2019) found that boas actively avoided open agricultural and pasture areas. Boas appear to prefer structurally complex forest habitat, with lots of sizes of trees and vines for moving between the canopy and ground (Koenig et al. 2007; Koenig 2019). At a smaller scale, boas prefer live broadleaf evergreen trees, trees overgrown with vines, large bromeliads, and rock piles for diurnal refugia or nocturnal foraging (Koenig 2019). Preferred tree species include *Adenanthera pavonina* and *Ficus membranacea* (Miersma 2010). Boas seem to prefer arboreal diurnal refugia, especially tank bromeliads (Miersma 2010), and they will also use human structures, such as roof eaves or hollow metal tubes, to rest during the day (Koenig 2019); they have even been found in bedrooms (Gosse 1851).

*Abundance*

Very little focused effort on abundance and detection of Jamaican boas has occurred, probably because even in areas with robust populations the boas are exceedingly difficult to find (Oliver 1982). All the work that has been done to date on this has come from the Windsor Research Centre in Cockpit Country, Trelawny Parish. Nevertheless, the generalization is that boas occur in low abundance in most areas of Jamaica except for Cockpit Country. Even when boas are being tracked, they can be difficult to see, as visual detection of radio-transmitted boas in Cockpit Country was as low as 20% despite the observer being within 1–2 m of the animal (Koenig 2019). The only reliable estimate of detectability was a multiyear survey conducted in Cockpit Country, Trelawny Parish, that relied upon 6768 p-h to find twenty-six boas, or 0.004 boas/p-h (Miersma 2010; Koenig 2019).

*Activity and Trophic Ecology*

Like all other West Indian boas, this species is mostly active at night, although it is known to bask openly in the morning after a heavy rain (Grant 1940). Jamaican boas likely employ both active and ambush foraging modes (Fig. 5.102), and they are trophic generalists (Reynolds et al. 2016a) in that they consume a wide variety of prey including lizards, birds, bats, introduced rodents, and mongooses (Bain and Hurst 1982). The diet of boas in Cockpit Country is primarily composed of introduced rats (Grant 1940; Miersma 2010). Additionally, *C. subflavus* has been documented to feed on Jamaican twig anoles (*Anolis valencienni*), hatchling Jamaican iguanas (*Cyclura collei*), domestic fowl

**Figure 5.101.** *Top*: Typical daytime resting position for many species of *Chilabothrus*. This *C. subflavus* is sleeping in a dense vine tangle in the karst region of Jamaica. *Bottom*: Tropical wet forest developed around limestone haystack hills in Trewlany Parish, Jamaica. This is a preferred habitat of *C. subflavus*. Photos by Brent Newman.

**Figure 5.102.** Actively foraging adult *Chilabothrus subflavus* with a more orange coloration. This species can range from very yellow to very orange, although no regional patterns are evident. Photo by Brent Newman.

(*Gallus gallus*), Olive-throated Parakeets (*Eupsittula nana*), Jamaican Woodpecker (*Melanerpes radiolatus*), Yellow-billed Parrots (*Amazona collaria*), Black-billed Parrots (*A. agilis*), Pallas's mastiff bats (*Molossus molossus*), and Jamaican fruit-eating bats (*Artibeus jamaicensis*) (Gosse 1851; Cruz and Gruber 1981; Koenig 2001; Koenig et al. 2007, 2015; Henderson and Powell 2009; Miersma 2010). Boas are considered the major predator of Black-billed Parrots and are responsible for a significant proportion of nesting failures (Koenig 2001). Probable predation of the nest of a Jamaican Crow (*Corvus jamaicensis*) has been reported (Schaefer et al. 2019). The species is also known to consume introduced cane toads (*Rhinella marina*), which are toxic and likely responsible for acute poisoning and death of the snake (Wilson et al. 2011).

Jamaican boas are one of four members of the genus *Chilabothrus* that are known to consume bats at the entrances of caves (the others being *C. angulifer*, *C. striatus*, and *C. inornatus*). A long-term study at Windsor Great Cave, Trelawny Parish, has documented this specialized boa feeding behavior (Prior and Gibson 1997; Koenig, 2019). This cave has one of the largest bat colonies in Jamaica, with over 100,000 bats of at least eleven species using the cave as a diurnal roost. Upon exiting the cave in the evening, bats encounter boas hanging around the mouth of the cave from strangler fig roots, grabbing bats in midair by means of touch sensing. This is apparently a behavior that improves with experience, as younger boas tend to strike more often, with more misses, than older boas (e.g., during 105 minutes, a 950 mm juvenile female *C. subflavus* made over 200 unsuccessful strikes at emerging bats; Prior and Gibson 1997). The foraging behavior of *C. subflavus* in caves is similar to that of *C. inornatus* but differs slightly from that of *C. angulifer*; this behavior, although documented, has not been studied in detail in *C. striatus* (see the respective species accounts). Cave-associated boas have much smaller home ranges than those found elsewhere (see below; Koenig 2019).

A radio-tracking study was conducted between 2008 and 2011 near the Windsor Research Centre in Cockpit Country, Trelawny Parish, Jamaica (Miersma 2010; Koenig 2019). The study implanted radio transmitters in eight females and six males, three of which were associated with a large cave at the site. Boas moved a mean distance of 20 m per day, with no difference in mean daily movement distance between males and females. Nevertheless, males did make very large movements relative to females from January to April and were much more sedentary relative to females from September to December. Interestingly, cave-associated boas had much smaller home ranges than surface-dwelling boas, presumably owing to their ability to exploit a ready food source at the cave. Nevertheless, boas around the cave did not associate with one another and were never found in close proximity, such as in the same tree. Adult home ranges (95% minimum convex polygon) of males not associated with Windsor Cave ranged from 16.28 ha to 70.11 ha and those of females from 2.16 ha to 19.56 ha. Core activity areas (50% minimum convex polygon) of males not associated with Windsor Cave ranged from 0.20 ha to 0.34 ha and those of females from 0.64 ha to 6.63 ha. For those males associated with Windsor Cave, home ranges ranged from 0.75 ha to 2.25 ha and core activity areas ranged from 0.20 ha to 0.34 ha; for a single cave-associated female, home range was 2.51 ha and core activity area was 1.12 ha (Koenig 2019). Boas were also found to move as far away as 1 km and return to the exact point of previous capture, suggesting excellent navigation skills and some evidence for philopatry and territoriality (Miersma 2010). There were no significant differences (cave and noncave) between males (6.8 ± 1.0 days) and females (7.2 ± 1.2 days) for the number of consecutive days they remained immobile. This apparent fidelity for or limited mobility around sites with abundance of trophic resources has been documented as well in the Cuban boa and the Puerto Rican boa (see the respective species accounts).

Newman et al. (2019) studied short-distance translocation in *C. subflavus* as a potential "conservation tool for species occurring in multipurpose landscapes where development may be necessary and human-wildlife conflict is unavoidable." Translocation distances for seven females ranged from 693 to 3545 m. Two boas returned to their original point of capture in the first 2 months following translocation; others were 500 m to >1000 m from their points of capture. The authors concluded that short-distance translocation had the potential as a possible management tool for *C. subflavus*.

*Predators and Defensive Behavior*

Boas will rarely strike when cornered, although juveniles will display much more aggressive behavior than adults. Defensive behaviors include hissing, biting, throat inflation, and voiding the contents of their cloaca when picked up (Newman et al. 2020).

Native predators of Jamaican boas are poorly known but presumably include birds of prey (hawks, owls) and other predatory birds such as herons (Newman et al. 2020). A single report from captivity suggested that a female boa killed and consumed a male cage mate (Tolson and Henderson 1993), although ophiophagy is unknown in the wild. Boa populations are likely mostly preyed upon by introduced vertebrate predators such as cats and mongooses (Grant 1940). Barbour (1910) observed that mongooses, which were introduced to Jamaica in 1872, could potentially be responsible for a perceived near-extirpation of boas on Jamaica, although he noted that the population on Great Goat Island was still intact and mongoose-free. Grant (1940) observed that the species declined in the area around Portland Point, the southernmost point on Jamaica, following the introduction of cats.

Although not potential predators, Jamaican Crows (*Corvus jamaicensis*) have on multiple occasions been observed to mob *C. subflavus* (Schaefer et al. 2019).

*Reproduction*

In the wild, reproduction occurs in the spring, and male and female boas have been found together in April (Koenig 2019). Males appear to move long distances during the first quarter of the year, presumably in search of mates (Koenig 2019). Females become sedentary when gravid in the summer, often thermoregulating in caves or large termite mounds (Koenig 2019). Like most members of the genus *Chilabothrus*, wild females likely reproduce every other year (biennially), although in captivity many are capable of annual reproduction. Females can become reproductive at 6–7 years of age (Huff 1979) and may continue reproducing until at least age 17 (Tolson 1991b). Lifespan is at least 24 years.

In captivity reproduction occurs from January to June, with copulation lasting 3–14 hours (Tolson 1980). Gestation can last up to 220 days, and young are born in the late summer to fall, with litter sizes in captivity of three to thirty-nine (Tolson 1980; Bloxam and Tonge 1981). Young were 12–19 g in mass and 360–530 mm SVL (Tolson 1980).

*Conservation*

*Chilabothrus subflavus* is listed on CITES Appendix I and is listed as Vulnerable on the IUCN Red List (Gibson 1996a; Gibson et al. 2021). The species was placed on the U.S. Endangered Species List in 1970. It is also protected by the Jamaican Wildlife Protection Act of 1945. The first IUCN Red List assessment (Gibson 1996a) found the species was experiencing population declines due to persecution, loss of habitat, and predation by introduced species such as mongoose (Henderson 1992). A revision of the Red List assessment (Gibson et al. 2021) finds that these threats have not

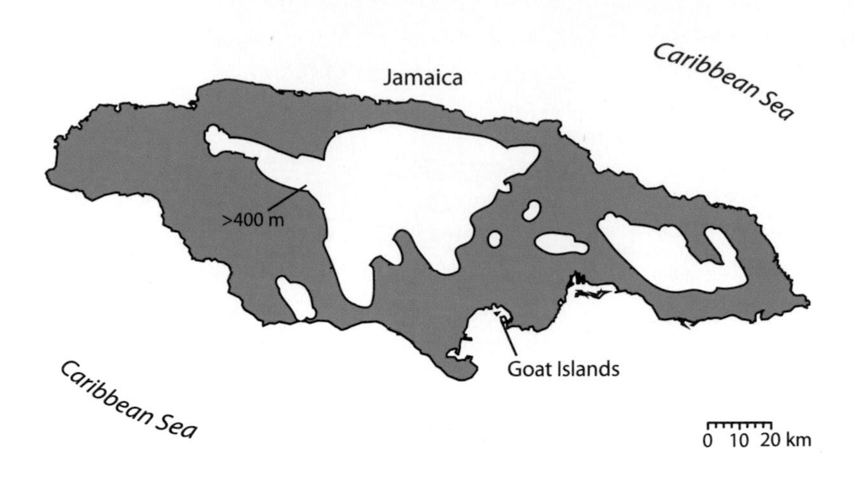


**Figure 5.103.** Range of *Chilabothrus subflavus*, endemic to Jamaica. The species was historically found below about 400 m island-wide, but the actual range is probably highly localized within this elevational zone (see text).

been mitigated and that the species is now also facing pressure from human collectors for the pet trade, eastern medicine, and consumption by foreign workers in Jamaica.

As for virtually every boa species in the West Indies, the two greatest threats to their conservation are habitat loss coupled with fragmentation and introduced species. Until the last decade, little was known about the extent of the remaining range of the species. Several recent studies, with the leadership of S. Koenig and the Windsor Research Centre in Cockpit Country, have collected sightings from across the island (Tzika et al. 2008; Miersma 2010; Newman et al. 2016). Together these studies revealed that the species occurs in twenty isolated localities, occupying less than 10% of the island (Newman et al. 2016). Commercial mining for bauxite contributes to fragmentation, which creates barriers to dispersal, a potential increase in mortality, and ease of access by humans to previously inaccessible areas of forest. Likewise, selective logging for construction materials, charcoal production, and firewood "progressively alters forest composition, structure, and function" (Newman et al. 2020). Introduced species in concert with habitat alteration have impacted *C. subflavus* abundance and distribution and have altered its prey base. Noteworthy invasives include mongooses, domestic cats and dogs, rodents (*Rattus* spp.), and a variety of ungulates (e.g., cattle, goats, pigs; Newman et al. 2020).

Local people frequently take pride in killing the snakes, and foreign workers from Asia with a tradition of snake consumption and use in traditional medicine are now buying snakes from local collectors, including from protected areas (Miersma 2010). In fact, this new demand for snakes is thought to have contributed to a 30% population size reduction since 2008, suggesting that exploitation is further contributing to extirpations on top of ongoing habitat loss, impact of invasive species, and human persecution. A final threat is collection for the pet trade, as interest in this species has grown over the last few decades (Wilson et al. 2011).

Human impacts on the species have been documented for over a century, beginning with Barbour (1910) observing that mongooses prey on boas. Indeed, invasive species have undoubtedly had a tremendous negative impact on populations of the Jamaican boa. Bauxite mining and timber harvesting are widely blamed for habitat loss and degradation for this species and other endemic Jamaican wildlife. A campaign to preserve and restore the Goat Islands has, as of this writing, been successful, and the long-term hope is to use Great Goat Island as a sanctuary to reintroduce iconic species such as the boas and the iguana *Cyclura collei*.

Other ongoing conservation measures include education campaigns based at the Windsor Research Centre and the Hope Zoo (Miersma 2010) and the creation of a robust captive colony at zoological institutions abroad (Tzika et al. 2009). Figure 5.103 shows the distribution of *Chilabothrus subflavus*.

## THE LESSER ANTILLES

As traditionally defined, the Lesser Antilles comprise an arc-like archipelago of oceanic islands that extends some 700 km from Sombrero in the north to Grenada in the south. Collectively, these islands form the eastern boundary of the Caribbean Sea. Sixteen island banks include from one (Sombrero, Saba, Redonda, Montserrat, Marie-Galant, Barbados) to more than thirty (Grenada) named islands ranging in size from barely emergent rocks to >1400 km$^2$ (Guadeloupe, when Basse-Terre and Grand-Terre are considered together). Extant native boa species occur on four of these banks (Dominica, St. Lucia, St. Vincent, and Grenada).

These islands developed along the border of the Caribbean and South American tectonic plates, where the latter is being subducted beneath the former (Mann 1999; Potter et al. 2004). The present Lesser Antillean Arc represents Eocene to Holocene eastward migration of the resultant volcanic activity (Malfait and Dinkelman 1972). Seventeen volcanoes remain active (Bouysse et al. 1990).

The first islands to form along this abutment arose over 36 MY and form what are called the "outer-arc" islands along the eastern (Atlantic) side of the archipelago. This outer arc includes the present-day islands of the Sombrero, Anguilla, Antigua, Marie-Galante, and Barbados Banks plus the Grande-Terre portion of Guadeloupe. The volcanoes responsible for forming these islands are extinct. Since their undersea birth some 36 to >50 MY, the outer-arc islands have been uplifted by tectonic activities, rising above sea level—only to be submerged again as a consequence of erosion and local subsidence of Earth's crust. During submerged periods, corals formed limestone caps over the previously exposed volcanic rock. Subsequent uplifts added areas to the islands that had never before been above sea level. Because the original volcanic surfaces of the islands in this outer, much older arc have become capped with limestone, they are sometimes called the "Limestone Caribees" (Martin-Kaye 1969). Extinct species of *Boa* are known from the outer arc islands of Antigua (*Boa* sp.), Martinique (*Boa* sp.), and Marie-Galante (*B. blanchardensis*).

The "inner-arc" islands or "Volcanic Caribees" are of much more recent volcanic origins (Martin-Kaye 1969). These islands form the Saba, St. Christopher, Redonda, Montserrat, Île des Saintes, Dominica, Martinique, St. Lucia, St. Vincent, and Grenada Banks plus the Basse-Terre portion of Guadeloupe. These islands continue to show varying degrees of volcanic activity, and threats of eruptions exist. Every major island has a max. elev. at least 50% higher than the highest peak in the Limestone Caribees. All extant boa species in the Lesser Antilles occur on these inner arc islands: species of *Boa* occur on the Dominica Bank and the St. Lucia Bank, and species of *Corallus* occur on the St. Vincent Bank and the Grenada Bank (Grenada + the Grenadines).

Inner-arc islands have sufficiently high elevations (above 600 m) to intercept moisture-laden air. Referred to in the English-speaking islands of the region as "snag islands" (because the peaks "snag" clouds), they actually hold clouds that have formed largely as a result of evaporation from lower areas. This results in the peaks receiving as much as twice the rainfall as dryer, lower elevations (Powell et al. 2015). In sharp contrast, the highest peaks of the outer-arc islands lie below the elevation required to form and arrest clouds. Consequently, those islands show less disparity in the amount of precipitation. The inner-arc islands, with greater heterogeneity of elevation and precipitation, generally support more diverse biotic communities.

The higher elevations on inner-arc islands provide conditions necessary to sustain perpetually moist forest formations. However, equally important are the transitional vegetative communities that vary according to elevation and orientation in regard to prevailing winds (generally speaking, windward slopes are moister and subjected to greater erosive forces than are leeward slopes). Consequently, inner-arc islands can support eight or more distinct vegetative communities. Such biotic diversity, combined with dramatic seasonal fluctuations in rainfall (approximately twice as much rain falls in August–November than in January–April), results in conditions that force terrestrial animals to be either ecologically versatile or able to sustain distributions restricted to the tiny fragments of the islands that provide suitable habitats.

**Figure 5.104.** *Boa nebulosa* resting on a rock, northern Forest Reserve, Dominica. Photo by Jeff Lemm.

## DOMINICA BOA

*Boa nebulosa* (Lazell 1964)

### *Taxonomy*

Originally described as a subspecies of *Constrictor constrictor*. Prior to Lazell's (1964) description, the boa population on Dominica was considered to belong to *Boa diviniloqua* (what is now *B. constrictor*) by Günther (1888) and then synonymized with the St. Lucia population (*Constrictor orophias*; e.g., Barbour 1930, 1937). It was elevated to species status by Henderson and Powell (2009) based on scale and pattern characters provided by Lazell (1964) as well as the geographic isolation of the species. Bezerra de Lima (2016) considered *B. nebulosa* a distinct lineage in the *B. constrictor* complex based on phylogenetic analysis of DNA sequence data.

### *Etymology*

The specific name *nebulosa* refers to the "extremely dark, clouded appearance" (Lazell 1964) of individuals of this species presumably relative to mainland *Boa constrictor*. Boas are known locally on Dominica as "tete'chien" (or "tet chyen," "tet chien," or "tête-chien"), a name that translates to "dog-head snake" (Ober 1899; Lazell 1964) or "dog head" in the Kwéyòl language (J. Brisbane, in litt. to RWH, 14 August 2018).

### *Description*

Relative to the other boas that occur in the Lesser Antilles (*Corallus* spp.), *Boa nebulosa* (along with *B. orophias*) has the potential to be much longer and more massive (Fig. 5.104). Our description here is based largely on Lazell (1964). *Boa nebulosa* has a prominent snout and a convex canthus (Fig. 5.105). It attains a maximum SVL of about 4.0 m (Malhotra et al. 2011) and is sexually dimorphic in size, with females attaining larger sizes than males. The number of dorsal scale rows at midbody is 58–69, ventrals 263–273, supralabials 19–21, and infralabials 20–22. The head has a dark temporal stripe extending from the eye to beyond the commissure of the mouth, and this stripe is distinctly darker along its ventral edge. The face is paler gray. The dorsal ground color varies from cloudy gray brown (with indistinct markings) to nearly black.

**Figure 5.105.** *Boa nebulosa* from Batalie Beach, near Coulibistrie (St. Joseph Parish). Photo by Robert Powell.

The number of transverse dark markings on the body ranges from 32 to 35; they may become more distinct posteriorly (black, bordered by dark brown), although can be variable in appearance among individuals depending on dorsal ground color (Fig. 5.106). The pattern on the tail consists of saddles of very dark brown with black borders (Fig. 5.107). The venter is ash gray anteriorly and slate gray to black posteriorly. Irregular dark blotches occur along the lateral edges of the ventrals anteriorly, and they become obliterated posteriorly. The chin is gray.

### *Distribution*

Endemic to the Lesser Antillean island of Dominica, West Indies. The species is widespread on the island and in a variety of habitats, "at least to elevations of circa 350 meters" (Lazell 1964). It might be restricted to wet ravines in drier parts of the island (e.g., around Scott's Head Village at the southern tip of the island; Lazell 1964).

**Figure 5.106.** *Top*: *Boa nebulosa* from Kalinago Territory (St. David Parish). *Bottom*: *Boa nebulosa* habitat along the Layou River (St. Joseph Parish). Photos by Ricky Lockett.

*Habitat*

This is an ecologically widespread species that occurs in virtually all habitats on the island, including from coastal xeric and littoral woodland to montane rainforest (Malhotra and Thorpe 1999; Fig. 5.106) as well as gardens (R. Thorpe, in litt. to RWH, 20 February 2018). Lazell (1964) encountered *B. nebulosa* along the edges of machine-cleared banana fields. Vandeventer (1992) encountered boas in rock piles, a root system, a shallow undercut in a stream bank in the vicinity of hot springs, and sulfur vents along mountain streams. According to R. Lockett (in litt. to RWH, 15 February 2018), boas occur in agouti burrows, tree trunks, under rocks, and in trees. Knapp et al. (2009) encountered *B. nebulosa* on the ground in a shallow cave along a rock wall 12 m from the ocean. RWH has encountered *B. nebulosa* sunning on a termite nest in the early afternoon, under a discarded piece of plywood in a small patch of woodland about 50 m from the sea, and foraging at ground level in the early evening. Younger (smaller) boas are likely more arboreal and will ascend into trees.

*Abundance*

Ober (1899) described this species as "rather abundant in the forests," and Lazell (1964) considered this species to be "amazingly abundant." He considered it "not at all uncommon" to encounter congregations of three to twelve individuals "denning in the same hollow log or tree stump . . . especially along the edges of machine-cleared banana fields where suitable den sites are often in profusion." During a 2-week span in February, Vandeventer (1992) encountered fifteen adult boas. Only one boa was alone; all others were found in small groups of two, five, and seven in rock piles, a root system, and a shallow undercut in a stream bank. Malhotra et al. (2011) considered *B. nebulosa* still "fairly common," although "larger individuals are becoming less commonly seen." Similarly, J. Brisbane (in litt. to RWH, 14 August 2018), a biologist and life-long resident of Dominica, has observed a decline in large boas and snakes in general.

*Activity and Trophic Ecology*

We have observed this species active by day and night. We have encountered individuals basking on a termite nest and under a plywood board by day and active on the ground at night. According to Malhotra and Thorpe (1999), these boas are more active by night than by day and are often found dead on roads in the morning.

*Boa nebulosa* employs both active and ambush foraging strategies (J. Brisbane, in litt. to RWH, 14 August 2018). The diet has not been well documented but includes both endothermic and ectothermic prey. A boa with an SVL of 1950 mm and a mass of 5.9 kg contained an agouti (*Dasyprocta leporina noblei*); a female with SVL of 1606 mm contained a *Rattus* (A. Schwartz field notes, 8 and 12 March1962). At the entrance to

**Figure 5.107.** *Boa nebulosa* from near Mahut (St. Paul Parish), illustrating the contrasting tail pattern. Photo by Joseph Burgess.

Stinking Hole Cave, a boa of ~3.0 m total length hung from an aerial root and struck at bats exiting the cave. The boa was first observed at 1600 hours, more than 2.5 hours before bat emergence began. It eventually captured a *Brachyphylla cavernarum* and, while constricting the bat, was simultaneously attempting to catch others (Fig. 5.108). The boa began ingestion of the bat after evening emergence of bats from the cave decreased (Angin 2014).

*Boa nebulosa* has been observed probing Lesser Antillean iguana (*Iguana delicatissima*) burrows while actively foraging (A. Mitchell in Henderson and Powell 2009). Knapp et al. (2009) documented two instances of predation on *I. delicatissima*, both by boas ~2.0 m TL; at least one of the predations occurred in a tree. Boas will enter coops in residential areas and prey on domestic fowl (Anonymous 2014; J. Brisbane, in litt. 14 August 2018). Additional (but undocumented with photographs or written accounts) prey animals include native birds, opossums (*Didelphis marsupialis*), domestic and feral cats, and small dogs (J. Brisbane, in litt. 14 August 2018). It is noteworthy that, with the exception of the iguana, bat, and (possibly) native birds, prey is composed largely of introduced species.

Although we do not believe the diet of *B. nebulosa* is restricted to the four species for which we have documentation, it nevertheless has access to a much smaller prey base than do mainland species of *Boa*. With little effort, we accrued literature records of forty prey species for *B. constrictor* (from throughout its range) and

**Figure 5.108.** This *Boa nebulosa* has just captured a bat (*Brachyphylla cavernarum*) at Stinking Hole (St. George Parish). Photo by Baptiste Angin.

**Figure 5.109.** *Top*: *Boa nebulosa* photographed at night in Kalinago Territory (St. David Parish). Photo by Ricky Lockett. *Bottom*: Resourceful Dominicans hoping to attract tourist dollars with photo opportunities starring *B. nebulosa*. The lack of body turgor in the boas suggests that they are unhealthy or dehydrated. Photo by Ricky Lockett.

twenty-one prey species for *B. imperator*. We are confident that both species take many additional prey species for which we do not have documentation. We are similarly confident that *B. constrictor*, for example, has the potential to prey on forty or more species at a given locality in portions of its range. Certainly, prey diversity on Dominica has been altered from pre-human settlement to the post-Columbian era. Although we are unaware of fossil evidence of now-extinct rodents on Dominica, their absence is attributed to "a sampling artifact rather than a genuine absence from Late Quaternary faunas" (Turvey et al. 2010). Therefore, we confidently assume there must have been *Rattus*-sized prey (e.g., *Megalomys*) on Dominica long before the introductions of *Rattus* and *Dasyprocta*.

*Predators and Defensive Behavior*

Humans are almost certainly the most important predator (Fig. 5.109). The boas may be killed for fear of snakes (although no snakes dangerous to humans occur on the island), because they apparently prey on livestock (e.g., chickens, rabbits) and dogs (J. Brisbane, in litt. 14 August 2018), or for the apparent pleasure or satisfaction some people get by killing a snake. Humans will sometimes eat large boas (R. Lockett, in litt. to RWH, 15 February 2018). Long (1974) notes that

**Figure 5.110**. Range of *Boa nebulosa*, endemic to Dominica up to an elevation of ~350 m.

some farmers introduce *B. nebulosa* to their plantations to "discourage" rats. Cars and trucks also contribute to significant mortality, and J. Brisbane (in litt. to RWH, 14 August 2018) witnessed drivers to "have gone out their way to ensure the snake was crushed." Lescure (1979) reported predation by the large leptodactylid frog *Leptodactylus fallax*.

When threatened, *Boa nebulosa* will try to escape; if unsuccessful it will hiss, strike, bite, and void its cloaca. People have inadvertently stepped on or even stood on boas without any aggressive (or defensive) response from the snake (J. Brisbane, in litt. 14 August 2018).

*Reproduction*

Little information on the reproductive habits of this species has been published. A 1606 mm SVL female had twelve enlarged eggs on 12 March; sixteen nearly full-term young were removed from a female on 25 April (Schwartz and Henderson 1991). Vandeventer (1992) reported three litters (seven, nine, and eleven neonates) from captive-held boas in late July and August. Whether the aggregations of *B. nebulosa* reported by Lazell (1964) and Vandeventer (1992) were reproductively driven is uncertain.

*Conservation Status*

This species has been assessed as Least Concern based on IUCN criteria (Daltry et al. 2018). Nevertheless, its distribution is limited to a small (790 km$^2$) island and could probably trigger a Vulnerable status (Fig. 5.110). According to Lazell (1964), the people of Dominica "enjoy being deathly afraid of the Tet'chien. . . . Children, once initiated, are not at all afraid of even the biggest ones, however, and their parents are often shamed into changing their minds about the danger of the Tet'chien simply because the children come to regard them as play-toys." These boas are killed for the "medicinal" oil rendered from their fat, which is used as a local remedy for joint pain and possibly as an aphrodisiac (J. Brisbane, in litt. to RWH, 14 August 2018). They are also killed because they prey on domestic chickens (Henderson and Powell 2009; Anonymous 2014). *Boa nebulosa* helps control agricultural pests (*Rattus* spp.), and some farmers allow the snakes to stay on their property to reduce rodent-inflicted damage to their crops. Some resourceful Dominicans will carry boas to areas of heavy tourism and present them as photo opportunities for a small fee (R. Lockett, pers. comm.to RWH, 22 February 2020; Fig. 5.109). Of course, some boas are killed for no reason other than for being a snake. *Boa nebulosa* is afforded some protection by the Forestry and Wildlife Act (1976), ostensibly for the protection and management of wild fauna and the management of their forest habitats (Malhotra et al. 2011); however, there are no laws that specifically protect snakes on Dominica. The National Parks Act will protect *Boa* habitats but not specifically for that reason.

A Dominica-based online photograph and article about a *B. nebulosa* that was killed in the process of swallowing a domestic chicken generated a great deal of online discussion (Anonymous 2014). Of eighty-two comments that could be judged as either positive boa (or snake) or negative boa, fifty-six (68.3%) were in favor of the boa not being killed and twenty-six (31.7%) were glad it had been killed. Most of the negative response was based on the false fear that boas will kill and eat children; most of the positive responses focused on boas keeping rat populations in check and stressing that Dominica has no snakes that pose a threat to humans (of any size). Based on 425 primary and secondary school respondents (urban and rural schools), J. Brisbane (in litt. 14 August 2018) found that *B. nebulosa* was the least favorite animal among seven choices (including a parrot, an opossum, and a wild pig). She found a general lack of proper awareness about the boas, leading to misinformation and subsequent mistreatment of these animals. Figure 5.110 shows the distribution of *Boa nebulosa*.

## ST. LUCIA BOA

*Boa orophias* Linnaeus 1758

### *Taxonomy*

Originally described as a species by Linnaeus (1758), it was placed in the synonymy of *Constrictor diviniloquus* (or *diviniloqua* [Tyler 1850; Boulenger 1893] or *diviniloquax* [Jan 1863]) by Laurenti (1768) and subsequently by Duméril and Bibron (1844). Barbour (1914) regarded it as full species (*Constrictor orophias*), but it was subsequently downgraded to subspecies by Amaral (1930). It was again elevated to species status by Stull (1935) as *C. orophias*, but Lazell (1964) considered it a subspecies of *C. constrictor*, as did Peters and Orejas-Miranda (1970; as *Boa c. orophias*). It was elevated to species rank by Henderson and Powell (2009) based on scale and pattern characters in Lazell (1964) as well as geography. Bezerra de Lima (2016) considered *B. orophias* a distinct lineage in the *B. constrictor* complex.

### *Etymology*

We are uncertain as to the derivation of the specific name *orophias*, but Long (1974) translated it as "tame." It could also refer to "mountain lover" from the ancient Greek oros (mountain). The local name for this species is "tet'chien" (Lazell 1964). Tyler (1850) used the spelling "Téte Chien" with the note that the head resembles that of a greyhound. See the *Boa nebulosa* account for additional spellings.

### *Description*

Our description is based largely on Lazell (1964) and from J. D. Lazell's field notes (July–August 1962). *Boa orophias* has a prominent snout and a convex canthus. It can reach a maximum SVL of about 3.0 m and is sexually dimorphic in size, with females attaining larger size than males. The number of dorsal scale rows at midbody is 65–75, and ventral scale counts range between 270 and 288. The head has a dark temporal stripe extending from the eye to beyond the commissure of the mouth; it is distinctly darker along its ventral edge (Fig. 5.111). The subocular stripe is distinct and complete; the loreal stripe is indistinct. The face is pale and suffused with pink. The dorsal ground color ranges from yellowish brown to golden brown to pale gray brown and bordered with yellow; it becomes paler and

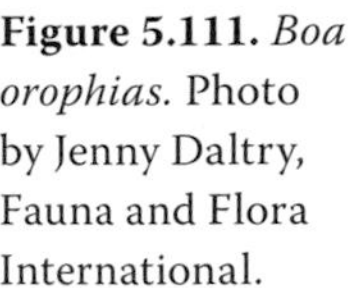

**Figure 5.111.** *Boa orophias.* Photo by Jenny Daltry, Fauna and Flora International.

grayer laterally and then pinkish close to the venter (Fig. 5.111). Twenty-seven to thirty-one distinct, subrectangular dorsal saddles are chestnut to chocolate brown. The pattern on the tail consists of saddles of very dark brown to slate black with yellow borders. The venter is white to yellow white with pronounced black or gray spotting. The chin and infralabials are marked with dark pigment ("not closely corresponding to the facial stripes"; Lazell 1964). Tyler (1850) described its size as eight to ten feet (2.4–3.0 m) but rarely longer than fourteen feet (4.3 m). We suspect that today encountering *B. orophias* of that size is unlikely.

*Distribution*

*Boa orophias* is endemic to the Lesser Antillean island of St. Lucia, West Indies, where it occurs to elevations of ~350 m. Its distribution is largely coastal (windward and leeward), but it could be absent from the relatively dry northern tip of the island as well as from the more southern coastal areas (Lazell 1964; Schwartz and Henderson 1991; Daltry 2009). Its range broadly overlaps that of the pit viper *Bothrops caribbaeus* (Lazell 1964).

*Habitat*

This species occurs in riparian forest and ravines through areas of dry forest, forest gardens, and banana plantations (Daltry 2009; Fig. 5.112). James D. Lazell (field notes, 29 July 1962, 6–23 August 1962) encountered them in a variety of situations: an 1880 mm TL male at 2.4 m in a cacao tree (Marquis Estate); a 1943 mm TL male at 7.6 m in a large tree (Praslin Estate); a 2365 mm TL male on the ground and a female in a cacao tree (Anse-la-Raye); a 2311 mm TL female in a hollow log (Praslin Estate); and coiled in a ball at ~12 m in a large tree at Fond Citron.

*Abundance*

Daltry (2009) considered *B. orophias* to be "still locally common in some areas" but that interviews "indicate it has declined in many parts of the island." Tyler (1850) found it "in great abundance" in sugarcane and that "it is highly valued as a means of destroying rats." During the 1930s, Barbour (1930, 1935, 1937) stated that this species was rare. Unfortunately, we again note that Barbour's assessments of whether or not a species was

**Figure 5.112.** *Left*: *Boa orophias* in the Millet Forest Reserve. *Right*: *Boa orophias* habitat near Louvet. Photos by Jenny Daltry, Fauna and Flora International.

**Figure 5.113.** *Boa orophias* from Grand Bois Forest, Anse La Raye. Photo by Joseph Burgess.

rare must be taken with a great deal of skepticism as they were often lacking in firsthand knowledge. Considering that this species is well camouflaged and easily overlooked, Daltry (2009) believed "the Saint Lucia population is relatively infrequently seen compared to [*Boa*] populations . . . on uninhabited islands." During 55 hours of standardized plot surveys, Daltry (2009) encountered one *B. orophias*, "equivalent to 2.08 per hectare," which should be taken as a very rough estimate given that only one individual was recorded. There seems to be some consensus that *Boa nebulosa* is more common/abundant than *B. orophias* (Barbour 1937; Lazell 1964; Malhotra and Thorpe 1999).

*Activity and Trophic Ecology*

*Boa orophias* is encountered active by day and by night (Fig. 5.113). Little is known about the natural diet of this species. The now-extinct rodent *Megalomys luciae* (Sigmodontinae: Oryzomyini) was the size of *Rattus*

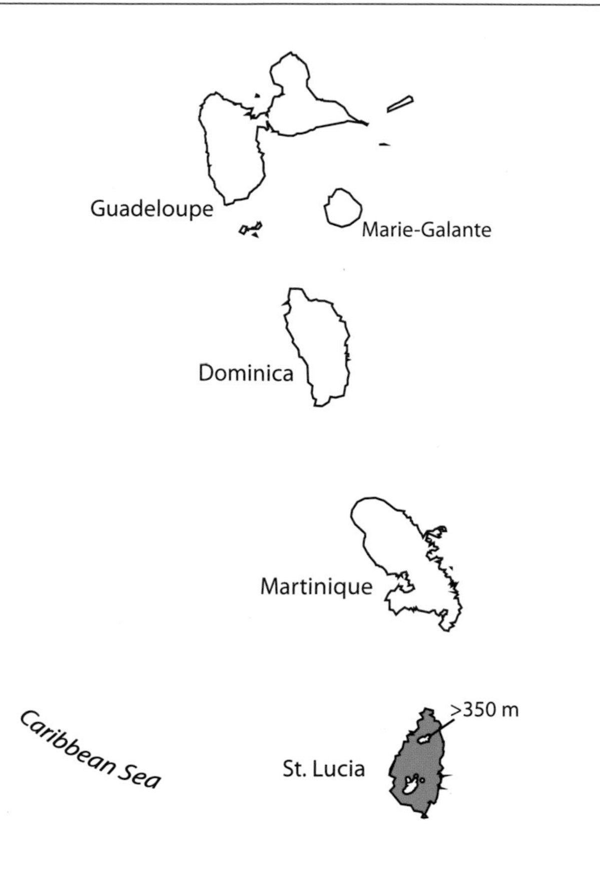


**Figure 5.114**. Range of *Boa orophias* on St. Lucia up to an elevation of ~350 m.

*rattus* and may have survived into the early years of the twentieth century (Morgan and Woods 1986; Turvey et al. 2010). We can assume it was a frequent prey species for *B. orophias*.

We can speculate that the diet of *B. orophias* likely parallels that of *B. nebulosa* on Dominica and that it preys on lizards (*Iguana iguana*), possibly birds, and introduced mammals (opossums, rats, agoutis, cats, dogs, and mongooses). Tyler (1850) described its diet as including fowl, rats, and cats. The only documented prey we have for this species is a bat. At about 1640 hours (fading daylight) in rainforest, a 1100 mm TL (397 g) *B. orophias* captured a bat (*Brachyphylla cavernarum*) at a roosting site situated in a tree (*Sloanea caribaea*) hollow. The boa had positioned itself at the cavity entrance. The bats were not exiting the hollow; rather, they moved higher up in the hollowed trunk. The bat was captured as it tried to ascend higher in the inner wall of the cavity. The bat (a gravid female; 67 g) was swallowed head-first; the predator:prey mass ratio was 16:9 (Arendt and Anthony 1986).

*Predators and Defensive Behavior*

Humans are almost certainly the most impactful predator. The boas may be killed for fear of snakes, as St. Lucia has both a native boa and a dangerously venomous pit viper (*Bothrops caribbaeus*). Some boas are almost certainly killed because people misidentify the harmless boa for the venomous *Bothrops caribbaeus*. Apparently, the fat of *Boa orophias* is harvested for local remedies, and sometimes the oil is extracted from live boas, a practice that is both inhumane and almost certain to be fatal to the snake (Daltry 2009). It is likely boas are persecuted for preying on domestic chickens and for simply being snakes. Cars and trucks may also contribute to significant mortality.

Like most members of the genus, *B. orophias* will try to escape when threatened. If escape is prevented, it will hiss, strike, bite, and void its cloaca. Referring to the boa's defensive reaction to humans, "fear is however perfectly unnecessary, as although it constantly leaves its teeth in the object of its attack, no result more than from the scratch of a thorn ensues" (Tyler 1850).

*Reproduction*

A breeding ball composed of one female and five males was observed in February in the Roseau River valley (Malhotra and Thorpe 1999). Upon being discovered, the boas appeared unperturbed and exhibited neither aggression nor alarm (R. Thorpe, in litt.to RWH, 20 February 2018). James D. Lazell (field notes, 15 August 1962) collected a *B. orophias* with a total length of 362 mm on 15 August. It is difficult to imagine neonates of this species being much smaller, and an August birth corresponds well with February mating.

*Conservation Status*

This species is listed as Endangered on the IUCN Red List (Daltry 2018) as it has a limited distribution on a small (604 $km^2$) island (Fig. 5.114). There is some evidence indicating *Boa orophias* is exported illegally, especially to Martinique (J. Daltry, in litt. 6 July 2018). These boas are often killed by people (see above). Just recently (Gaillard 2020) it has been reported that *B. orophias* is being killed for the purpose of human consumption, and the Forestry Division on St. Lucia does "not want this spiraling out of control." Persecution of *B. orophias* carries a penalty fine of $5000 in Eastern Caribbean currency (~ US$1850 on 22 May 2021). Figure 5.114 shows the distribution of *Boa orophias*.

## ST. VINCENT TREEBOA

*Corallus cookii* Gray 1842

*Taxonomy*

Originally described as *Corallus cookii*, it was relegated to subspecies rank by Stull (1935) as *Boa enydris cookii*; when Forcart (1951) resurrected *Corallus*, it became *Co. e. cookii*; Roze (1966) recognized it as *Co. hortulana cookii*. Henderson (1997) elevated it to its former full species status as *Co. cookii* and confined its range to the island of St. Vincent. Recent molecular evidence (Colston et al. 2013; Reynolds et al. 2014a) shows *Co. cookii* to be a recently diverged lineage in *Co. hortulana* (Fig. 0.2). Based on morphological characters and geography, Henderson (2015), Henderson and Powell (2018), and Reynolds and Henderson (2018) continued to recognize it as a valid species.

*Etymology*

The species name is a patronym honoring E. W. Cooke, an English artist and naturalist (Gray 1842). Beolens et al. (2011) correctly noted that the vernacular name should be "Cooke's," (i.e., Cooke's treeboa), but the terminal "e" has been dropped in error. Hedges et al. (2019) recommend the name "St. Vincent Treeboa." Vernacular names include "congo snake" and "lazy snake." "Congo snake" is also used to refer to *Corallus grenadensis* on many of the Grenadine Islands.

*Description*

These slender snakes have a laterally compressed body, a prehensile tail, and a somewhat chunky head with conspicuous labial pits attached to a very slender neck. The maximum known SVL is 1374 mm (Henderson 1997). The number of dorsal scale rows at midbody is 39–48, ventrals 257–278, subcaudals 100–122, scales between supraoculars 7–13, and infraloreals 0–4.

In a sample of forty-seven specimens, forty-one (87.2%) were taupe and the others were gray or brown. The main element of the dorsal pattern is best described as an hourglass or dumbbell shape, rarely like a stout spade; the dorsal-most portion of the shape is open (i.e., dorsal ground color is visible). The main elements are usually edged in black, with or without white margins. Areas between the main elements sometimes include a dorsoventrally elongated blotch.

**Figure 5.115.** *Corallus cookii* from the Botanic Garden in Kingstown, St. Vincent, displaying a fairly typical color pattern. Photo by Father Alejandro J. Sanchez Muñoz.

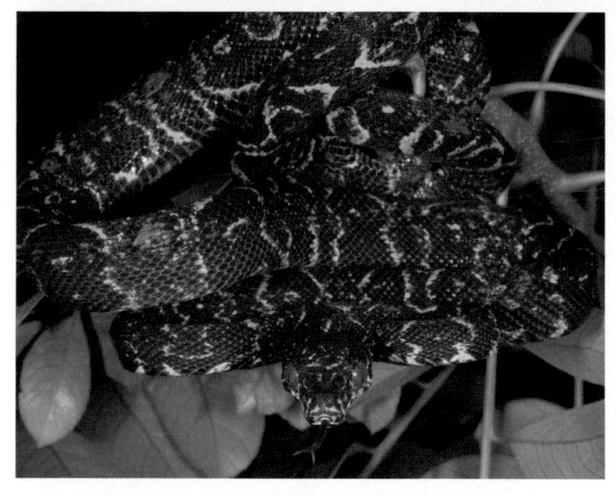

**Figure 5.116.** An especially dark *Corallus cookii* from the Vermont Nature Reserve (320 m asl), St. Andrew Parish, St. Vincent. Photo by Robert Powell.

**Figure 5.117.** A *Corallus cookii* from near the Vermont Nature Reserve; it retains some of the pink, salmon, or orange juvenile coloration. Photo by Robert Powell.

The dorsal pattern may become a series of longitudinally elongated middorsal blotches with another longitudinally elongated series situated more laterally (Fig. 5.115). One individual from near the Vermont Nature Reserve was very dark brown with white reticulations between the dark pattern elements, somewhat reminiscent of *Co. grenadensis* at elevations >400 m asl on Grenada (Fig. 5.116). The underside of the head is immaculate, but the ventrals may be lightly to heavily patterned, and lateral encroachment of the dorsal pattern onto the ventrals is common. The top of the head has rounded blotches or a vermiculate pattern. Mental and gular regions are white but marked (often heavily) with brown. Juvenile coloration is not dramatically different from that of adults (Fig. 5.117). The most conspicuous difference is that in neonates and juveniles the pale color surrounding elements of the dorsal pattern is pink to coral; during ontogeny, that color becomes white, off-white, or cream (Henderson 1997, 2015; Henderson and Powell 2018).

**Figure 5.118.** *Left*: Habitat at the Vermont Nature Reserve (320 m asl), St. Andrew Parish, was borderline habitat for *Corallus cookii*. Photo by Robert Powell. *Right*: The Botanic Garden in Kingstown, St. Vincent (70 m asl) provided excellent habitat for *Co. cookii*. Photo by Mike Treglia.

*Distribution*

*Corallus cookii* is restricted to St. Vincent. It occurs from at or near sea level up to at least 425 m at Hermitage (St. Patrick Parish) and in relatively intact habitats as well as those that are highly altered. Our knowledge of the geographic and ecological distribution of *Co. cookii*, however, is inadequate. We feel certain that its distribution is more widespread than currently documented and that it occurs in more ecological situations than have been recorded to date.

*Habitat*

Like most islands in the Lesser Antilles, St. Vincent's landscape has been substantially altered over the past 400–500 years, with most of the forest sacrificed to agriculture. Nevertheless, *Co. cookii* can be amazingly abundant in certain habitats. In the mid-1980s, *Co. cookii* was abundant in an area of second growth and mixed agriculture interspersed by human residences in the Layou Valley (St. Patrick Parish). It was collected or observed foraging in mango (*Mangifera indica*), coconut (*Cocos nucifera*), nutmeg (*Myristica fragrans*), and cacao (*Theobroma cacao*) trees. In the Vermont Nature Reserve (about 320 m asl; Fig. 5.118), an area of lush rainforest, *Co. cookii* foraged in strangler fig (*Ficus* sp.) and tree ferns (*Cyathea* sp.). At Hermitage, individuals were loosely coiled on a dead *Cyathea* frond and on the trunk of a Caribbean pine (*Pinus caribaea*); others foraged in bamboo (*Bambusa* sp.) in the Orange Hill area and the Vermont Nature Reserve. At Campden Park (150 m, St. Andrew Parish), *Co. cookii* occurred above a heavily disturbed residential area. By day, a snake was roosting in a mango tree at 1430 hours on a road cut along the Chateaubelair River (St. David Parish). We have only one observation of *Co. cookii* alive or dead on a road, and we have seen only one moving at ground level (at the top of a road-cut in viny vegetation in the Vermont Nature Reserve). No treeboas were encountered on a descent from La Soufrière volcano (1234 m) at dusk and into the night, but it is likely boas do occur at lower elevations on the volcano's slopes (Henderson 2015; Henderson and Powell 2018).

Powell et al. (2007) surveyed eight sites for *Co. cookii* and made 130 observations during 78.1 p-h of searching. The sites included (1) upland rainforest in a nature reserve (320 m asl); (2) an area of mixed orchard trees and native vegetation (150 m; Fig. 5.119); (3) a small area of introduced orchard trees (240 m); (4) a banana (*Musa*) monoculture (50 m); (5) an urban (in Kingstown) botanical garden with native and introduced trees and shrubs (70 m; Fig. 5.118); (6) a site along a dirt road bordered primarily by the introduced tree *Gliricidia sepium* (locally known as "quick stick"; 10–25 m); (7) a paved road bordered predominantly by *G. sepium* (30 m); and (8) a residential area in Kingstown with streetlights, numerous houses, frequent vehicular and pedestrian traffic, and numerous dogs; interspersed between the houses were numerous trees (orchard and native; 140 m). On any given night, encounter rates ranged from 0 in the *Musa* monoculture to 3.2 in the Kingstown residential area, but mean rates for the eight sites ranged from zero (site 4, banana plantation) to 2.83 ± 0.11/p-h (site 2, mixed orchard trees and native vegetation). Encounter rates between sites differed significantly (ANOVA, $F = 14.79$, $P < 0.0001$). In addition, Powell et al. (2007) used a suite of ten binary variables (presence/absence) to determine increasing degrees of relative disturbance of a survey site: (1) streetlights, (2) vehicular traffic at night, (3) pedestrian traffic at night, (4) introduced orchard trees, (5) houses, (6) livestock, (7) obvious evidence of deforestation, (8) paved roads, (9) clear-cuts, and (10) a monoculture habitat. The numbered sites above are listed from least (upland rainforest) to most disturbed (Kingstown residential area). The number of disturbance variables and mean encounter rates at the eight sites were not significantly correlated, nor were mean encounter rates and elevation. Although encounter rates were not significantly correlated with rankings

**Figure 5.119.** Probably close-to-ideal habitat for *Corallus cookii* at Ferret, St. Patrick Parish, St. Vincent (150 m asl). Photo by Mike Treglia.

of disturbance or elevation, the results of Powell et al. (2007) do indicate that human-mediated alterations to habitat can affect relative abundance. Apparently, the type of disturbance is more important than the cumulative number of different disturbances in determining the impact on abundance in a given area. Although infrequently encountered in dry forest, Mallery et al. (2007) observed *Co. cookii* in that habitat at about 200 m asl at Akers in St. Patrick Parish.

*Abundance*

Although Barbour (1937) suggested that *Corallus* had been eliminated from St. Vincent and A. Schwartz (pers. comm. to RWH, in 1987) considered it rare, multiple visits to the island by RWH between 1987 and 2006 revealed *Co. cookii* to be locally abundant depending on habitat. Powell et al. (2007) did a series of surveys in a variety of habitats and recorded encounter rates of 0.0–2.78 ± 0.32/p-h. In all instances except one (a banana plantation where the encounter rate was 0.0/p-h), observers worked in close proximity. If we assume one person would have encountered the same number of boas in the same amount of time, encounter rates ranged from 0.0 to 8.59/p-h. The highest rates were at a site with introduced orchard trees mixed with natural vegetation (Ferret, St. Patrick Parish) and two sites where the impact of humans was blatantly obvious (the Botanic Garden and a residential area). The lowest rates occurred where the impact of humans was least obvious (a nature reserve) and also very conspicuous (a *Musa* monoculture and along roadways lined with introduced quick stick trees; Powell et al. 2007; Henderson 2015; Henderson and Powell 2018).

*Activity and Trophic Ecology*

*Corallus cookii* is nocturnal. Encounters during daylight hours are rare. Snakes have been observed during the day on the roof of the St. Vincent parrot enclosure in the Botanic Garden in Kingstown (St. George Parish), and another was resting on branches near the Chateaubelair River in St. David Parish. Yet another was 2.0 m above ground level on a shaded branch at 1415 hours, its body resting on but not around the branch, and one was coiled at 2–3 m in a mango tree along a road cut at 1430 hours (Henderson 2002, 2015).

*Corallus cookii* has been observed foraging at heights from ground level to more than 20 m. Based on a sample of only twenty-two observations, most foraging occurred at heights <5.0 m. Although the number of observations available is much smaller than for *Co. grenadensis*, most *Co. cookii* were actively foraging on the distal ends of tree branches in a manner similar to that observed in *Co. grenadensis*.

On 6 June at Ferret in St Patrick Parish (~160 m asl), an approximately 755 mm SVL *Co. cookii* was observed at 200 cm in a tree and about 15 cm from a sleeping adult *Anolis trinitatis*. At 2220 hours, the boa slowly approached the lizard from behind and slightly above. By 2230 hours, the snake was within 2.0 cm of the anole and, for the next 15 minutes, movement by the boa was almost imperceptible. It was virtually touching the anole before it grabbed it in the nuchal region, threw a single coil around it, and subdued it. Swallowing was completed at 2256 hours, and the snake ascended into the crown of the tree (Henderson et al. 2007).

We have documentation of only nine prey items (five *Anolis*, one bird, three mammals; Henderson and Pauers 2012; Henderson 2015), thereby preventing an accurate assessment of diet composition and a possible ontogenetic shift in diet (as seen in other species of *Corallus*). Anoles (*Anolis griseus* and *A. trinitatis*) were taken by snakes 396–910 mm SVL, rodents were taken by snakes 655–1374 mm SVL, and the bird was recovered from a snake that was 655 mm SVL. The anole taken by the 910 mm *Co. cookii* was *A. griseus* (SVL 110 mm). All other anoles were taken by snakes 396–752 mm SVL, suggesting that *Co. cookii* likely undergoes an ontogenetic shift in diet similar to that exhibited by *Co. grenadensis*.

Prior to the introduction of *Mus* and *Rattus* (which arrived with Europeans), the primary endothermic prey for *Co. cookii* was probably a recently extinct species of small rice rat (*Oligoryzomys victus*; Turvey et al. 2010). This *Mus*-sized rat was extant until at least 1892, so it coexisted with introduced rodents for several hundred years. Turvey et al. (2010) noted that the species has not been recorded from pre-Columbian archaeological sites on St. Vincent, in contrast to rice rat species on other Lesser Antillean islands. They attributed this absence to the possibility that it was a relatively recent human-assisted translocation to the Windward Islands from an unknown source population on mainland South America, or, given its small size, pre-Columbian Amerindians did not perceive it as a valuable food item. Undescribed small-bodied rice rats reported from Amerindian middens on nearby Carriacou and Grenada may be conspecific with *O. victus* (Turvey et al. 2010; but see Mistretta 2019).

*Predators and Defensive Behavior*

We have no documentation of predation on *Co. cookii*. We suspect that hawks, falcons, and owls prey on this species, as well as opossums and possibly mongooses. The Barn Owl (*Tyto alba*) is known to prey on *Co. hortulana* in Brazil (da Costa Silva and Henderson 2013) and could potentially prey on West Indian *Corallus* as well. Humans, of course, regularly kill this species for no other reason than being a snake. Scars and incomplete or stubbed tails are not unusual on adult boas. These might well be the results of unsuccessful attacks from would-be predators, or they might have been inflicted by a prey species (e.g., a rat) during an attempted predation episode.

A suite of defensive strategies allows *Co. cookii* to function in a potentially dangerous environment. Crypsis is a passive strategy, and the rather dull, neu-

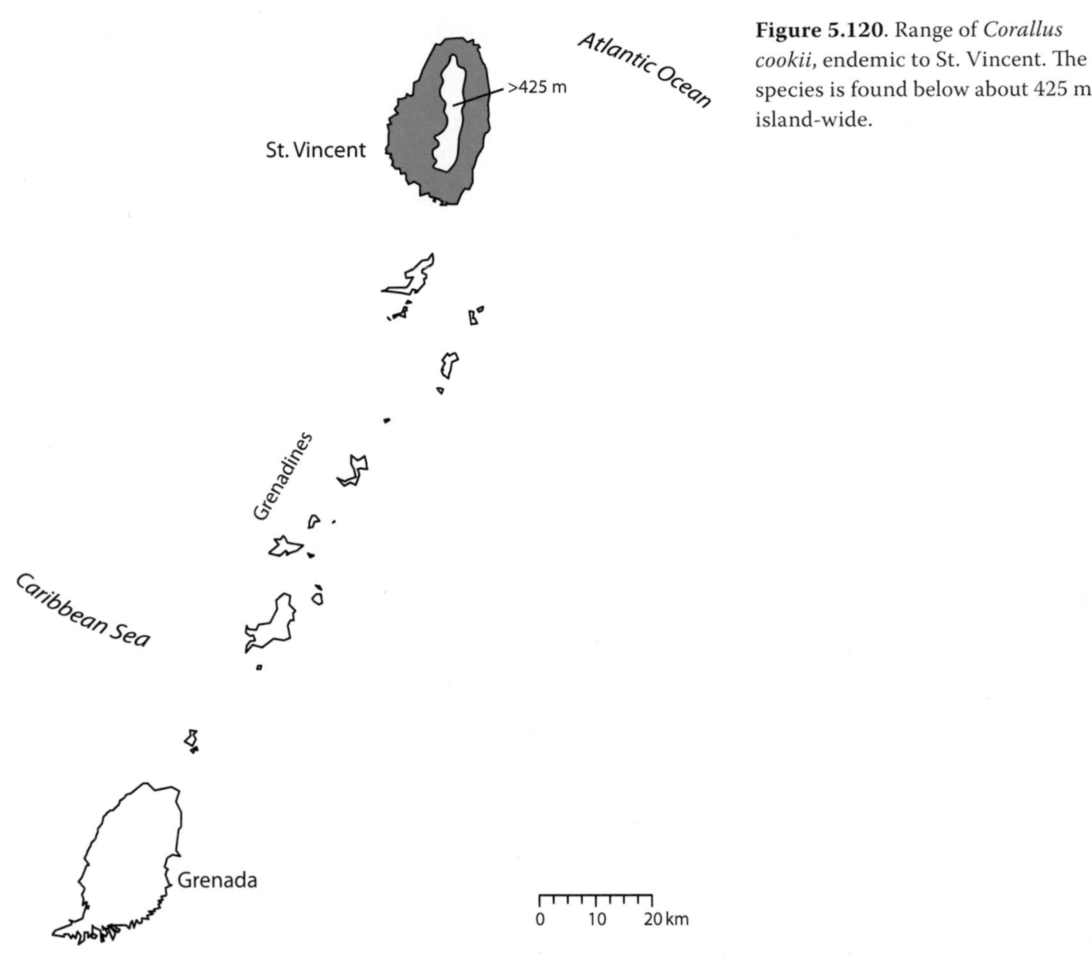


**Figure 5.120**. Range of *Corallus cookii*, endemic to St. Vincent. The species is found below about 425 m island-wide.

tral color pattern of *Co. cookii* renders it inconspicuous in its arboreal, foliage-filled environment. When capture of these snakes is attempted, they will usually try to escape to higher perches or into the interior of the tree crown. If captured, boas will assume a balled-up posture, usually with the head hidden, although the efficacy if this strategy is unknown. More proactive behaviors include voiding the contents of the cloaca, rapid rotation of the body, and biting (Henderson 2015; Henderson and Powell 2018).

*Reproduction*

We have no data regarding reproduction in this species, aside from those fortuitously accumulated via collecting. Based on collection dates of boas <500 mm SVL, births occur between August and January. RWH has personally collected *Co. cookii* <500 mm SVL in October (451 mm SVL), November (337–375 mm SVL), and March (372–492 mm SVL). October and November are wet-season months, but rainfall is diminished in December–March. Most birth data coincide with the wet season in St. Vincent. Prey (*Anolis* spp.) of appropriate size for small treeboas are available throughout the year (Henderson 2015; Henderson and Powell 2018).

On 6 June 2006, six small *Co. cookii* were encountered in close (<1.0–1.5 m) proximity to one another at Ferret (St. Patrick Parish). One with an SVL of 365 mm and mass of 11.5 g was captured (and later released at the capture site). These babies likely were recently born and had not yet dispersed. If correct, this would indicate a prolonged breeding season (Henderson 2015), and recent observations of courtship in *Co. grenadensis* (Henderson et al. 2015) suggest that reproductive activity is more prolonged than previously thought.

*Living with Humans*

Although this species has the smallest range of any species of *Corallus*, it, along with *Co. grenadensis*, has apparently adapted to the presence of human activity with greater success than some continental congeners. In only a few places (e.g., Vermont Nature Reserve) is *Co. cookii* found far from human activity (roads, residences, structures housing small businesses, or agricultural sites). This is the only species of *Corallus* that could be considered an "urban" treeboa (e.g., Powell and Henderson 2008). *Corallus cookii* has been observed to be common in residential areas of Kingstown, the largest and most densely populated city in St. Vincent. It was abundant in the country's botanical garden, and it was observed crawling through a chain-link fence along a busy thoroughfare. One evening in a busy residential area, as people were just returning home from their places of business, two people dressed in business suits walked along a sidewalk in animated conversation. Not 2 m away was a 1.0-m *Co. cookii*, wending its way slowly along a metal fence at the edge of someone's residential yard. The people were oblivious to its presence (RWH, pers. obs., June 2006). The ability of *Co. cookii* to use a variety of habitats and abundant trophic resources in the proximity of human activity renders this species robustly tolerant of anthropogenic development.

*Conservation*

*Corallus cookii* has been assessed as Near Threatened on the IUCN Red List (Henderson and Powell 2021a). This species is afforded some protection under CITES Appendix II. In May 2019, live individuals of this species were advertised for sale on the internet from a U.S.-based location. To the best of our knowledge, no legally collected and exported boas are coming out of St. Vincent and the Grenadines. *Corallus cookii* is afforded some additional protection as it shares a portion of its range with the national bird of St. Vincent, *Amazona guildingii*, which has an IUCN Red List assessment of Vulnerable. *Corallus cookii* will reap some benefit if *A. guildingii* habitat is protected. Figure 5.120 shows the distribution of *Corallus cookii*.

## GRENADA BANK TREEBOA

*Corallus grenadensis* (Barbour 1914)

*Taxonomy*

Originally described as *Boa grenadensis*, Barbour (1935) subsequently relegated it to a subspecies of *B. cookii*; meanwhile, Stull (1935) synonymized it with *B. enydris cookii*; and Barbour (1937) continued to recognize it as a subspecies of *B. cookii*. After Forcart (1951) untangled *Boa*, *Constrictor*, and *Corallus*, and McDiarmid et al. (1996) did the same for *Co. enydris*/*hortulana* (but see also Gonzalez et al. 2021), Henderson (1997) resurrected *Corallus grenadensis* to full species status. Recent molecular genetic evidence (Colston et al. 2013; Reynolds et al. 2014a) shows *Co. grenadensis* to be a recently diverged lineage in *Co. hortulana* (Fig. 0.2). Based on morphological characters and geography, Henderson (2015, 2019), Henderson and Powell (2018), and Reynolds and Henderson (2018) continued to recognize it as a valid species.

*Etymology*

The name *grenadensis* refers to the island of Grenada, the site of the type locality. Local common names include "serpent" ("sarpint"; Groome 1970), "Lazy Snake" (on Grenada and Carriacou), "tèt-chen" (possibly restricted to Carriacou; Kephart 2019), and "congo snake" (on the rest of the Grenadines).

*Description*

These slender snakes have a laterally compressed body, a prehensile tail, and a somewhat chunky head with conspicuous labial pits attached to a very slender neck. The maximum known SVL is 1625 mm (Henderson 1997). Based on snakes from Grenada and the Grenadines, neonate SVL is 295–393 mm, with mass 7–12 grams. The number of dorsal scale rows at midbody is 37–46, ventrals 251–278, subcaudals 100–119, scales between supraoculars 3–9, and infraloreals 0–4.

Dorsal ground color is variable, ranging from yellow, orange, red, gray, taupe, and pale brown to dark brown. The predominant element of the dorsal pattern is spade shaped, and this pattern occurs throughout the Grenada Bank, with the greatest variation being whether the spade has sharp angles (53.5%) or is rounded (26.5%). Although the rounded spades may

**Figure 5.121.** Color and pattern variation in *Corallus grenadensis*. *Top left*: Les Avocats, St. David Parish, Grenada; *Top right and bottom left*: Grand Bras, St. Andrew Parish, Grenada; *Bottom right*: Petite Martinique, Grenada Grenadines. Photos by Robert W. Henderson.

appear in any habitat, they predominate at higher elevations on Grenada. A dorsal ground color of gray or taupe predominates on most of the Grenadine islands (Fig. 5.121). Boas with a yellow dorsal ground color are occasionally encountered on Bequia, Union, and Carriacou (where it accounts for ~1.0 % of all boas). An exception to the dominance of the gray-taupe dorsal ground color is Mayreau; there red orange is predominant (Fig. 5.122). The primary element in the dorsal pattern in snakes from Mayreau also is consistently different from other populations on the bank (more of a balloon shape rather than a spade; a similar pattern occurs on the other islands on the bank but not to the exclusion of the more typical spade shape) (Fig. 5.122).

Based on the examination of living snakes, dorsal ground color often is correlated with iris and tongue colors (i.e., snakes with a yellowish dorsal ground color usually have yellowish irises and a pale-colored

**Figure 5.122.** *Corallus grenadensis* from (*top*) Carriacou, Grenada Grenadines, and (*bottom*) Mayreau, St. Vincent Grenadines. Photos by R. Sajdak.

tongue; snakes that are taupe to dark brown have very dark irises and dark-brown to black tongues; Henderson 2004). The ventral ground color usually is dull yellow, but it may be white or cream. It may be immaculate (in snakes with a yellowish dorsal ground color), marked with flecks, spots, or large blotches, or almost completely covered with dark brown. The ventral pattern usually becomes denser posteriorly. The predominant dorsal ground color at any locality is associated with elevation, rainfall, and percent possible sunshine (e.g., Henderson 1990, 2002, 2015; Fig. 5.123), with yellow dorsal ground color predominating in open, sun-drenched habitats at low elevations (Fig. 5.121), and very dark boas occurring at higher elevations (>400 m asl) with cooler temperatures, frequent cloud cover, and frequent precipitation (Fig. 5.123).

As in *Co. cookii*, juvenile coloration is not dramatically different from that of adults. The most conspicuous difference is that in neonates and juveniles the pale color surrounding elements of the dorsal pattern (if present) is pink to coral; during ontogeny that color becomes white, off-white, or cream. Patternless *Co. grenadensis* have a dorsal ground color that is usually "yellowish," which in neonates and juveniles can be a quite bright. With ontogeny, however, the yellow becomes rather dingy, and large adults are often pale brown rather than yellowish.

*Distribution*

*Corallus grenadensis* is widely distributed on the Grenada Bank, occurring on Grenada (up to about 530 m asl) and at least ten of the Grenadines (Bequia, Baliceaux, Isle à Quarte, Mustique, Canouan, Mayreau, Union, Petite Martinique, Petit St. Vincent, Carriacou; Henderson and Breuil 2012; Henderson 2015; Henderson and Powell 2018). It is likely *Co. grenadensis* occurs on additional islands in the Grenadines archipelago.

*Habitat*

This species exhibits impressive ecological plasticity on Grenada. Of the eighteen land cover and forest formation categories assessed by Helmer et al. (2008), *Co. grenadensis* occurs in ten (55.6%) categories. If the three "no vegetation" categories are eliminated, it occurs in ten of fifteen (66.7%). These land use/forest formation categories include a wide range of environmental variables (e.g., temperature, rainfall, insolation, vegetation; Henderson 2002, 2015; Henderson and Henderson 1995). About 72% of the land area of Grenada is devoted to agriculture, whereas the remaining native vegetation covers only about 23% (the remaining 5% comprises urban habitat and pastures). Because *Co. grenadensis* often is common in agricultural areas with a contiguous canopy of orchard trees (Henderson and Winstel 1995; Henderson et al. 1996, 1998), the large amount of land devoted to agriculture is not necessarily detrimental to the distribution or population densities of this species, although some agricultural

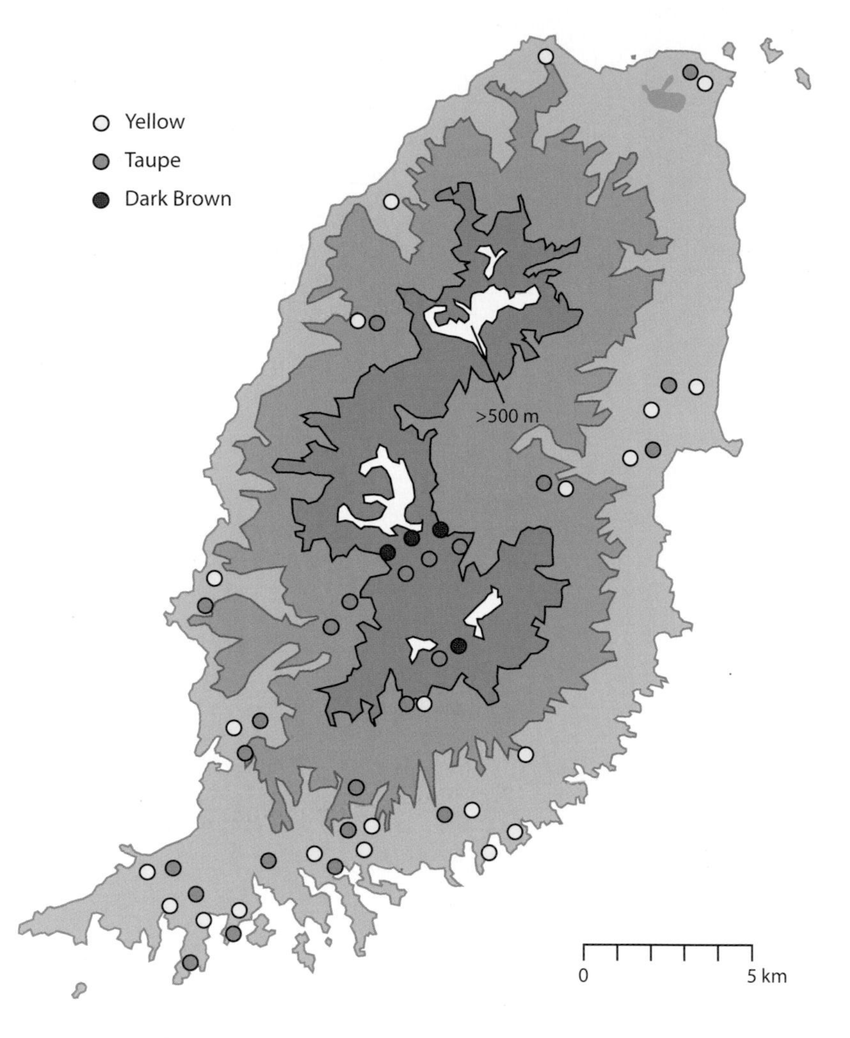


**Figure 5.123.** Map of Grenada showing the distribution of three color varieties of *Corallus grenadensis* across elevational gradients. The species is found up to about 530 m in elevation.

practices (e.g., use of pesticides, clear-cutting) obviously have serious ramifications. Many people have observed that *Co. grenadensis* is not shy about entering houses.

Treeboas are nowhere more common on Grenada than in areas of mixed agriculture between sea level and about 150 m asl, including stretches of trees along open areas (Fig. 5.124). In such areas *Co. grenadensis* forages among passion fruit vines and in mango, nutmeg, coconut, breadfruit, citrus, cacao, and banana trees.

We have never encountered *Co. grenadensis* at elevations above about 530 m (Grand Etang). Since trees occur above 530 m and, therefore, potential treeboa habitat exists, some other factor or factors must limit their elevational distribution; the two most likely agents are temperature and rainfall. Kubalal, at 670 m, is the only meteorological station on Grenada that is at a higher elevation than Grand Etang. The only temperature data available for Kubalal are for January–March 2004. Average minimum and maximum temperature for those three months are 17.2 °C and 23.7 °C, respectively; the same data for Grand Etang are 17.8 °C and 24.5 °C, and Grand Etang receives more rain than Kubalal (Henderson 2015). Other factors that may contribute to its exclusion at the higher elevation are elevated wind speed, reduced solar radiation, and vegetational structure.

Using the same ten variables to assess the potential impact of various human-mediated disturbances

**Figure 5.124.** Desirable habitat for *Corallus grenadensis* at (*top left*) Pearls and (*bottom left*) Grand Bras, both in St. Andrew Parish, Grenada. *Right*: *Co. grenadensis* habitat at Les Avocats (400 m asl), St. David Parish, Grenada. This site is part of the Grand Etang forest. Photos by R. Sajdak (top) and Robert W. Henderson (bottom and right).

on *Co. cookii* (see that account), Powell et al. (2007) did a similar assessment for *Co. grenadensis* at six sites listed in order from least to most disturbed: (1) Grand Etang National Park, tropical forest (~530 m); (2) Mt. Hartman Bay, dense stands of acacia and mangrove (primarily *Rhizophora mangle*; ~1 m); (3) Mt. William, an area of active cacao and nutmeg production mixed with native trees (~400 m); (4) Pearls, an area of orchard trees mixed with native vegetation (10 m); (5) Westerhall Estate, a site of mixed agriculture, orchard trees, and native vegetation (~45 m); and (6) St. George's (Grand Anse), a stretch of paved road illuminated by streetlights with hotels and ornamental and native vegetation (~5 m). A total of 157 p-h of searching resulted in 323 observations. Nightly encounter rates ranged from zero (Pearls, Grand Etang) to 11.3/p-h (Pearls), and mean encounter rates ranged from 0.23 (Grand Etang) to 5.03/p-h (Mt. William). Encounter rates between sites differed significantly (ANOVA, $F = 8.04$, $P < 0.0001$). The number of disturbance variables ranged from one (Grand Etang) to six (St. George's), but as for *Co. cookii*, no significant correlation existed between number of disturbance variables and mean encounter rates at the six sites (Spearman correlation, $Z = 0.83$, $P = 0.41$). Likewise, mean encounter rates were not significantly correlated with elevation.

*Corallus grenadensis* requires only a narrow corridor of trees in which to move and forage. At Westerhall, and prior to Hurricane Ivan in 2004, concentrations of treeboas were observed along small plots of sugarcane in tree lines only one tree wide, usually with a high incidence of mango and breadfruit trees and contiguous with other trees (either native or introduced). Similarly, at Mt. Hartman, *Co. grenadensis* occurred in a strip of mangroves often <5.0 m wide. At both sites, however, narrow avenues of vegetation could be used to reach broad expanses of trees.

In the Grenadines, undisturbed habitat is virtually nonexistent (Nichols 1891; Howard 1952; RWH, pers. obs.), and treeboas on these islands generally are encountered in the widespread dry scrub vegetation (e.g., *Vachellia* spp.) typical of the archipelago (Beard 1949; Howard 1952). From 3 to 20 June 2010 on Union Island, Quinn et al. (2011) conducted visual surveys during which they encountered thirty-seven *Co. grenadensis* at eight sites at times from 2000 to 2345 hours. With the exception of mangrove habitat, all sites were variants of dry seasonal forest, although dominant species of trees (e.g., *Hippomane mancinella*, *Vachellia* sp., *Conocarpus erectus*) varied. Elevations ranged from sea level to 130 m (mean ± SD = 50.8 ± 46.1 m). Snakes always were on tall vegetation (≥3 m), although perch heights varied (0.5–15 m; mean = 4.0 ± 2.9 m). Most snakes (78%) were young-of-the-year or subadults. In thirty-one of the thirty-seven encounters, boas were actively foraging, moving slowly through vegetation with bodies extended. When boas were not foraging, they were stationary and coiled on a branch or branches.

On Mayreau, with a max. elev. of only 61 m, *Co. grenadensis* has a wide distribution in the dry scrub habitat that occurs throughout the island (Fig. 5.125). During author visits in June 2012 and November 2015, treeboas were encountered in roadside dry scrub vegetation at heights of 1.0–3.2 m. Five adults were at 1.0–2.1 m, with four of the five at 1.0–1.5 m. Vegetation in which we encountered boas included *Pisonia fragrans*, *Leucaena leucocephala*, *Pithecellobium ungis-cati*, *Randia aculeata*, and *Cordia curassavica*. By day, M. de Silva (pers. comm., June 2010) observed treeboas in bromeliads and outhouses. On Canouan, like Mayreau, the habitat is largely dry scrub with a high incidence of thorny vegetation. Treeboas forage in flamboyant (*Delonix regia*) and tamarind (*Tamarindus indica*) trees, as well as on prickly pear cactus (*Opuntia dillenii*). Eight *Co. grenadensis* were foraging at 0.5–2.2 m in roadside vegetation, but we found no boas in mangrove or sea grape (*Coccoloba uvifera*) habitats.

**Figure 5.125.** *Corallus grenadensis* habitat on Mayreau, St. Vincent, Grenadines. Photo by R. Sajdak.

Contiguous crown vegetation is the single most important criterion in habitat selection by *Co. grenadensis* (and perhaps any arboreal snake). It provides both arboreal habitat for foraging and other basic activities and also a corridor for movement between adjacent habitats.

*Abundance*

*Corallus grenadensis* might attain higher population densities than any member of the genus (Henderson 2015), although *Co. cookii* and *Co. hortulana* can also be relatively abundant in some situations. On Grenada, nightly encounter rates ranged from <1.0/p-h to >10.0/p-h. The highest encounter rates occurred in areas of mixed agriculture and moderate rainfall at elevations <100 m asl. Many of the encounter rates fell between five and six snakes/p-h, including a sample for one of the Grenadines. Four persons searching on Carriacou encountered fifty *Co. grenadensis* at three sites in just under 3 hours (Henderson et al. 2013a), a rate of 4.2 boas/p-h. (See also Demography below.)

On Union Island, encounter rates of *Co. grenadensis* averaged 1.32 ± 0.86 snakes/h (Quinn et al. 2011). However, in ideal habitat on Grenada (i.e., natural vegetation mixed with orchard trees), encounters could

**Figure 5.126.** *Corallus grenadensis* conspicuously basking in the open on Carriacou, Grenada Grenadines; an unusual behavior. Photo by Esau Pierre.

exceed twenty boas per hour (Henderson 2002; Henderson et al. 2009). That habitat is lacking on Union (or on any of the Grenadines).

*Activity and Trophic Ecology*

During the day, *Co. grenadensis* will usually (~90% of observations) coil on the distal end of a branch with its prehensile tail sometimes wrapped under the branch. The coils are more or less concentric, but often with some overlap (i.e., a portion of one coil resting on a portion of another), and the head resting on top of one of the coils. Inactive snakes are almost invariably under an umbrella of leaves shielding them from full sunlight. Choosing such situations might provide protection from avian predators while also minimizing heat loading. The canopies often were little more than a small sprig of vegetation but sufficient to conceal the snake and to create a sun-shade mosaic. Inactive snakes observed in mangrove habitat (usually *Rhizophora mangle*) invariably selected perches that were over water. Aside from slight shifts of posture, we have not personally observed activity during daylight hours. However, at the St. Rose Nursery (St. George Parish; 215 m asl), a snake coiled in vegetation adjacent to a greenhouse routinely shuttled between shade and sunlight during the day (Henderson 2015). At Grand Bras (St. Andrew Parish) beginning on 3 May 2013 and for the next 20 days, I. Monah (pers. comm. to RWH, 12 September 2013) monitored an adult *Co. grenadensis* that used a cavity in a papaya tree (*Carica papaya*). The boa exited the cavity at dusk but would again be in the tree the next morning; it would retreat into the cavity soon after the sun's rays found it. An adult yellow *Co. grenadensis* was photographed coiled near the hollow of a tree trunk on Carriacou; it was presumably (and conspicuously) basking (Fig. 5.126).

Activity begins at dusk. The earliest time at which foraging behavior has been observed is 1755 hours. The onset of activity is associated with declining air temperature and light intensity. Based on 528 observations, peak activity in *Co. grenadensis* occurs from astronomical dusk (headlamp required) until approximately midnight, a span of about 6 hours (Henderson and Winstel 1997; Henderson 2015); activity tends to taper off after midnight. At a site in Lower Woburn (St. George Parish), two young-of-the-year boas were still apparently foraging in acacia vegetation shortly before 0500 hours (astronomical dawn). At about the same time, a subadult was moving into the interior vegetation of an acacia tree, possibly signaling the termination of that snake's foraging. By 0520 hours, a headlamp was no longer necessary, and locating boas by the reflection of their eyeshine was impossible. No activity has been observed between 0535 and 1755 hours.

Lizards (*Anolis aeneus* and *A. richardii* and the iguanid *Iguana iguana*), small passerine birds (Bananaquit, *Coereba flaveola*, Tropical Mockingbird, *Mimus*

**Figure 5.127.** *Corallus grenadensis* that has captured and is swallowing a Bananaquit (*Coereba flaveola*) on Carriacou. Photos by R. Sajdak.

*gilvus*; Sajdak and Henderson 2017, 2019), introduced murid rodents (*Mus musculus* and *Rattus rattus*; Henderson and Pauers 2012; Henderson 2015), introduced frogs (*Eleutherodactylus johnstonei*; Rush et al. 2013), and a marsupial (*Marmosa robinsoni*; Rush and Henderson 2014) are documented prey items. Records for eighty-four prey items include two (2.4%) frogs, fifty-four (64.3%) lizards, twenty-two (26.2%) mammals, and six (7.1%) birds. Of the lizards, one was an *I. iguana*, and the other fifty-three were anoles (98.1% of lizard prey; Henderson 2015). Mammalian and arboreal lizard diversity on the Grenada Bank are extremely limited but nevertheless provides more than 90% of the prey taken by *Co. grenadensis*. The avian fauna is much richer but is largely underexploited.

Like *Corallus cookii* on St. Vincent, prior to the introduction of *Mus* and *Rattus* with Europeans, the principal endothermic prey of *Co. grenadensis* was probably rice rats (Oryzomyini; Lippold 1991; Pregill et al. 1994; Turvey et al. 2010; Mistretta 2019). The overexploitation of native taxa (including rice rats) by Ceramic Age peoples on certain West Indian islands is well documented (Newsom and Wing 2004; Fitzpatrick and Keegan 2007), and rice rats may have become extinct on Grenada long before the arrival of alien *Mus* and *Rattus*. The apparent limited prey diversity available to West Indian *Corallus* is intriguing relative to that available to their close relative *Co. hortulana*, which has a wide distribution in South America. With little effort, we were able to identify a minimum of fifty documented prey species taken by *Co. hortulana* (many birds and mammals could not be identified to species). In sharp contrast, we have documentation of a known maximum of nine prey species taken by *Co. grenadensis* and five by *Co. cookii*. It is likely that both West Indian species may take more bird species than is currently documented, and likely bats as well, but the same can certainly be said for *Co. hortulana*.

*Corallus grenadensis* exhibits an ontogenetic shift in diet from *Anolis* lizards to rodents. Lizards (*Anolis* spp.; $n = 37$) were taken by boas with sizes between 347 and 1025 mm SVL (mean = 650.9 ± 30.1 mm) and constituted nearly 96% of the prey items in the size class 300–699 mm SVL; no mammals were represented. The transition from lizards to mammals occurred at 700–800 mm SVL. Although anoles were still an important prey component (64.7%), the boas were now large enough for mammals to be prey targets (29.4%). Snakes >800 mm SVL, although still taking anoles (21.0%), took a much higher percentage of mammals (nearly 74%). Mammals (all rodents; $n = 19$) were taken by boas 705–1170 mm SVL (mean = 914.4 ± 31.7 mm), and birds ($n = 3$) were taken by boas 655–1220 mm SVL (mean = 878.3 ± 173.5 mm). Snake sizes associated with different prey groups differed significantly ($P \leq 0.001$; Henderson 2015). Coincident with an increase in SVL and the shift in prey is an apparent change in frequency of feeding. In a sample of seventy-six preserved *Co. grenadensis* <750 mm SVL, 44.7% had food in their stomachs, 94.1% of which was composed of *Anolis* lizards. Of fifty-nine boas 750–950 mm SVL, 27.1% contained prey (25% anoles and 75% endotherms); only 16.9% of fifty-nine boas >950 mm SVL had prey, and 80% was mammalian (Henderson 2002).

When hunting sleeping lizards, snakes actively forage on the distal portions of branches on the peripheries of trees and bushes, investigating most thoroughly those branches and leaf surfaces below the branch on which they are supported, presumably because they are less likely to disturb a sleeping anole if they are not actually on the same branch. The snakes move slowly with their bodies sprawled over several branches (sprawl-crawl posture) and almost constantly with frequent tongue flicking on the surfaces of leaves and branches, sometimes stopping and pressing their snouts against branch surfaces. Yorks et al. (2003) described foraging behavior and videotaped two predation events at Westerhall Estate. They designated three SVL size classes of *Co. grenadensis* for purposes of comparison (<700 mm, 700–950 mm, >950 mm) and found that the smallest size class (i.e., that which preys most heavily on sleeping lizards) perched higher in the vegetation, moved the slowest, but spent the most time moving in a sprawl-crawl fashion. Their heads were oriented below the perch that supported their bodies, and they more frequently tongue flicked the branch they were on rather than a branch above or below that which supported their bodies. They found that the largest size class perched lower in the vegetation, moved at a faster rate (cm/min), moved vertically at a faster rate, and spent more time stationary than boas in the smaller size classes. In the first of the two videotaped predation events, the boa (750 mm SVL) was observed initially about 7.0 m from a sleeping *Anolis richardii*. It spent about 30 minutes tongue flicking a tree trunk, vines, a branch, and leaves and pressing its snout against leaf surfaces. By then it was 20 cm away from the anole and tongue flicks stopped in an apparent awareness of the lizard's presence. Another 24 minutes passed before the boa attacked the lizard, after moving at a rate of 0.8 cm/min over the last 20 cm. Twenty minutes passed between anole capture and the start of swallowing; ingestion took 7.0 minutes.

After about 4.0 minutes into the second predation episode (that included over 3.0 minutes of no movement), an 85-cm SVL *Co. grenadensis* moved its head toward a sleeping *A. richardii* on a branch 60 cm away. The boa approached slowly and, as in the first episode, tongue flicks stopped when the boa and anole were 20 cm apart. The boa took 14 minutes to cover the last 20 cm and attack the lizard (rate of movement was 1.4 cm/min). From capture to start of swallowing was 13.5 minutes and ingestion required 5.8 minutes. The prolonged approaches to sleeping lizards may seem counterproductive in that the likelihood of prey escape would increase with time. This possibility, however, is offset by the extreme stealth with which *Co. grenadensis* approaches prey in concert with using branches adjacent to the prey. Albeit based on few observations, this strategy presumably results in a high incidence of successful captures once a sleeping lizard is located.

Thirty boas collected for marking on Carriacou had recently ingested prey items, and in twenty-eight cases Henderson et al. (2021c) were confident they could identify the prey item as either an ectotherm (almost surely an anole) or an endotherm (likely an introduced rodent). They found that twelve of fifteen boas (SVLs 430–819 mm, mean = 635.7 ± 131.3 mm; a boa with an SVL of 865 mm was observed killing and ingesting an anole) identified as having consumed an ectotherm were still actively foraging (i.e., in the sprawl-crawl activity); the other three boas were in a coiled posture. Among boas that were identified as having ingested an endotherm, however, eleven of thirteen (SVLs 598–1206 mm, mean = 907.1 ± 159.5 mm) were found coiled in a tree; one of the other two boas was in sprawl-crawl mode and the other was sprawled but motionless. These observations suggest that boas that have recently consumed an anole are frequently still in foraging mode and searching for additional prey. Boas that have recently ingested a more massive endotherm will usually terminate foraging and sequester themselves in arboreal retreats.

An actively foraging *Co. grenadensis* likely is attempting to locate the scent trail of an anole that has gone to roost on the distal end of a branch (sleeping either on a leaf surface or on the branch itself). If a scent trail is detected, it is followed until either the anole is encountered or until the trail is lost. This strategy probably leads to many unsuccessful trials, as the scent trail of an anole that was merely moving across a branch or leaf hours before going to roost may lead the snake on the proverbial wild goose (or lizard) chase. Of course, foraging treeboas might be able to discriminate the chemical intensity of a trail, thereby discounting many scent trails that were made long before anoles would be seeking roost sites. Short-term observations suggest that the snakes are thorough in their searches for the scent trails, tongue flicking many branch and leaf surfaces. Nevertheless, most trail following probably does not terminate with a feeding episode.

In a sample of 443 observations made at four localities (Henderson 2002), 72.6% of the treeboas were foraging at heights <5.0 m and 40.8% were at heights between 2.0 and 3.9 m. Only 7.2% of the observations were of snakes foraging at heights <1.0 m, 12.2% were higher than 6.9 m, and only 0.7% were at heights >20.0 m. Foraging heights differed between localities, most likely reflecting variations in vegetative structure. Based on another sample of thirty-seven encounters on Union Island (Quinn et al. 2011), perch heights ranged from 0.5 m to 15.0 m (mean = 4.0 ± 2.9 m); most of the boas were young-of-the-year or juveniles. Although large (>1.0 m) treeboas may occur anywhere in the vertical structure at a given site, they frequently occur close to ground level, employing an ambush foraging mode to capture nocturnally active rodents (Fig. 5.128) moving at or near ground level (Fig. 5.129). Henderson (1993), in a sample of thirty-three observations of both West Indian taxa (then both under the taxonomic umbrella of *Co. enydris*), found that the mean SVL of treeboas foraging <1.0 m above the ground (940 ± 95 mm) was significantly greater than that of boas foraging between 1.0 and 2.9 m (662 ± 51 mm; $P < 0.01$, *t*-test). At the Mt. Hartman site, mean SVL for treeboas foraging at <1.0 m was 815.7 mm (range = 375–1332 mm SVL), at 1.0–1.9 m was 603.7 mm (range = 389–1150 mm SVL), at 2.0–2.9 m was 582 mm (350–1150 mm SVL), and at 3.0–3.9 m was 680 mm (range = 400–1130 mm SVL). Similarly, Yorks et al. (2003) found that the largest size class of boas (i.e., that most likely to prey on mammals) perched lower in the vegetation. Recent fieldwork on Carriacou, however, showed no significant differences in perch heights. Based on 448 observations of three size classes (<700 mm SVL, $n = 171$; 700–900 mm SVL, $n = 162$; and >900 mm SVL, $n = 115$), there were no significant differences in foraging heights (ANOVA, $P = 0.529$). Although foraging heights ranged from 0.2 m to 8.0 m, the mean height for the three size classes was 2.3 ± 1.5 m, 2.3 ± 1.4 m, and 2.5 ± 1.6 m, respectively (RWH and collaborators, unpubl. data).

*Impact of Hurricane Ivan*

On 7 September 2004, Hurricane Ivan, a Category-4 storm with winds in excess of 225 km/h made a direct hit on Grenada. Estimates indicated that more than 90% of the homes and buildings on the island sustained some damage and many were completely destroyed. Henderson and Berg (2005) arrived in Grenada on 10 November 2004 and visited several study sites. On 12 November at Mt. William (St. David Parish; 400 m asl), treeboas were observed foraging in the crowns of fallen trees, and others were moving through leafless trees and in low shrubs above a roadcut. Along a 600-m transect, seven *Co. grenadensis* were encountered in 65 minutes, including young-of-the-year and large adults. On 14 November at the same site, four boas were encountered in 52 minutes. Hurricane Ivan essentially destroyed nutmeg production at Mt. William, and what orchard trees and forest that remained were subsequently cleared for other agriculture. Fieldwork at Pearls was essentially abandoned as trails became impassable, although a sharp decline in *Co. grenadensis* numbers had been witnessed prior

**Figure 5.128.** *Corallus grenadensis* that has captured a rat (*Rattus*; arrow) on Carriacou. Photo by Robert W. Henderson.

to Hurricane Ivan (Henderson et al. 2009). At Westerhall, orchard trees and native vegetation were largely leveled, and only a single treeboa was encountered in a brief (30 minute) search. Hurricanes are a recurring phenomenon in the West Indies, and over many thousands of years, *Co. cookii* and *Co. grenadensis* have had to "weather the storm." Certainly, treeboas are killed during violent weather and their habitats severely altered, but that a hurricane could completely eradicate a *Corallus* population is unlikely.

Westerhall update: post–Hurricane Ivan fieldwork at Westerhall (2010–2011) emphasized the importance of the composition and structure of plant communities in treeboa habitat. Seventeen years had passed since a previous project at Westerhall (1993) addressed habitat use (Henderson and Winstel 1995; Henderson et al. 1998). On 19 February and 6 June 2002, fourteen and fifteen boas, respectively, were encountered. The site was not visited again until post-Hurricane Ivan. The hurricane leveled many trees and what was once considered ideal *Co. grenadensis* habitat was transformed into submarginal habitat with little contiguous crown vegetation along the transect worked in 1993. By 2010, however, the transect was again wooded along its entire length, but the composition and structure had changed dramatically. Fifty 10-m transect sections (out of a possible 122) harbored mango trees in 1993, yet only four did so in 2010. Additionally, twenty-nine sections had evidence of sugarcane cultivation in 1993 compared with none in 2010, and thirty-one sections had breadfruit trees in 1993 but only one in 2010. In 1993, only thirty-three sections were 100% uncultivated, whereas all 122 sections were uncultivated in 2010. Although anoles were still common throughout

**Figure 5.129.** *Left*: *Corallus grenadensis* in an ambush posture at Grand Bras, St. Andrew Parish, Grenada. *Right*: A male *Co. grenadensis* exiting a tree hollow at Grand Bras, St. Andrew Parish, Grenada, in order to pursue a female for possible courtship. Photos by Robert W. Henderson.

the transect in 2010 and rodents were frequently observed, the mean number of treeboa observations per night was 1.6 ± 0.2 (range = 0–3) compared with 9.5 ± 0.7 (5–15) in 1993. Subsequently, in the course of thirteen visits to the site between November 2012 and November 2018, there has been an encouraging increase in abundance: the mean number of boas during those visits was 9.2 ± 1.8 (range 5–12). We are unable to attribute the increase in numbers to anything in particular confidently, although time and plant succession are at the forefront. Additionally, human activity at the site appears to have increased with some agricultural activity evident.

*Movements*

Records for movements by *Co. grenadensis* are available from three sites. At Mt. Hartman Bay (St. George Parish, Grenada) a subadult male originally captured on 8 October 1996 was recaptured on 2 Feb 1998 (482 days elapsed) 10–20 m from the original site and in the same stand of *Vachellia nilotica*. An adult female originally captured on 4 December 1995 was recaptured on 17 April 1996 (134 days elapsed) along the same trail and 140–150 m from the original site. That same female was recaptured again on 30 September 1996 (166 days elapsed), again along the same trail and in the same stand of *Rhizophora mangle*, 50–60 m from the April recapture site. She was within 100 m of where originally captured on 4 December 1995 (Henderson 2002).

Thirteen movement records at the Pearls site (St.

Andrew Parish, Grenada) were made between 2002 and 2004. The mean number of days between recaptures was 211.7 ± 28.7 days, range 111–349 days; distances moved ranged from 17 to 64 m (mean = 29.9 ± 4.4 m); snakes ranged in size from 670 mm to 1228 mm SVL (mean = 877.0 ± 49.0 mm). The longest distance traveled (adult male) between original capture and last recapture was 102 m over a span of 379 days; the shortest distance recorded between recaptures for a long-term movement was 17 m over 120 elapsed days for an adult female. Of course, both of these snakes could have moved hundreds of meters in the interim between recaptures.

Based on 197 movement records (77–1100 elapsed days) on Carriacou, mean distance moved by boas <700 mm SVL was 27.2 ± 46.5 m (range = 0.0–338 m; *n* = 53); those boas 700–900 mm SVL moved 39.6 ± 63.3 m (2.6–468 m; *n* = 89); and those >900 mm SVL moved 48.5 ± 51.6 m (2.5–243 m; *n* = 55); there was no relationship between distance moved and elapsed time (days) between recaptures (RWH and collaborators, unpubl. data). The smallest size class is composed of anole predators, the middle size class is boas in transition from anoles to rodents, and the largest size class feeds almost exclusively on rodent prey. As anole densities can be very high, small boas can stay in relatively small areas to actively forage for prey. Some records of movement showed boas returning to GPS points that were within 0–4 m of the previous capture (mean = 2.2 ± 0.3 m; *n* = 10), even after, in some cases, hundreds of days between captures (range = 78–480 days; mean = 216.4 ± 39.7 days); the mean SVL of those snakes was 701.6 ± 47.4 mm (range = 596–1081 mm). The mean of thirty-one movements between 5 and 10 m was 7.5 ± 0.3 (range = 5.2–9.9 m) over 81–475 elapsed days (mean = 162.7 ± 17.7 days); the mean SVL of the boas that made those movements was 787.3 ± 30.6 mm (range = 488–1136 mm) (Henderson et al. 2021a; RWH and collaborators, unpubl. data). With few exceptions, these boas were in the size class that fed almost exclusively on anoles. Although these data might suggest *Co. grenadensis* is relatively sedentary, many marked boas were not recaptured, perhaps indicating they were transients and had moved out of the study area. It is possible, also, that radio telemetry would present a very different picture of movements in *Co. grenadensis*.

*Predators and Defensive Behavior*

African mona monkeys (*Cercopithecus mona*; Cercopithecidae) were introduced to Grenada during the slave trade in the late-seventeenth and eighteenth centuries (Glenn 1998). These monkeys have been observed molesting *Co. grenadensis* at Grand Etang National Park (Henderson 2002), where treeboas are uncommon and frequent treeboa–monkey encounters are unlikely. Henderson et al. (1996) reported a small *Co. grenadensis* found in the stomach of an opossum (*Didelphis marsupialis*) on Grenada. Didelphid opossums are notorious snake predators (including venomous species; Voss and Jansa 2012), and *D. marsupialis* is, like *Co. grenadensis*, arboreal and nocturnal. Certainly, they encounter each other with regularity, and predation on young *Co. grenadensis* probably is a common occurrence. We have observed up to eighteen *D. marsupialis* along a 1.0-km stretch of road on Carriacou. Isidore Monah (pers. comm., 2 February 2014) witnessed predation on a treeboa by a Broad-winged Hawk (*Buteo platypterus*) on Grenada. Likely, such events also are not rare.

Like *Co. cookii*, *Co. grenadensis* has a relatively wide array of defensive behaviors. Retreat is the first avenue of defense, but if physical contact is made, striking and biting ensue (even boas <500 mm SVL can deliver a surprisingly effective bite). Voiding of the cloaca is frequently employed if the boa is either partially (i.e., still on a tree branch) or completely (i.e., captured) restrained. Once in hand, balling behavior is not uncommon behavior in boas of all sizes. Among West Indian boas, rapid rotation of the body apparently appears only in *Corallus*; it is a startling behavior when it occurs.

*Demography*

Henderson et al. (2021a) provided a demographic analysis of *Corallus grenadensis*. From 2015 to 2019, 254 boas were marked along a 1 km stretch of road on Carriacou. One hundred and twenty-seven of the 242 boas (52.4%) that had the potential to be recaptured were recaptured in the ~4.48-ha study site. Capture probabilities increased with sampling effort and decreased with increasing SVL. Estimates of abundance ranged from 96 to 141 individuals, but confidence intervals overlapped among years. Although the authors did not determine population densities, based on the size of their study site and the estimates of abundance, densities ranged from 21.4 to 31.5 *Co. grenadensis*/ha, with the two most precise estimates being 25.0 and 27.9/ha. Annual survival of residents was 0.71, 95% confidence interval = 0.54–0.83. The proportion of transients (i.e., boas captured and marked but never recaptured) increased with increasing SVL, with the estimate being distinguishable from zero starting at ~810 mm SVL, coinciding with the size at which a shift from ectothermic to endothermic prey begins (Fig. 5.130). Although Henderson et al.'s (2021a) study site was capable of supporting a substantial resident population of *Co. grenadensis* that relied on *Anolis* lizards as prey, it supported a much smaller resident population of larger boas that subsisted largely on introduced rodents. Most large boas that were encountered in the site were transients and quickly (or eventually) abandoned it (i.e., permanent emigration), ostensibly to search for more trophically promising habitat that would maximize foraging efficiency and energy intake. Ontogenetic dietary shifts are widespread among snakes (Shine and Wall 2007), and at least eleven of eighteen species of West Indian boas exhibit an ontogenetic dietary shift, either from ectotherms (mostly *Anolis*) to introduced rodents or from one type of lizard (e.g., diminutive *Sphaerodactylus* geckos) to larger species of lizards (*Anolis, Leiocephalus, Cyclura, Iguana*). The prominence of transients in the study site may be indicative of their demographic and ecological importance among other snake species.

*Reproduction*

Relatively little is known about social behavior in species of *Corallus*, but some evidence suggests that *Co. grenadensis* engages in aggregating behavior. Over several nights at the Westerhall site, eight observations were made of three or more boas in the same tree in close proximity (or sometimes in contact); in at least two instances, one of the boas was a male.

On 9 June 2015 (moon in its third quarter, wet season with rain earlier in the day) at a site in central Grenada, we observed what we consider courtship in *Co. grenadensis*. At 2100 hours, we observed a large (1470 mm SVL on 23 January 2015) female resting on the branch of a *Senna alata* tree and close to the top rail of a chain-link fence bordering a cemetery. At 2103 hours, an adult male (~1250 mm SVL) crawled rapidly into a hollow in the trunk of an allspice tree (*Pimenta*

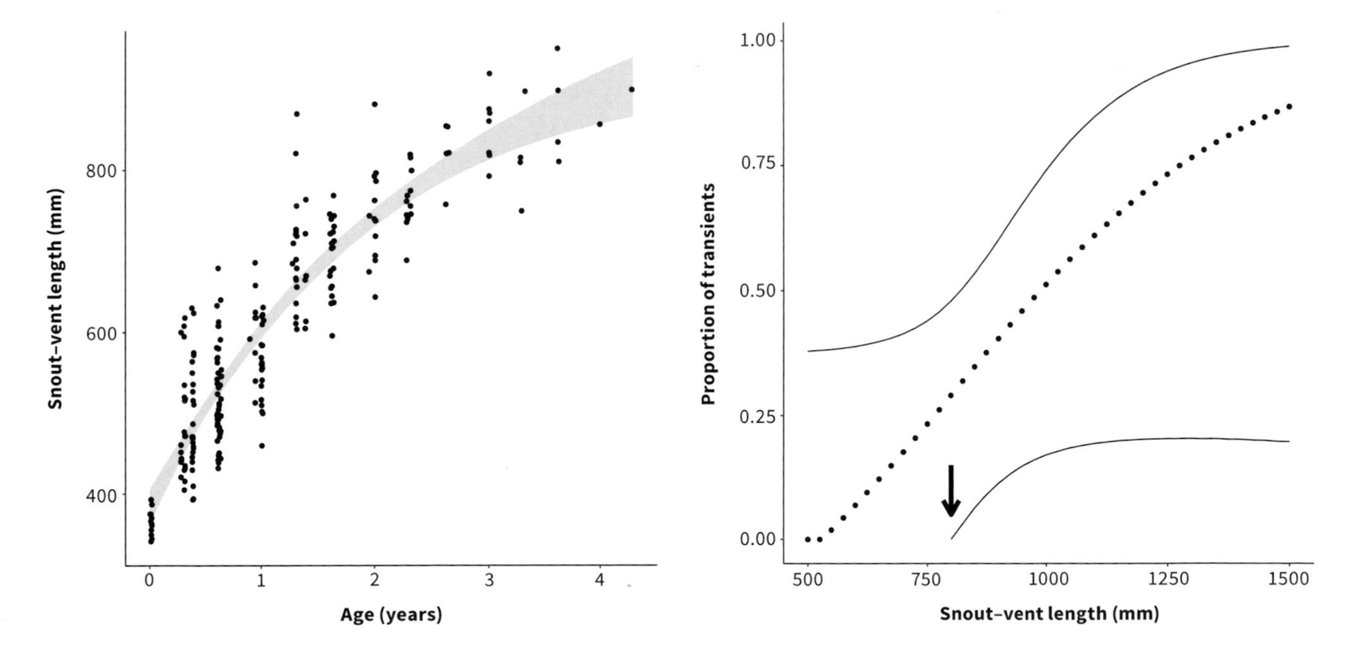


**Figure 5.130.** *Left*: Estimated growth trajectory for *Corallus grenadensis* on Carriacou, Grenada; the shaded area is the 95% confidence interval. *Right*: Larger *C. grenadensis* are more likely to be transient, as shown in these data from Carriacou, Grenada. After Henderson et al. (2021a).

**Figure 5.131.** *Corallus grenadensis* (*top*) foraging in an out building at Grand Bras (St. Andrew Parish, Grenada) and (*bottom*) on a chain-link fence at Les Avocats, St. David Parish, Grenada. Photos by R. Sajdak (top) and Robert W. Henderson (bottom).

*dioica*). At 2105 hours, a smaller adult (unsexed) was coiled on the upper rail of the chain-link fence and about 7 m from the female. At about 2107 hours, the male emerged from the tree hollow (Fig. 5.129) and moved down to the chain-link fence and the proximity of the female. With very slow and intermittent advances, the male approached the female and by 2120 hours they were in contact. The male displayed frequent tongue flicks to the female's body, but at 2150 hours their heads were facing away from one another and their tails were not close. At 2215 hours, the male had his tail wrapped around the female's mid-section and their heads were in virtual contact. At 2221 hours, we were forced to terminate observations and quickly sexed the two snakes. At 2225 hours, the male had retreated into the hollow of the allspice tree, but the female remained on the fence. These behaviors presumably are portions of a courtship sequence (Henderson et al. 2015), with the frequent tongue flicks, dorsal body looping, dorsal advance, and possible chin rubbing serving as elements of the tactile-chase, described by Gillingham (1987), Tolson (1992b), and Tolson and Henderson (1993) as the initial phase of courtship. We suspect that the male had been making advances toward the female when we arrived and saw him retreat into the tree hollow. We also suspect that he resumed his advances after we departed.

Derek Yorks (in Henderson et al. 2009) had been watching an apparently pregnant female for several days and found newborn boas on 25 August at Westerhall Estate. The neonates ($n$ = 6) had SVLs of 295–310 mm (mean = 305.8 ± 2.4 mm), and all were in close proximity to one another. Another group of neonates ($n$ = 11) collected at La Sagesse (St. David Parish) on 1 September ranged in SVL from 320 to 355 mm (mean = 333.6 ± 3.0), and a third group ($n$ = 8), also from Westerhall but not found in close proximity to one another, ranged in size from 300 to 380 mm SVL (mean = 331.3 ± 10.6 mm). The mean SVL of all twenty-five neonates is 326.2 ± 4.23 mm. Similarly, a young snake enthusiast on Grenada captured a pregnant *Co. grenadensis*, and she produced twenty-three young in August. On Canouan, M. de Silva (pers. comm., 9 June 2012) encountered a fresh litter of *Co. grenadensis* on the ground (Henderson 2015). Neonates have been encountered in the Grenadines in mid-July but not in June. It is likely births occur from July–October, possibly depending on island, and on Grenada, elevation likely has some role in determining when parturition occurs.

Little is known concerning litter size. A 1587 mm SVL snake had twenty-nine undeveloped eggs; a boa 1400 mm SVL had seventeen nearly full-term young. The young snake enthusiast mentioned above provided the best data with the litter of twenty-three. Based on captive animals, litters were produced on 30 June and 31 July, and each litter was composed of eight live babies and two stillborn. Mass of the live neonates ranged from 6.8 g to 8.6 g (J. Murray, in litt. 30 April 2020). On Carriacou, we have encountered babies with masses as small as 5 g (SVL 312 mm) and 6 g (SVL 337 mm) in October (RWH and collaborators, unpubl. data).

*Living with Humans*

Because *Corallus grenadensis* is active when most humans are not, and because it lives in trees, it has, for the most part, successfully coexisted with humans. Almost from day one of working in Grenada, we were told that humans and treeboas lived in close association with one another, and our observations supported that. In October 1989, in the True Blue section of St. George's, we collected *Co. grenadensis* about 4.0 m from a bright street light at 0400 hours. That same year and again in November 2007, boas were observed in trees bordering the parking lot of a luxury hotel in Grand Anse. In April 2000 at Pearls, a large treeboa was in a tree next to the only edifice within several hundred meters, and a large male was within 2.0 m of the same house after it was abandoned. In June 2002, a foraging boa was observed within 10 m of a house in a Beausejour development.

More recently, at a home in St. Andrew Parish in November 2012 and during several visits in 2013, we observed boas crawling on vines covering a small chicken coop, in the wire mesh of the coop and, in an outbuilding (Fig. 5.131), both about 20 m from the property owner's house. Others were seen within 1.5 m of a large chicken coop and foraging in a cacao tree above a concrete outbuilding. This property was adjacent to a cemetery, and *Co. grenadensis* was common in trees (mostly mango) surrounding the cemetery. In June 2002, at St. Rose Nursery (St. George Parish), an adult *Co. grenadensis* was observed as it slept/rested in foliage adjacent to a greenhouse. Workers at the nursery told us it was a long-time resident that shuttled between shade and sunlight during the course of a day, and it remained unmolested by the nursery staff.

This species was observed using power lines in Birch Grove (St. Andrew Parish) to get from one side of the street to the other (and from one group of trees to another); that practice sometimes got the snakes electrocuted. At the waterworks at Les Avocats, boas have been observed foraging in a chain link fence sometime used by rats to travel above ground level (Fig. 5.131). Birch Grove residents also told us that treeboas often entered the roofs of houses and, if the house were very warm, would enter the house itself. Groome (1970) noted that *Co. grenadensis* often enters "the roofs of houses and keeps them free of rodents in exchange for an occasional bath—a habit which should not be discouraged."

On Princess Margaret Beach, a popular tourist beach on Bequia (St. Vincent Grenadines), *Co. grenadensis* was found coiled in a tree at the edge of the beach "almost directly above [one of the] beach chairs" (Bennett 2013). On Mayreau, M. de Silva (pers. comm. to RWH, June 2013) encountered this species in human-made shelters, including outhouses. *Corallus grenadensis* is apparently rare on Petite Martinique.

One of two that we (Henderson and Berg 2012) encountered was a young-of-the-year observed 2.0 m high in a tree along a paved road in the Kendace (or Kendeace) area, 36 m from a streetlight and across the street from a house where people were sitting on a porch. On Carriacou in September 2013, we observed a 654 mm SVL treeboa foraging in a tree across the road from a large house; another was partially illuminated as it foraged 27 m from a streetlight. Similarly, it was not unusual to find treeboas in trees around the Union Island salt pond and very close to homes on the opposite side of the road (RWH, pers. obs.).

Not all human–treeboa encounters end well for the snake. Despite successfully living in proximity to human activity, and despite many human inhabitants on the Grenada Bank being aware that *Co. grenadensis* poses no threat to human safety, these boas are encountered and killed every day. Some are killed in the erroneous belief that they are dangerous, but others are killed only because they are snakes.

On J'ouvert morning during Grenada's annual carnival, masqueraders (jab jabs) carry live *Co. grenadensis* in order to frighten carnival-goers. These boas have likely been deprived of food for prolonged periods with either their mouths sewn shut or their formidable teeth removed. In either case, it is ultimately fatal for the boa. As noted by a forestry officer, "The jab jabs once again brought the snakes out of the forests and used them to enhance their acts . . . and I am concerned . . . because these snakes will not be brought back to the forest but will be left to die on the side of the road in the hot sun" (Straker 2008). Although it is illegal to use the boas during carnival, the police do not enforce the law. According to the assistant commissioner of police "there are times when certain things are allowed to happen not necessarily that there is no law to deal with it, but because there are elements that have to be taken into consideration as well" (Campbell 2018). In other words, there are times when it is okay to overlook animal abuse.

*Conservation*

Although restricted to the Grenada Bank, this species falls under the jurisdiction of two countries, as most of the Grenadine Islands are governed by St. Vincent. That this species has a multi-island distribution is certainly a plus for its future. Like *Co. cookii*, where *Co. grenadensis* occurs, human activity is likely to be nearby. Although this species, like any tree- and shrub-dwelling species, can be affected by significant reductions in arboreal habitat, *Co. grenadensis* uses habitats and includes in its diet prey species that are the direct result of human activities. It can be found in a wide variety of habitats over many of the islands and it should be safe during the foreseeable future with minimal management—if commercial exploitation and hunting for use in traditional medicine or cuisine by any foreign presence on the island can be avoided. This species has received an IUCN Red List assessment of Least Concern (Henderson and Powell 2021b). Taking into consideration its area of occupancy (<400 km$^2$), its fragmented distribution, and the decline in habitat quality that has been well documented, a Near Threatened assessment seems more apropos. Like all boids that are not listed on CITES Appendix I, this species is listed on CITES Appendix II.

Although *Co. grenadensis* has apparently been exploited commercially, at least occasionally for at least 50 years (Progscha 1972), it has, seemingly, become more prevalent in recent years. Based on data from CITES and reported in Outhwaite (2019), 106 *Co. grenadensis* were exported between 2013 and 2017, and they (along with *Chilabothrus angulifer*) accounted for the majority of reptiles in the trade in endemic Caribbean reptiles. Nearly all of the *Co. grenadensis* were exported either by the Netherlands or Dutch Sint Maarten and were described as captive bred (CITES data, courtesy of W. Outhwaite, in litt. 14 April 2020). We assume boas were smuggled off their native islands (most likely one or more of the St. Vincent Grenadines) and then shipped off to Europe or the United States to be sold in the pet trade. Based on internet surfs, we are aware of live individuals being illegally taken from Grenada and from St. Vincent and the Grenadines. In 2017, live individuals of this species ostensibly collected on Mayreau and on Union Island were offered for sale on the internet from a U.S.-based site; in 2018, live individuals from Union were advertised on the internet (from the same site) with an asking price of US$750 for one or $1400 for a pair. A post on Facebook, 2 August 2020, announced the birth of twelve *Co. grenadensis* with the information that the parents originated from "Union Isle" (= Union Island, St. Vincent Grenadines). Union is the site of the type lo-

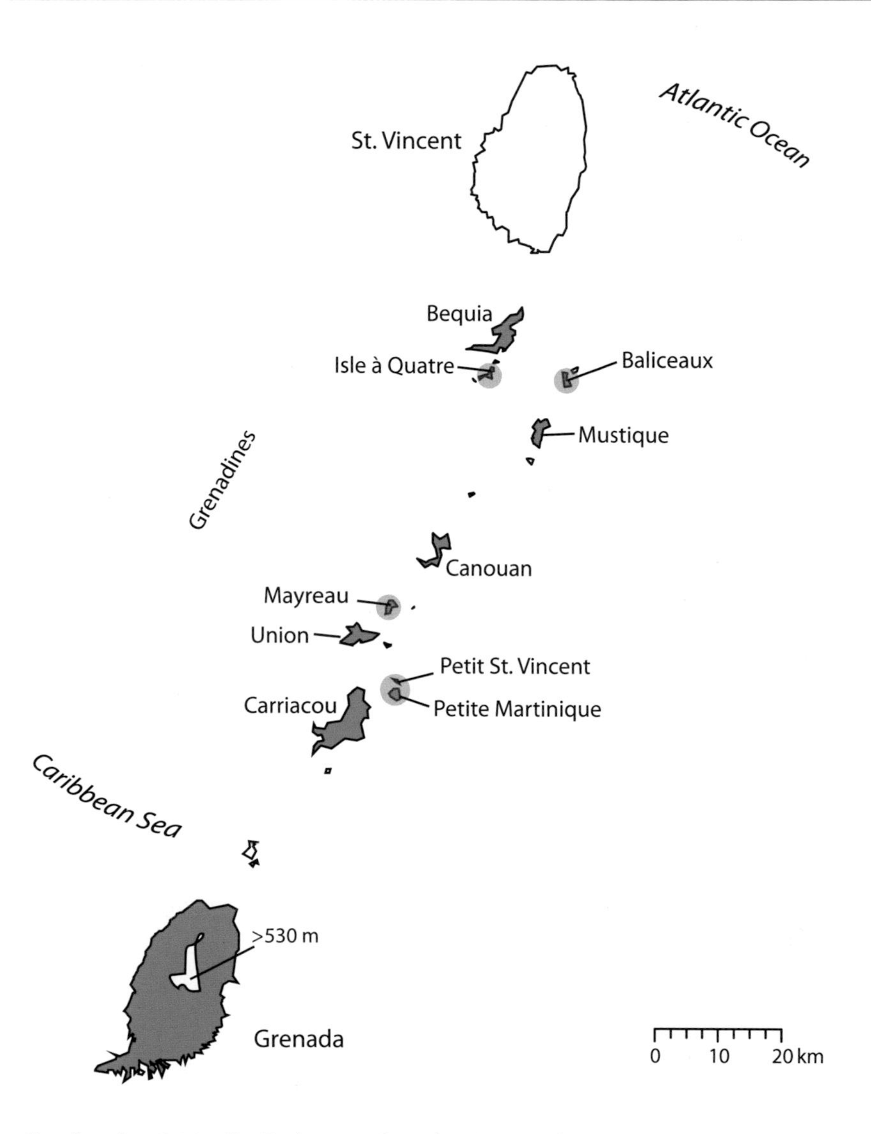


**Figure 5.132**. Range of *Corallus grenadensis* in the Grenadines and on Grenada up to an elevation of 530 m. The species is far more common below 150 m.

cality for the Critically Endangered gecko *Gonatodes daudini*; its wildlife is closely guarded, and the legal collection and exportation of any reptile species is highly unlikely. On the German website for Terraristik, Union Island *Co. grenadensis* were advertised for sale in February 2018, and individuals of the St. Vincent endemic *Co. cookii* were advertised for sale on the internet as recently as November 2018. Additionally, five *Co. grenadensis* (no locality given) were offered for sale in September 2016 at €600 (= US$680) each. Baby *Co. grenadensis* were purchased for US$250 each from importers in the United States and elsewhere. Apparently, no CITES permits have been issued by the Forestry Department in St. Vincent and the Grenadines (I. Vique, in litt. 27 April 2020). This echoes information from G. Gaymes of the same department on St. Vincent (in litt. 15 January 2019); that is, they have never issued permits (including CITES) for the exportation of species of *Corallus*. Figure 5.132 shows the distribution of *Corallus grenadensis*.

## EXTINCT SPECIES OF WEST INDIAN BOAS

Recently, in-depth analyses of fossil boa material have revealed some exciting finds: specifically, that three extinct species of boa from the greater Caribbean region are now recognized, although not formally described. These fossils expand our understanding of the historical biogeography of boas in the region, and they also corroborate some previous biogeographic and evolutionary hypotheses regarding West Indian boas. Below are two species, both described in 2018, that represent one extinct member of the genus *Boa* in the Lesser Antilles and one extinct species of *Chilabothrus* from peninsular Florida. Both are significant in that they greatly expand the known range of these genera in the region. *Boa blanchardensis* extends the historical range of boa northward towards Guadeloupe (Marie-Galante Island; also see Introduction) while *Chilabothrus stanolseni* represents a lineage that likely colonized the mainland United States from the proto-Antilles during the Miocene. A candidate species of *Boa* is suggested from material from Guadeloupe (Basse Terre and La Désirade islands); we include it here as *Boa* sp. Other fossil material from Antigua likely represents another *Boa*, although the material found thus far does not reliably allow it to be assigned to a species, and we treat it below as the Antigua boa (*Boa* sp.). A fifth potential species might have occurred on Martinique. We include these last two herein (also as *Boa* sp.) even though the fossil material, which is thought to exist, has not yet been published (see Lescure et al. 2020). The exciting possibility of upcoming discoveries and analyses will likely continue to expand our understanding of historical range of boas in the region in the near future. Figure 5.133 shows the distributions of extinct boas in the Caribbean.

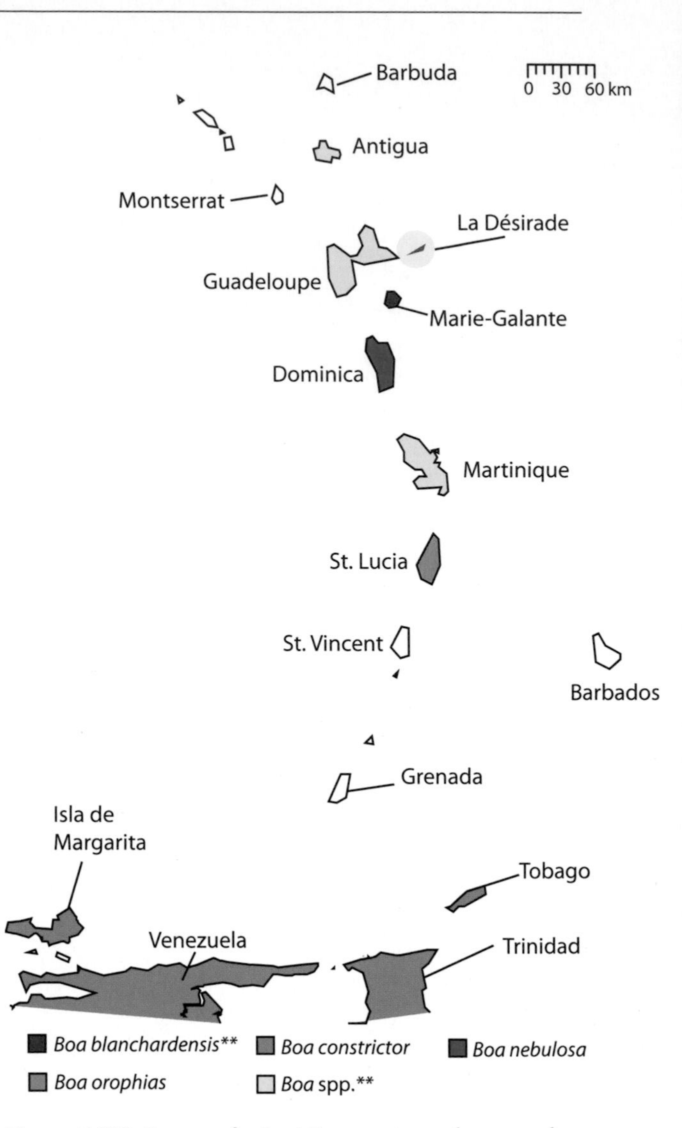


**Figure 5.133**. Ranges of extant *Boa* species and proposed ranges of extinct species of *Boa* in the Lesser Antilles. A double asterisk indicates extinct species.

## MARIE-GALANTE BOA

*Boa blanchardensis* (Bochaton and Bailon 2018)

*Taxonomy*

The Pleistocene subfossil material recovered from Marie-Galante forming the basis for the description of this species had not been previously assigned to a species, although it had been diagnosed as belonging to the genus *Boa* (Bochaton et al. 2015). The species was described based on a reevaluation of additional subfossil material from three different deposits (Bochaton and Bailon 2018; Bochaton et al. 2021).

*Etymology*

This species is named for Blanchard Cave on Marie-Galante Island, where the fossil material was found. The French name is "boa de Marie-Galante."

*Description*

Subfossil material including vertebrae and skull fragments representing the Marie-Galante boa has been found in at least three different deposits on the island of Marie-Galante (Bochaton et al. 2021), including Grotte Blanchard ($n$ = 254 bone fragments), Grotte Cadet 2 ($n$ = 16), and Abri Cadet 3 ($n$ = 4). This material is from the late Pleistocene (34,000–15,000 years BP; Stouvenot et al. 2014; Bailon et al. 2015; Bochaton et al. 2015, 2021), and no evidence presently suggests that the species survived into the Holocene (Bochaton and Bailon 2018). Reconstruction of body sizes suggests that this was a small species, with a total length of 73–139 cm (this is represented as "73–139 mm" in Bochataon and Bailon 2018; we consider that a likely typographical error). This species is diagnosable from *Boa constrictor* and *B. nebulosa* based on at least six autapomorphies (Bochaton and Bailon 2018).

*Distribution*

*Boa blanchardensis* is known from only Marie-Galante Island in the French West Indies. Additional material from Guadeloupe (see below) could be conspecific but comes from later Holocene deposits and is not detailed enough to ascribe to *B. blanchardensis*.

## GUADELOUPE BOA

*Boa* sp. Bochaton et al. 2021

*Taxonomy*

This species has not been officially described but is recognized as *Boa* sp. by Bochaton et al. (2021). This designation is based on two pieces of archaeological evidence, that of Holocene pre-Columbian ornamental "beads" fashioned from *Boa* vertebrae recovered from Basse-Terre and La Désirade islands (Bochaton 2020; Bochaton et al. 2021).

*Description*

This boa is described from Holocene materials that are not likely to yield much descriptive information; nevertheless, if additional material is recovered in the future this might provide more insight. Two archeological pieces, presumably beads fashioned from vertebrae, are mentioned in Bochaton (2020) and Bochaton et al. (2021), one from Cathédrale de Basse-Terre (Basse-Terre Island) and one from Pointe Gros Rempart 6 (La Désirade Island). Recently described Holocene (pre-Columbian) material from an owl deposit ~550–350 BP on La Désirade (Bochaton 2020) might also represent this taxon, although the author does not describe the material to species because of the degradation of key diagnostic features.

*Distribution*

Archeological evidence suggests that a *Boa* sp. might have been present during the pre-Columbian Holocene on Basse-Terre and La Désirade islands (Bochaton 2020; Bochaton et al. 2021). Bochaton et al. (2021) suggest that this indicates that a boa was present across the Guadeloupe islands, which would have been connected during periods of lower sea level. Nevertheless, because some of the records are from beads fashioned from vertebrae, that suggests that they were presumably decorative or of cultural significance, and hence it is unclear whether they originated on the islands on which they were found (but also see Bochaton 2020). Also see a discussion of historical biogeography in the Introduction.

## MARTINIQUE BOA

*Boa* sp. Lescure et al. 2020

*Taxonomy*

This species has not been officially described but is recognized as *Boa* sp. 2 by Lescure et al. (2020). A boa was described from Martinique by l'Anonyme (Anonymous) de Carpentras (1618–1620) in a work by Pere Labat (1724), the latter of which has provided support for the assertion by Benito-Espinal (1978, 1990) that boas occurred on Martinique prior to the introduction of the mongoose between 1880 and 1850 (Breuil 2002). Numerous authors (Lescure 2001; Lorvelec et al. 2007; Breuil 2009) also included *Boa* on species lists from Martinique. Dewynter et al. (2019) have also included *Boa* sp. on their species list for Martinique and suggested that one of the authors (C. Bochaton) is working to describe the fossil material that will demonstrate the historical presence of *Boa* on the island.

*Description*

The original description was traced to the locality of Anse Couleuvre on the northwest coast (Dewynter et al. 2019). No fossil material is yet formally described, although Dewynter et al. (2019) state that this is a study in progress ("étude en cours") by C. Bochaton.

*Distribution*

Unknown but presumably would have occurred islandwide up to 350 m in elevation if the species was similar to *B. orophias* and *B. nebulosa*. There are currently no *Boa* on Martinique, and none has been reported since at least the nineteenth century. Pre-Columbian beads fashioned from *Boa* vertebrae have been reported from Dizac Beach in southern Martinique (Bochaton 2020), which could suggest either the historical presence of the species or arrival via trade.

## ANTIGUA BOA

*Boa* sp. Lescure et al. 2020

*Taxonomy*

Steadman et al. (1984) provide an assertion that fossil material from Antigua represents the species *Boa constrictor*. There is no description of these fossils in that publication, other than that they were found at the Indian Creek Site, which is a ceramic culture archeological site on the southern coast of Antigua dating to between 1915 and 845 years BP (Table 1.1). Pregill et al. (1988) describe another subfossil, this a single vertebra, from the Burma Quarry site on the north coast of Antigua dating to 4300–2560 years BP (Table 1.1). These authors conclude that the specimen is too worn to determine which species it might belong to but placed it into the genus *Boa* owing to a lack of diagnostic characters that would represent other possible lineages of snakes in the region (*Chilabothrus*, *Tropidophis*, and West Indian xenodontines). Lescure et al. (2020) recognize this material as representing *Boa* sp. 1.

*Description*

No description is offered of the material from Indian Creek, while the single vertebra from Burma Quarry is sufficiently worn as likely to preclude assignment to species. Nevertheless, it lacks characters typical of other snakes from the region and most closely resembles the genus *Boa* (Pregill et al. 1988).

*Distribution*

Subfossil material is known from only two archeological sites on Antigua—Burma Quarry, near the modern-day international airport, and Indian Creek, nearly due south near the coast (see illustration of sites in Steadman et al. 1984).

## FLORIDA BOA

*Chilabothrus stanolseni* (Vanzolini 1952)

### *Taxonomy*

Fossils representing this species were originally described as two species of the genus *Neurodromicus* Cope 1873: *N. stanolseni* and *N. barbouri* (Vanzolini 1952). Auffenberg (1963) merged fossils from these two species into a new genus *Pseudoepicrates stanolseni*. Kluge (1988b) carved the original *N. barbouri* material back out and placed it into the genus *Boa*. Albino (2011) then assigned all of the original material to the extant species *Boa constrictor*. Onary and Hsiou (2018) carefully reevaluated the fossil material and concluded that it represented an extinct species of *Chilabothrus*.

### *Etymology*

The name *stanolseni* was given by Vanzolini to honor S. Olsen, the individual who "assembled the material studied" (Vanzolini 1952).

### *Description*

This is one of the few known fossils of the genus *Chilabothrus* (although some subfossil material exists, see Introduction) and the only extinct species for which fossil or subfossil material is available. A number of autapomorphic features of the vertebrae distinguish this species from extant boa taxa (Onary and Hsiou 2018). The presence of a "weakly developed median lobe" on the zygosphene of the vertebrae is a diagnostic character that distinguishes this species from extant *Chilabothrus* (Onary and Hsiou 2018).

### *Distribution*

This species is known from a single fossil deposit—the Thomas Farm site located in Gilchrist County, Florida, near the city of Gainesville. This site dates to 18.5 MY, and thus the species is near the root of the insular West Indian *Chilabothrus*, which is thought to have evolved around 22 MY (Reynolds et al. 2013a). Onary and Hsiou (2018), based on an interpretation of Reynolds et al. (2013a), argued that this species plausibly colonized Florida while other members of the genus were colonizing the proto-Antilles—as the age of the species is younger than the initial inferred colonization of the West Indies by this genus. Also see a discussion of historical biogeography in the Introduction.

*Chilabothrus gracilis* from Higüey, Domincan Republic. Photo by Luis M. Díaz.

## *Chapter 6*

# A BRIEF HISTORY OF THE STUDY OF WEST INDIAN BOAS

Four hundred years before it was recognized as a unique taxon (Lazell 1964), and nearly two hundred years before Linnaeus (1758) described its closely related congener, *Boa orophias*, from St. Lucia, a French explorer had a run-in with *B. nebulosa* on Dominica. In 1564, René Laudonniere and his men, out for an excursion, "spied two serpents of exceeding bignes. . . . My soldiers went before them thinking to let them from going into the woods; but the serpents nothing at all astonied at these gestures glanced into the bushes with fearful hissings; yet for all that, my men drew their swords and killed them, and found them afterward 9 great foote long, and as big as a mans leg" (Hakluyt 1904). On Dominica in 1876, 312 years after Laudonniere and his men had their *B. nebulosa* encounter, Frederick Ober had his. "It was twelve feet in length and looked capable of crushing a sheep to death—as indeed I was told it could. . . . It is a terror to the negroes and Indians, who fear contact with its slimy skin. . . . Fortunately, although rather abundant in the forests, they do not willfully attack man, and seldom do harm more than to pay occasional visits to the hen-roosts of sequestered settlements" (Ober 1899). In the late 1950s, 83 years after Ober's observations, James "Skip" Lazell (1964) found *B. nebulosa* to be "amazingly abundant. . . . The people of Dominica, unlike most in Saint Lucia, have no *Bothrops* to fear and thus enjoy being deathly afraid of [*B. nebulosa*]; sometimes they even kill them. . . . Children, once initiated, are not at all afraid of even the biggest ones, however, and their parents are often shamed into changing their minds about the danger . . . simply because the children come to regard them as play-toys." About 50–60 years later, Malhotra et al. (2011) considered *B. nebulosa* still "fairly common," although "larger individuals are becoming less commonly seen." Similarly, J. Brisbane (in litt.to RWH, 14 August 2018), a biologist and life-long resident of Dominica, has observed a decline in large boas and snakes in general.

Almost certainly similar scenarios occurred on virtually every West Indian island on which humans and boas either currently or at one time co-occurred. However, long before Columbus set foot on Dominica in 1493 (or on any other of his landfalls), Taíno, Lucayan, Arawak, Carib, or members of other Amerindian cultures likely had frequent encounters with boas on any one of dozens of Antillean islands, and, we suspect, those boas were routinely killed. Indeed, in the early sixteenth century (1535) Captain Gonzalo Fernández de Oviedo y Valdes (Fernández de Oviedo y Valdés 1851a, 1851b) described very large snakes encountered on Cuba and Hispaniola, some as stout as a man's thigh and as long as 25 to 30 feet (7.6 m to 9.1 m) but very tame and nonvenomous; he stated that the Amerindians used to eat them. We can only imagine what those islands were like 500–600 years ago and share the lament of M. Graham Netting and Coleman J. Goin (1944), regarding the events of the "morning of October 12, 1492, when Columbus made landfall at Watling's Island. This event was not immediately prejudicial to the snake population . . . but it sent in motion a long chain of destructive factors. . . . We can only regret that no 'Cinquecento' herpetologist had opportunity to wander through the then extensive stands of tropical hardwoods and to observe the large boas which, we believe, may then have fattened on the richer avifauna in the cool glades on many of the islands."

We view the history of the study of boas in the West Indies in three phases: (1) Those persons who were engaged in describing and categorizing the natural world but who, in most cases, had no firsthand experience with the animals they were describing. In regard to West Indian boas, this includes (among others) C. Linnaeus, G. Bibron, J. E. Gray, A. C. Günther, E. D. Cope, L. Stejneger, G. A. Boulenger, and O. G. Stull. (2) The second group includes biologists who traveled to West Indian islands and encountered boas where they live, but their efforts were directed toward making taxonomically broad collections or general observation on native wildlife. Among those in this category are P. H. Gosse, J. Gundlach, T. Barbour, M. G. Netting, and A. Schwartz. (3) Members of the last group have conducted dedicated research on one or more species of West Indian boas, often for many years. They, more than their predecessors, have contributed to our current knowledge of boid natural history and evolutionary biology and have addressed conservation concerns for many of the species covered in this book.

## THE EARLY DESCRIBERS

Although this is not an exhaustive list of early contributors, it is apparent that, with the exceptions of Linnaeus and Stull, most taxonomic activity occurred in the nineteenth century when efforts to describe and catalog the natural world were at a frenzy, including for descriptions of new reptiles; those numbers were not to be surpassed until the twenty-first century (Uetz and Stylianou 2018). Numbers of described reptile taxa by various authors noted below are from Uetz and Stylianou (2018).

**Carl Linnaeus (1707–1778).** Although responsible for naming several of the most iconic boas (e.g., *Boa constrictor, Corallus caninus, Eunectes murinus*) and numerous snakes worldwide, Linnaeus described only one West Indian boa, the St. Lucian endemic *Boa orophias.*

**John Edward Gray (1800–1875).** Working out of the British Museum, Gray was a prolific describer and published over 1100 papers on a wide taxonomic range of plants and animals. He described 335 reptile taxa but only one West Indian boa, *Corallus cookii.*

**Gabriel Bibron (1805–1848).** Although best known for his collaboration with A.-M.-C. Duméril and their multivolume *Erpétologie Générale ou Histoire Naturelle Complète des Reptiles* (1834–1854), Bibron described the Cuban endemic *Chilabothrus angulifer* in de la Sagra's multivolume work on Cuba (*Historia física, política, y natural de la isla de Cuba*).

**Johannes T. Reinhardt (1816–1882).** A native of Denmark, Reinhardt circumnavigated the globe on the Danish corvette *Galathea* (1845–1847), which included stops in South America. He returned to Brazil on two more occasions (1850–1852 and 1854–1856). Based on collections in the Royal Natural History Museum and the University of Copenhagen and in collaboration with Christian F. Lütken, he produced a monograph on the amphibians and reptiles of the Caribbean (Adler 2014). He described the Puerto Rican endemic, *Chilabothrus inornatus.*

**Johann Gustav Fischer (1819–1889).** A zoologist at the Naturhistorisches Museum in Hamburg, Germany, Fischer concentrated on fishes as well as reptiles. He described two currently recognized West Indian boas: *Chilabothrus gracilis* and *C. striatus.*

**Albert C. L. G. Günther (1830–1914).** Born in Germany, Günther received an appointment to work at the British Museum in 1857. Although his primary focus was ichthyology, he was a prolific describer of reptiles as well; he succeeded J. E. Gray as Keeper of Zoology at the museum. Among the many (364) reptilian taxa he described was the Hispaniolan endemic *Chilabothrus fordii.*

**Edward Drinker Cope (1840–1897).** Long associated with the Academy of Natural Sciences in Philadelphia, only Boulenger described more herpetological taxa (659) than Cope (385). He is responsible for describing the Lucayan species *Chilabothrus chrysogaster* and *C. strigilatus.*

**Leonhard Stejneger (1851–1943).** Although he also worked on birds and seals, the Norway-born Stejneger was the long-time curator of herpetology at the U.S. National Museum (Smithsonian Institution). He did fieldwork in Puerto Rico, but his primary contribution to West Indian boa taxonomy was the description of the Jamaican endemic *Chilabothrus subflavus* in 1901.

**George Albert Boulenger (1858–1937).** Although he did not describe any West Indian boas, Boulenger, the most prolific reptile taxonomist of all time (659 named taxa), was responsible for taxonomic updates and new combinations that included West Indian boas.

**Olive G. Stull (1905–1969).** Stull produced a checklist of the Boidae in 1935; it might have introduced as much taxonomic confusion as it clarified. She described *Chilabothrus granti*, but as a subspecies of *C. inornatus* (Stull 1933). She had intended a major revision of the boas and pythons of the world, but the specimens she had accumulated and was studying were accidently incinerated; her manuscript was never published (Adler 2007).

## TAXONOMISTS WITH WEST INDIAN EXPERIENCE

Like the early describers noted above, persons listed below were also primarily taxonomists and cataloguers. The difference, however, is that members of this group of biologists and naturalists traveled to West Indian islands and, in many cases, personally collected the specimens they would later describe; they had firsthand experience with the animals, saw the habitats in which they lived, and witnessed aspects of their behavior. Their publications provided information about the animals that was lacking in those of the earlier describers. They all had experience with live examples of one or more West Indian boas.

**Philip H. Gosse (1810–1888).** Gosse spent 18 months in Jamaica (1844–1846) and likely was "the first [European] naturalist to see the full diversity of Jamaica's habitats, to observe its fauna and flora. . . . And he did so before the introduction of the mongoose in 1872" (Adler 2012). Although primarily interested in birds, his classic *A Naturalist's Sojourn in Jamaica* (1851) contains many useful observations on frogs and reptiles, including the endemic boa, *Chilabothrus subflavus*.

**Juan C. Gundlach (1810–1896).** Gundlach was born in Germany but immigrated to Cuba in 1839 and spent the rest of his life there. He was particularly interested in birds and insects (as well as mollusks and mammals), but his *Erpetología Cubana* (1894) contains extensive observations on the natural history of frogs and reptiles, including the endemic *Chilabothrus angulifer*. He is responsible for the (exaggerated?) historical size record in *C. angulifer* of about "7 varas" total length (i.e., 6.4 m).

**Thomas Barbour (1884–1946).** Barbour was the most prolific contributor to West Indian herpetology during the first half of the twentieth century. Although he made numerous trips to the Antilles, he often relied on locals to do the collecting for him, thus precluding firsthand experience with the herpetofaunas. Barbour spent long periods at the then Atkins Institution of the Arnold Arboretum at Soledad, property of Harvard University (today Jardín Botánico de Cienfuegos), in Cienfuegos, Cuba, where he gained extensive experience with much Cuban wildlife. Barbour provided some interesting observations on *Chilabothrus angulifer* (Barbour and Ramsden 1919; Barbour 1943). He described *Corallus grenadensis* and coauthored (with Benjamin Shreve in 1935) the description of *Chilabothrus relicquus* (now a subspecies of *C. chrysogaster*).

**Chapman Grant (1887–1983).** In addition to being the founder of the journal *Herpetologica* and the Herpetologists' League, Grant was a prolific contributor to the literature on the West Indian herpetofauna. He was especially active in the 1930s and 1940s, with an emphasis on the Puerto Rico Bank but with major papers on the Cayman Islands and Jamaica as well. His publications included notes on boas, and *Chilabothrus granti* (Stull) was named in his honor.

**M. Graham Netting (1904–1996).** Netting had a long career as a herpetology curator and, eventually, as director of the Carnegie Museum of Natural History. Although most of his publications focused on eastern North America, he collaborated (with Coleman J. Goin) on the description of *Chilabothrus exsul*; they collected several of the specimens upon which the description was based.

**Albert Schwartz (1923–1992).** Considered the "dean" of West Indian herpetologists (Fig. 6.1), Schwartz has 299 reptilian taxa to his credit, most at the subspecies level. His major contribution to boa taxonomy is the 1974 review of *Chilabothrus* (then *Epicrates*) with Bruce R. Sheplan. Therein they described five new subspecies of *C. striatus* and one each for *C. fordii* and *C. gracilis*; Schwartz (1979) subsequently named another subspecies of *C. fordii*. Schwartz's publications always contained detailed descriptions of color and pattern variation, as well as extensive meristic data.

**Figure 6.1.** Some twentieth century describers of West Indian boas. *Top left*: Olive Griffith Stull. *Top right*: Chapman Grant. *Bottom left*: Thomas Barbour. *Bottom right*: Albert Schwartz at a microscope. Photos courtesy of Kraig Adler.

**James D. Lazell.** Lazell was primarily interested in *Anolis* lizards while collecting reptiles in the Lesser Antilles in the 1960s. Nevertheless, based on his field notes and a publication (Lazell 1964), he had memorable encounters with species of *Boa* and *Corallus*, and he published the original description of *B. nebulosa* on Dominica.

**Donald W. Buden.** Buden contributed to describing the boas of the southern Bahamas (Buden 1975) and spent some time on Crooked-Acklins Bank, Great Inagua, and the Caicos Islands. His important work describing the boas of the *Chilabothrus chrysogaster* "complex" included the first reports of boas from the Crooked-Acklins Bank, now recognized as *C. schwartzi* (Reynolds et al. 2018). He provided some accounts of these boas and islands for the revision of the genus (Sheplan and Schwartz 1974).

## DEDICATED BOA RESEARCH

This group contains researchers who actively study boas where the animals live, often doing long-term studies that may continue for many years or even decades. Also included here are those biologists who might not work directly with the animals on the islands but use molecular tools to determine phylogenetic relationships and historical biogeography in the Booidea, thus providing a high-resolution view of boa evolution in the West Indies.

### *Boa*

The two West Indian representatives of this famous genus have received surprisingly little attention since Lazell's (1964) descriptions and notes on behavior and ecology. Largely anecdotal observations appear in a few scattered publications. Alberto R. Puente-Rolón and RGR (Reynolds et al. 2013b) have done a genetic analysis of invasive *Boa constrictor* on Puerto Rico and are monitoring the population and its potential impact and proliferation on the island. Other researchers have conducted genetic and population analyses of *Boa* introductions elsewhere in the Caribbean (see Chapter 3).

*Chilabothrus*

**Peter J. Tolson.** Tolson has devoted decades of field and laboratory research to species of *Chilabothrus*. He is a recognized authority on several species in this genus, with particular long-term focus on *C. angulifer* at Guantanamo Bay, Cuba, *C. granti* (Tolson 1996a) in the U.S. Virgin Islands, and *C. monensis* on Isla de Mona (Tolson et al. 2007). Tolson's contributions are numerous and include examining evolutionary relationships and morphology (Tolson 1987), reproduction (e.g., Tolson 1980), and many other topics related to the biology and natural history of *Chilabothrus*, such as captive breeding and conservation.

**Alberto R. Puente-Rolón.** Puente-Rolón has spent most of his professional career working with the Puerto Rican boa, *C. inornatus* (e.g., Puente-Rolón and Bird-Picó 2004). Through his dissertation work and years of field experience with the species he has contributed greatly to our understanding of feeding ecology, reproduction, and movement of this species. He has also worked on the Mona boa, *C. monensis*, and the Virgin Islands boa *C. granti*, and has likely seen more individuals of these species than most other researchers combined. Over the last decade, Puente-Rolón and Reynolds have collaborated on numerous studies of the genus *Chilabothrus*, and they authored descriptions of three newly recognized species (*C. strigilatus*, *C. argentum*, and *C. schwartzi*).

**R. Graham Reynolds.** Reynolds has worked for over 15 years on this genus, starting with work on the Turks Island boa *C. chrysogaster* for his doctoral dissertation. Reynolds specializes in molecular phylogenetics, population genetics, and on-island study of this genus, with particular focus on species in the Bahamas, Turks and Caicos, and Puerto Rico (e.g., Reynolds et al. 2013a, 2016a). His 15-year-long study on *C. chrysogaster* on the Caicos Bank (with Glenn Gerber) has included capturing and marking over 1200 individuals, and his field and molecular work with Puente-Rolón and others has led to the description of four species in the genus, including the discoveries of *C. argentum* (Reynolds et al. 2016c) and *C. ampelophis* (Landestoy T. et al. 2021b).

**Tomás M. Rodríguez-Cabrera.** Based in Cuba, Rodríguez-Cabrera has made significant contributions to our knowledge of the Cuban endemic *C. angulifer* based on field research with various collaborators (e.g., Rodríguez-Cabrera et al. 2015, 2016a, 2016b, 2022). He is actively collecting the most comprehensive data set to date on the species, which includes ecological studies, geographic records, natural history observations, photographs, and historical records.

**Others.** Because Puerto Rico has for many years been home to an active group of biologists, the Puerto Rican endemic *C. inornatus* has received noteworthy attention from several researchers. They did not, however, as individuals, make multiple or long-term commitments to *C. inornatus* biology. This group includes Juan A. Rivero, Fernando J. Bird-Picó, Armando Rodríguez-Durán, Javier A. Rodríguez-Robles, James W. Wiley, and Joseph M. Wunderle Jr. Each of these persons usually worked in collaboration with other biologists. On Jamaica, Susan E. Koenig has overseen at least a decade of work on *C. subflavus* by students and researchers. In the Dominican Republic, Miguel A. Landestoy T. has probably seen more individual boas of all four species than almost any other naturalist, and he discovered and described the species *C. ampelophis* (Landestoy T. et al., 2021b).

*Corallus*

**Robert W. Henderson.** Henderson has devoted 30 years to studying various aspects of treeboa biology, including taxonomy, ecology, and conservation. Most of his efforts have focused on the two West Indian endemics, *Corallus cookii* and *Co. grenadensis*, with the latter species receiving the most attention. His work is summarized in four publications (Henderson 2002, 2015, 2019; Henderson and Powell 2018), although he has published dozens of other papers on these species. With Peter Tolson, he coauthored the first book dedicated to West Indian boas in 1993.

**George B. Pendlebury.** In the 1960s, Pendlebury collected specimens of *Corallus grenadensis* in Grenada and the Grenadines, examined stomach and intestine contents, and was the first to document the sharp dichotomy between the diets of West Indian and mainland *Corallus* (Pendlebury 1974).

*Phylogenetics*

**Arnold G. Kluge.** Kluge used morphological data (Kluge 1988a, 1989) to study phylogenetic relationships among species of *Chilabothrus*—thus laying the framework for those who followed. His 1991 publication presented a cladistic analysis of boine taxa based on morphological characters.

**Frank T. Burbrink.** Burbrink provided one of the early molecular phylogenetic assessments of the boas (Burbrink 2004) and thus provided a foundation for most subsequent molecular genetic studies of the family Boidae. His later collaborative work has improved our understanding of squamate evolutionary relationships, including that of the boas (Pyron et al. 2014).

**Barry N. Campbell.** Campbell's 1997 dissertation (Campbell 1997) provided one of the first broad molecular genetic analyses of the boas. He characterized priming regions for the mitochondrial gene cytochrome *b*, now a workhorse in boa phylogenetics, and provided a phylogenetic hypothesis for the group that differed from those based on morphology (Kluge 1991) or allozymes (Tolson 1987). His dissertation was the beginning of the extensive use of molecular data to characterize the evolutionary history of boids.

**Timothy J. Colston.** Colston is a tropical herpetologist and accomplished field biologist specializing in snake phylogenetics and the evolution of snake gut and venom microbiomes. He, along with coauthors, has contributed important papers on the phylogenetics, morphological evolution, and evolutionary relationships of the genus *Corallus* (Colston et al. 2013; Henderson et al. 2013b).

**Brice P. Noonan.** Noonan is an evolutionary Neotropical biologist who produced some important earlier works on boa molecular phylogeny and evolution (Noonan and Chippindale 2006; Noonan and Sites 2010) that provided an important foundation for numerous subsequent and ongoing studies.

**R. Alexander Pyron.** Pyron is a molecular evolutionary systematist whose work usually focuses on large-scale systematic treatments of entire genera or families. His work includes revising the phylogeny of the squamates (Pyron et al. 2013) and collaborative work with Reynolds and Burbrink (Pyron et al. 2014) revising higher-level taxonomy in boas.

**Athanasia C. Tzika.** Tzika, along with her dissertation advisor M. Milinkovitch, contributed important molecular studies of the Jamaican boa, *Chilabothrus subflavus*, including an analysis of genetic diversity across the island as well as in captive populations (Tzika et al. 2008).

Numerous other researchers have contributed important work on Caribbean boas, and although we do not list everyone here, we are grateful for their contributions that have moved the field forward.

# EPILOGUE

With few (if any) exceptions, boas in the West Indies today live on islands that must look far different from those inhabited by their long-ago ancestors. For the past 7000 years, humans have wrought changes to boa environments, and they go largely unchecked to this day. Through Archaic and Ceramic Ages, forest habitat was reduced by indigenous peoples in order to construct shelters and to make room for agriculture in small villages and settlements. Historically, with the arrival of European explorers and then settlers, vast areas of land were cleared, sometimes deforesting an entire island for the cultivation of sugarcane, cotton, bananas, orchard trees, and other commercial produce. Besides dramatic alterations to island landscapes, humans introduced chickens, goats, pigs, cattle, horses, cats, dogs, mice, rats, and mongooses. Yet, through it all, from the Bahamas, through the Greater Antilles, and as far south as Grenada in the Lesser Antilles, boa species have managed to persist on nearly one hundred islands.

Today, species of *Boa*, *Chilabothrus*, and *Corallus* face a wide variety of threats on their native islands, including habitat loss, invasive predators and competitors, illegal collecting for the pet trade, and, although they pose no threat to human safety, intentional killing for no other reason than for being a snake. Looking to the future, climate change will pose additional survival complications. All West Indian boas are afforded some legal protection on their respective islands, but whether that protection is enforced is sometimes in question. All are listed under either Appendix I or II by CITES, and IUCN Red List assessments for the eighteen species range from Least Concern to Critically Endangered.

## THE FUTURE

Despite their iconic status among wildlife in general and snakes in particular, as well as more than 150 years of fascination in West Indian boas by lay persons regardless of how they are perceived, there have been dedicated field studies of only seven of the eighteen species: *Chilabothrus angulifer*, *C. chrysogaster*, *C. granti*, *C. monensis*, *C. inornatus*, *C. subflavus*, and *Corallus grenadensis*. Long-term study on *C. argentum* is just now beginning. The abundance of some species, their multi-island distributions, and the proximity of the islands to North and South America, should make West Indian boas attractive subjects for long-term field projects. Below are some areas that we think are of high importance.

### *Distribution*

A number of species and subspecies of West Indian boas have poorly characterized distributions. *Chilabothrus gracilis* and *C. fordii* are examples. We know where the "hotspots" are to find them, but we do not know much about the extent of the ranges, the microclimatic or environmental variables they prefer, or how fragmented the populations are. This will require dedicated research to look for these cryptic species in areas where they might not have been previously reported, work that will be difficult, frustrating, and possibly dangerous. But such data are major gaps in our understanding of the biology of West Indian boas, and hence such studies are very important.

*Focused Study*

Seven of the eighteen species of West Indian boas have had dedicated, multiyear studies carried out on them by the same researcher or team of researchers. These studies are immensely valuable to our collective understanding of West Indian boas, but they are time-consuming and expensive. Nevertheless, these studies have shown that it is possible to secure funding to study boas annually for years or even decades. We encourage future researchers to identify species on which they might be able to initiate such studies and to go for it! We know little about population biology, movement ecology, details of foraging behavior, or in-depth habitat analyses for most West Indian boas.

*Natural History*

When writing this book, it struck us, in so many cases, how little we know about the natural history of these species. What does *C. exsul* eat? Does *C. exsul* live on Grand Bahama? What does *C. fordii* eat? Does *C. fordii* climb often? What are the average sizes of *C. striatus*? Are female *C. subflavus* consistently longer than males? Does tail length accurately predict sex? How impactful are hurricanes on boa populations? How do parasites and fungal disease impact boas? What is the status of boa populations in Haiti? The list is extensive. We especially encourage researchers to publish their data or otherwise make it accessible. Every published study, whether a short note in a specialized journal or a big paper in *Nature*, gets us closer to a more comprehensive understanding of West Indian boas, and we encourage naturalists, whether amateur or professional, to contribute observations of boa natural history to the literature. While publication is the preferred way to communicate information to the literature, there are other ways to make information available, such as through media or film.

*Evolutionary Biology*

Although some of us are working on unraveling the evolutionary history of West Indian boas, important discoveries are out there waiting to be found. Suppose there are additional fossils somewhere on Hispaniola or Cuba? Or evidence of additional species of *Boa* in the Leeward Antilles? Or new information about the paleohistory of Jamaica? Such findings could add immensely to our reconstruction of the origins of boa biodiversity. We encourage collaboration of boa researchers with geologists, paleontologists, and other scientists to fill in our understanding.

*Applied Conservation*

Conservation is challenging and expensive and can often feel unappreciated. Nevertheless, we stand to lose some magnificent species of West Indian boas without focused intervention. Six of the eighteen species are vulnerable to extinction or worse (Table 4.1). *Chilabothrus argentum* is the most immediately imperiled, but others are in serious jeopardy. We know that conservation intervention can work to reduce the loss of populations or even prevent or delay extinction, but we need dedicated people to implement conservation, monitoring, and biosecurity on islands in the West Indies. It takes vision, hard work, creative fundraising, and relentless effort to save species, but the more people who engage in this work, the easier it becomes for everyone. We hope that some readers will take up this call, either by directly assisting or by helping with fundraising.

## HOW YOU CAN SUPPORT BOA RESEARCH AND CONSERVATION

Our field of study has been small for decades, with just a few handfuls of people working on boas in the Caribbean over the past century. Nevertheless, we hope that this book has convinced you that, while we know a fair amount about these boas, there are heaps of unanswered questions. We welcome dedicated, enthusiastic, and collaborative newcomers, and we hope that this volume will inspire the next generation of boa biologists to work with us. We especially hope to encourage young students from the Caribbean to become scientists and conservationists, be it as researchers, policy makers, directors of nongovernmental organizations, or in other roles.

*Organizations Dedicated to West Indian Boa Research and Conservation*

The following list, although far from comprehensive, provides some examples of nonprofit or nongovernmental organizations that have designated programs for protecting and learning more about boas in the West Indies. There are many other organizations that indirectly, or through collaborations, support boa research.

**The Silver Boa Trust.** Founded by RGR, the Silver Boa Trust is dedicated to supporting research and conservation activities for boas in the Caribbean.

**The Bahamas National Trust.** A nongovernmental agency based in Nassau, the Bahamas National Trust relies on financial support from people who care about nature in the Bahamas. They do excellent conservation and education work, but their most important and crucial role is managing the country's system of National Parks. https://www.bnt.bs.

**The North Carolina Zoo.** The North Carolina Zoo is engaged in boots-on-the-ground boa conservation work in the Bahamas, Puerto Rico, and the U.S. Virgin Islands. Curator Dustin Smith is leading an effort to propagate and head start critically endangered boas for reintroduction. https://www.nczoo.org.

**Iniciativa Herpetológica, Inc.** This Puerto-Rico-based company, founded in part by Alberto R. Puente-Rolón, works to increase research and education on endangered Puerto Rican wildlife, such as the Puerto Rican crested toad and the Puerto Rican boa.

**Windsor Research Centre.** This center is engaged in research, conservation, and training in an important biodiversity area in Jamaica. Many studies of *Chilabothrus subflavus* are owing to work done at or in collaboration with the Windsor Research Centre.

**The San Diego Zoo Wildlife Alliance.** Glenn Gerber of this alliance has been directly involved in one of the longer-running studies of a population of boas; that of *C. chrysogaster* on Big Ambergris Cay.

Finally, **West Indian Boas** is an excellent website dedicated to providing information about West Indian boas, managed by Jeff Murray and Michael Saina. We encourage readers to visit it.

# REFERENCES

Aaron, J.M. 1974. Geology and mineral resources of Mona Island. *In* Mona and Monito Islands: an assessment of their natural and historical resources 2:B1–B7. Puerto Rico: Environmental Quality Board.

Abbad y Lasierra, F.I. 1788. Historia geográfica, civil y natural de la Isla de San Juan Bautista de Puerto Rico. San Juan: Editorial Universitaria.

Abreu-Rodríguez, E., and S. Moyá. 1995. Culebrón de Puerto Rico, *Epicrates inornatus* (Reptilia: Boidae), un nuevo hospedero de *Ornithodoros puertoricensis* (Acari: Argasidae) en Puerto Rico. Journal of Agriculture of the University of Puerto Rico 79:1–2, 91–92.

Acevedo González, M. 1983. Geografía física de Cuba. Havana: Editorial Pueblo y Educación.

Acevedo-Torres, M., N. Ríos-López, and M. Del Carmen Ruiz-Jaen. 2005. *Epicrates inornatus* (Puerto Rican boa). Cannibalism. Herpetological Review 36:195.

Acosta, M., and L. Mugica. 2006. Aves en el ecosistema arrocero. *In* L. Mugica, D. Denis, M. Acosta, A. Jiménez, and A. Rodríguez, eds., Aves acuáticas en los humedales de Cuba, pp. 108–135. Havana: Editorial Científico-Técnica.

Acosta Chavez, V., E. Ballesteros, A. Batista, A. García Rodríguez, A. Ines Hladki, G. Köhler, M. Ramírez Pinella, et al. 2016. *Corallus ruschenbergerii*. IUCN Red List of Threatened Species 2016:e.T203211A2762201. http://dx.doi.org/10.2305/IUCN.UK.2016-3.RLTS.T203211A2762201.en.

Adler, K., ed. 2007. Contributions to the history of herpetology, vol. 2. Contributions to Herpetology 21. Ithaca, NY: Society for the Study of Amphibians and Reptiles.

Adler, K., ed. 2012. Contributions to the history of herpetology, vol. 3. Contributions to Herpetology 29. Ithaca, NY: Society for the Study of Amphibians and Reptiles.

Adler, K., ed. 2014. Contributions to the history of herpetology, vol. 1. Contributions to Herpetology 30. Ithaca, NY: Society for the Study of Amphibians and Reptiles.

Alberts, A.C., ed. 2000. West Indian iguanas: status survey and conservation action plan. Gland, Switzerland: International Union for Conservation of Nature (IUCN).

Alberts, A.C., T.D. Grant, G.P. Gerber, K.E. Comer, P.J. Tolson, J.M. Lemm, and D. Boyer. 2001. Critical reptile species management on the U.S. Naval Base, Guantanamo Bay, Cuba. Report to the United States Navy for project 62470–00-M-5219.

Albino, A.M. 2011. Morfología vertebral de *Boa constrictor* (Serpentes: Boidae) y la validez del género mioceno *Pseudoepicrates* Auffenberg, 1923. Ameghiniana 48:53–62.

Alfonso Álvarez, E. Morell Savall, R. Díaz Aguiar, R. Carbonell Paneque, F. Morera, and V. Berovides Álvarez. 1998. *Epicrates angulifer* (Majá de Santa María). *In* E. Pérez, E. Osa, Y. Matamoros, and U. Seal, eds., Taller para la conservación análisis y manejo planificado de una selección de especies Cubanas II, pp. 1–4. Apple Valley, MN: IUCN/SSC Conservation Breeding Specialist Group.

Allen, G.M. 1942. Extinct and vanishing mammals of the Western Hemisphere with the marine species of all the oceans. Washington, DC: American Committee for International Wild Life Protection, Special Publication 11.

Amaral, A. do. (1929) 1930. Estudos sôbre ophidios neotrópicos. XVIII. Lista remisiva dos ophidios da região neotrópica. Memoirs Instituto Butantan 4:129–271.

Angeli, N.F., W. Coles, S. McKinley, and D.G. Mulcahy. 2019. Introduction of an exotic constricting snake

(*Boa constrictor*) and its establishment on St. Croix, U.S. Virgin Islands. Herpetological Conservation and Biology 14:288–296.

Angin, B. 2014. Bat predation by the Dominican Boa (*Boa nebulosa*). Caribbean Herpetology 51:1–2.

Anonymous. 2014. Tet shyen (*Boa constrictor*) in de back yard. Dominica News Online, September 6, 2014.

Aranda Pedroso, E. 2019. Systematics of Quaternary Squamata from Cuba. MS thesis, Universidade de São Paulo, Museu de Zoologia, São Paulo.

Aranda [Pedroso], E., J.G. Martínez López, O. Jiménez [Vázquez], C. Alemán Luna, and L.W. Viñola[-]López. 2017. Nuevos registros fósiles de vertebrados terrestres para las Llanadas de Sancti Spíritus, Cuba. Novitates Caribaea 11:115–123.

Aranda [Pedroso], E., L.W. Viñola-López, and L. Álvarez-Lajonchere. 2020. New insights on the quaternary fossil record of Isla de la Juventud, Cuba. Journal of South American Earth Sciences 102:1–16.

Arendt, W.J., and D. Anthony. 1986. Bat predation by the St. Lucia boa (*Boa constrictor orophias*). Caribbean Journal of Science 22:219–220.

Armas, L.F. de, R. Armiñana, J.E. Travieso, and L.O. Grande. 1990. Breve caracterización de la artropofauna de tres cuevas calientes de la provincia de Villa Clara, Cuba. Poeyana 394:1–14.

Arredondo, O. 1971. Nuevo genero y especie de ave fosil (Accipitriformes: Vulturidae) del Pleistoceno de Cuba. Memorias de la Sociedad de Ciencias Naturales La Salle 29:415–431.

Arredondo, O. 1976. The great predatory birds of the Pleistocene of Cuba. *In* S.L. Olson, ed., Collected papers in avian paleontology honoring the 90th birthday of Alexander Wetmore. Smithsonian Contributions to Paleobiology 27:169–187.

Arredondo Antúnez, C. 1997. Composición de la fauna de vertebrados terrestres extintos del Cuaternario de Cuba. Revista Electrónica Orbita Científica 2:i, 1–13.

Arredondo Antúnez, C., and V.N. Chirino Flores. 2002. Consideraciones sobre la alimentación de *Tyto alba furcata* (Aves: Strigiformes) con implicaciones ecológicas en Cuba. El Pitirre 15:16–24.

Arredondo Antúnez, C., and R. Villavicencio. 2004. Tafonomía del depósito arqueológico Solapa del Megalocnus en el noroeste de Villa Clara, Cuba. Revista Biología 18:160–172.

Auffenberg, W. 1963. The fossil snakes of Florida. Tulane Studies in Zoology 10:131–216.

Augstenová, B., S. Mazzoleni, A. Kostmann, M. Altmanová, D. Frynta, L. Kratochvíl, and M. Rovastsos. 2019. Cytogenetic analysis did not reveal differentiated sex chromosomes in ten species of boas and pythons (Reptilia: Serpentes). Genes 10:934.

Aungst, E.R., A.R. Puente-Rolón, and R.G. Reynolds. 2020. Genetic diversity in U.S. captive populations of the endangered Puerto Rican boa, *Chilabothrus inornatus*. Zoo Biology 39:205–213.

Aya-Cuero, C.A., C.H. Cáceres-Martínez, and D.A. Esquivel. 2019. First record of predation on Greater sac-winged bat, *Saccopteryx bilineata* (Chiroptera: Emballonuridae), by the Colombian rainbow boa, *Epicrates maurus* (Serpentes: Boidae). Herpetology Notes 12:815–817.

Ayes, C.M. 1995. Mitigacion parcial del monticulo 'A' del yacimiento de Angostura, Barrio Florida afuera, Barceloneta, Puerto Rico. San Juan: Autoridad de Carreteras y Transportacion.

Bailon, S., C. Bochaton, and A. Lenoble. 2015. New data on Pleistocene and Holocene herpetofauna of Marie-Galante (Blanchard Cave, Guadeloupe Islands, French West Indies): insular faunal turnover and human impact. Quaternary Science Reviews 128:127–137.

Bain, J.R., and L. Hurst. 1982. Life history notes: *Epicrates subflavus* (Jamaican boa). Food. Herpetological Review 13:18.

Barbour, T. 1910. Notes on the herpetology of Jamaica. Bulletin of the Museum of Comparative Zoology 52:272–301.

Barbour, T. 1914. A contribution to the zoögeography of the West Indies, with especial reference to amphibians and reptiles. Memoirs of the Museum of Comparative Zoology 44(2).

Barbour, T. 1930. A list of Antillean reptiles and amphibians. Zoologica 11:61–116.

Barbour, T. 1935. A second list of Antillean reptiles and amphibians. Zoologica 19:77–141.

Barbour, T. 1937. Third list of Antillean reptiles and amphibians. Bulletin of the Museum of Comparative Zoology 82:77–166.

Barbour, T. 1941. A new boa from the Bahamas. Proceedings of the New England Zoological Club 18:61–65.

Barbour, T. 1943. Naturalist at large. Boston, MA: Little, Brown. Reprint, London: Robert Hale, 1950.

Barbour, T., and C.T. Ramsden. 1919. The herpetology of Cuba. Memoirs of the Museum of Comparative Zoology 47(2): 69–213, 15 plates.

Barbour, T., and B. Shreve. 1935. Concerning some Bahamian reptiles, with notes on the fauna. Proceedings of the Boston Society of Natural History 40:347–366.

Beard, J.S. 1949. The natural vegetation of the windward & leeward islands. Oxford: Clarendon Press.

Bénito-Espinal, E. 1978. La faune I (reptiles, mammifères

et amphibiens). *In* Antilles d'hier et d'aujourd'hui 2. Fort-de-France: Emile Désormeaux.

Bénito-Espinal, E. 1990. Grande encyclopédie de la Caribe, tomo 4. Pointe-à-Pierre, Trinidad: Sanoli.

Bennett, S. 2013. Snakes on a beach!: photo of the day. Uncommon Caribbean, https://www.uncommon caribbean.com/bequia/photo-of-the-day-snakes-on -a-beach/.

Beolens, B., M. Watkins, and M. Grayson. 2011. The eponym dictionary of reptiles. Baltimore: Johns Hopkins University Press.

Beovides-Casas, K., and C.A. Mancina. 2006. Natural history and morphometry of the Cuban iguana (*Cyclura nubila*) in Cayo Sijú, Cuba. Animal Biodiversity and Conservation 29:1–8.

Berensten, A.R., J.G. Garcia-Cancel, E.M. Diaz-Negron, A.R. Puente-Rolon, R.N. Reed, and K.C. Vercauteren. 2015. Boa constrictor (*Boa constrictor*). Distribution. Herpetological Review 46:572.

Berman, M.J., and P.L. Gnivecki. 1995. The colonization of the Bahama Archipelago: a reappraisal. World Archaeology 26:421–441.

Bernarde, P.S., and A.S. Abe. 2010. Hábitos alimentares de serpentes em Espigão do Oeste, Rondônia, Brasil. Biota Neotropica 10:1–10.

Berovides [Álvarez], V., and A. Comas. 1997a. Abundancia de la jutía conga *Capromys pilorides* (Rodentia, Capromyidae) en varios hábitats de Cuba. Revista Biología 11:25–30.

Berovides [Álvarez], V., and A. Comas G[onzález]. 1997b. Densidad y productividad de la jutía conga (*Capromys pilorides*) en mangles cubanos. Caribbean Journal of Science 33:121–123.

Berovides Álvarez, V., and R. Carbonell Paneque. 1998. Morfometría y abundancia del majá de Santa María *Epicrates angulifer* (Ophidia, Boidae). *In* E. Pérez, E. Osa, Y. Matamoros, and U. Seal, eds., Taller para la conservación análisis y manejo planificado de una selección de especies Cubanas 2, pp 1–4. Apple Valley, MN: IUCN/SSC Conservation Breeding Specialist Group.

Bezerra de Lima, L.C. 2016. Filogenia e delimitação de espécies no complexo *Boa constrictor* (Serpentes: Boidae) utilizando maracadores moleculares. PhD dissertation, Instituto de Biociências de Universidade de São Paulo, Brazil.

Bibron, G. 1843. *In* J.-T. Cocteau and G. Bibron, Reptiles, pp. 120–143. Vol. 4., Historia física, política y natural de la Isla de Cuba: reptiles y peces, edited by R. de la Sagra. Paris: Arthus Bertrand.

Biegler, R. 1966. A survey of recent longevity records for reptiles and amphibians in zoos. International Zoological Yearbook 6:487–493.

Bird-Picó, F.J. 1994. Final report on *Epicrates inornatus* survey throughout Puerto Rico. Cooperative agreement between the U.S. Department of the Interior, U.S. Fish and Wildlife Service, and the Department of Biology, University of Puerto Rico, Mayagüez campus. Cooperative Agreement 14-16-0004-92-958.

Birdsey, R.A., and L.P. Weaver. 1987. Forest area trends in Puerto Rico. U.S. Forest Service, Southern Forest Experimental Station, Asheville, North Carolina, Research Note SO-331:1–5.

Bloxam, Q.M.C., and S. Tonge. 1981. A comparison of reproduction in three species of *Epicrates* maintained at the Jersey Wildlife Preservation trust. Dodo, Journal of the Jersey Wildlife Preservation Trust 18:64–74.

Boback, S.M. 2005. Natural history and conservation of island boas (*Boa constrictor*) in Belize. Copeia 2005:880–885.

Bochaton, C. 2020. First records of modified snake bones in the pre-Columbian archeological record of the Lesser Antilles: Cultural and paleoecological implications. Journal of Island and Coastal Archeology. https://doi.org/10.1080/15564894.2020.1749195.

Bochaton, C., and S. Bailon. 2018. A new fossil species of *Boa* Linnaeus, 1758 (Squamata, Boidae), from Marie-Galante Island (French West Indies). Journal of Vertebrate Paleontology. https://doi.org/10.1080 /02724634.2018.1462829.

Bochaton, C., S. Grouard, R. Cornette, I. Ineich, A. Tresset, and S. Bailon. 2015. Fossil and subfossil herpetofauna from Cadet 2 Cave (Marie-Galante, Guadeloupe Islands, F. W. I.): evolution of an insular herpetofauna since the Late Pleistocene. Comptes Rendus Palévol 14:101–110.

Bochaton, C., E. Paradis, S. Bailon, S. Grouard, I. Ineich, A. Lenoble, O. Lorvelec, A. Tresset, and N. Boivin. 2021. Large-scale reptile extinctions following European colonization of the Guadeloupe Islands. Science Advances 7:1–13.

Bolívar, N. Aróstegui, C. González Diaz de Villegas, and N. del Río Bolívar. 2013. Ta makuende yaya y las reglas de Palo Monte: mayombe, brillumba, kimbisa, shamalongo. Havana: Editorial Jose Marti.

Borroto-Páez, R. 2009. Invasive mammals in Cuba: an overview. Biological Invasions 11:2279–2290.

Borroto-Páez, R. 2011. Los mamíferos invasores o introducidos. *In* R. Borroto-Páez and C.A. Mancina, eds., Mamíferos en Cuba, pp. 221–241. Vaasa: UPC Print.

Borroto-Páez, R. 2013. Nidos y refugios de ratas negras

(*Rattus rattus*) en Cuba (Mammalia, Rodentia). Solenodon 11:109–119.

Borroto-Páez, R., and C.A. Mancina. 2006. Importancia del mangle rojo (*Rizophora* [sic] *mangle*) para la conservación de las jutías (Rodentia: Capromydae). *In* L. Menéndez Carrera and J. M. Guzmán Menéndez, eds., Ecosistema de manglar en el archipiélago Cubano: Estudios y experiencias enfocados a su gestión, pp. 170–177. Havana: Editorial Academia.

Borroto-Páez, R., M. Tejeda, F. Lewis, and M.A. Rodríguez. 1990. Fluctuación poblacional de *Mus musculus* (L) y *Rattus rattus* (L) en el cultivo de la caña de azúcar. Revista Biología 4:121–132.

Boulenger, G.A. 1893. Catalogue of the snakes in the British Museum (Natural History) 1. London: Taylor and Francis.

Bouysse, P., D. Westercamp, and P. Andreieff. 1990. The Lesser Antilles island arc. Proceedings of the Ocean Drilling Program, Scientific Results 110:29–44.

Bowler, J.K. 1977. Longevity of reptiles and amphibians in North American collections. Society for the Study of Amphibians and Reptiles Herpetological Circular 21.

Brace, S., S.T. Turvey, M. Weksler, M.L.P. Hoogland, and I. Barnes. 2015. Unexpected evolutionary diversity in a recently extinct Caribbean mammal radiation. Proceedings of the Royal Society B. 282:20142371.

Bradnam, K.R., J.N. Fass, A. Alexandrov, P. Baranay, M. Bechner, I. Birol, S. Boisvert, et al. 2013. Assemblathon 2: evaluating de novo methods of genome assembly in three vertebrate species. GigaScience 2(1): 2047–217X-2–10.

Branch, W.R. 1981. Hemipenis of the Madagascar boas *Acrantophis* and *Sanzinia*, with a review of hemipenial morphology in the Boinae. Journal of Herpetology 15:91–97.

Brattstrom, B.H. 1958. More fossil reptiles from Cuba. Herpetologica 13:278.

Brash, A.R. 1987. The history of avian extinction and forest conversion on Puerto Rico. Biological Conservation 39:97–111.

Breuil, M. 2002. Histoire naturelle des amphibiens et reptiles terrestres de l'archipel Guadeloupéen. Guadeloupe, Saint-Martin, Saint-Barthélemy. Paris: Muséum National d'Histoire Naturelle, Institut d'Écologie et de Gestion de la Biodiversité, Service du Patrimoine Naturel, Patrimoines Naturels 54.

Breuil, M. 2009. The terrestrial herpetofauna of Martinique: past, present, future. Applied Herpetology 6:123–149.

Breuil, M., 2011. The terrestrial herpetofauna of Martinique: past, present, future. *In* A. Hailey, B. Wilson, and J. Horrocks, eds., Conservation of Caribbean island herpetofaunas. Vol. 2, Regional accounts of the West Indies, pp. 311–338. Leiden, Netherlands: Brill.

Buckner, S.D., R. Franz, and R.G. Reynolds. 2012. Bahama Islands and Turks & Caicos Islands. *In* R. Powell and R.W. Henderson, eds., Island lists of West Indian amphibians and reptiles. Bulletin of the Florida Museum of Natural History 51:85–166.

Buden, D.W. 1975. Notes on *Epicrates chrysogaster* (Serpentes: Boidae) of the southern Bahamas, with description of a new species. Herpetologica 31:166–177.

Buide, M.S. 1966. Reptiles de la península de Hicacos. Poeyana 21:1–12.

Buide, M.[S.] 1985. Reptiles de Cuba. Havana: Editorial Gente Nueva.

Bullen, R.P. 1964. The archaeology of Grenada, West Indies. Contributions of the Florida State Museum, Social Sciences 11:1–67, plates 1–25.

Burbrink, F.T. 2004. Inferring the phylogenetic position of *Boa constrictor* among the Boinae. Molecular Phylogenetics and Evolution 34:167–180.

Burbrink, F.T., F.G. Grazziotin, R.A. Pyron, D. Cundall, S. Donnellan, F. Irish, J.S. Keogh, R.W. Murphy, B. Noonan, and C.J. Raxworthy. 2020. Interrogating genomic-scale data for Squamata (lizards, snakes, and amphisbaenians) shows no support for key traditional morphological relationships. Systematic Biology 69:502–520.

Bushar, L.M., R.G. Reynolds, S. Tucker, L.C. Pace, W.I. Lutterschmidt, R.A. Odum, and H.K. Reinert. 2015. Genetic characterization of an invasive *Boa constrictor* population on the Caribbean island of Aruba. Journal of Herpetology 49:602–610.

Callaghan, R.T. 2003. Comments on the mainland origins of the Preceramic cultures of the Greater Antilles. Latin American Antiquity 14:323–338.

Callaghan, R.T. 2007. Prehistoric settlement patterns on St. Vincent, West Indies. Caribbean Journal of Science 43:11–22.

Callaghan, R.T. 2010. Crossing the Guadeloupe Passage in the Archaic Age. *In* S.M. Fitzpatrick and A.H. Ross, eds., Island shores, distant pasts: archaeological and biological approaches to the pre-Columbian settlement of the Caribbean, pp. 127–147. Gainesville: University Press of Florida.

Campbell, B.N. 1997. Hic sunt serpentes—molecular phylogenetics and the Boidae (Serpentes Booidea). PhD dissertation, Queen's University, Kingston, Ontario.

Campbell, C. 2018. Use of snakes during J'ouvert concerns Forestry Division. NOW Grenada, https://

www.nowgrenada.com/2018/08/use-of-snakes-during-jouvert-concerns-forestry-division/.

Campbell, H.W., and F.G. Thompson. 1978. Observations on a captive Mona Island boa, *Epicrates monensis monensis* Zenneck. Bulletin of the Maryland Herpetological Society 14:98–99.

Card, D.C., D.R. Schield, R.H. Adams, A.B. Corbin, B.W. Perry, A.L. Andrew, G.I. Pasquesi, et al. 2016. Phylogeographic and population genetic analyses reveal multiple species of *Boa* and independent origins of insular dwarfism. Molecular Phylogenetics and Evolution 102:104–116.

Card, D.C., R.H. Adams, D.R. Schield, B.W. Perry, A.B. Corbin, G.I. Pasquesi, K. Row, et al. 2019. Genomic basis of convergent island phenotypes in boa constrictors. Genome Biology and Evolution 11:3123–3143.

Carlson, L.A. 1994. Analysis of the vertebrate fauna from MC-32 on the north coast of Middle Caicos, Turks and Caicos Islands, B.W.I. Gainesville: Florida Museum of Natural History.

Carlson, L.A. 1999. Aftermath of a feast: human colonization of the southern Bahama Archipelago and its effects on the indigenous fauna. PhD dissertation, University of Florida, Gainesville.

Cassimiro, J., C.F. de Souza Palmuti, and J. Bertoluci. 2010. *Epicrates cenchria*: Diet. Herpetological Review 41:501.

Castro-Prieto, J., W.A. Gould, C. Ortiz-Maldonado, S. Soto-Bayó, I. Llerandi-Román, S. Gaztambide-Arandes, M. Quiñones, M. Cañón, and K.R. Jacobs. 2019. A comprehensive inventory of protected areas and other land conservation mechanisms in Puerto Rico. U.S. Department of Agriculture Forest Service, International Institute of Tropical Forestry, San Juan, General Technical Report IITF-GTR-50.

Černý, V. 1966. Nuevas garrapatas (Ixodoidea) en aves y reptiles de Cuba. Poeyana 26:1–10.

Černý, V. 1969. The tick fauna of Cuba. Folia Parasitologica 16:279–284.

Chandler, C.R., and P.J. Tolson. 1990. Habitat use by a boid snake, *Epicrates monensis*, and its anoline prey, *Anolis cristatellus*. Journal of Herpetology 24:151–157.

Cisneros-Heredia, D.F. 2016. *Corallus blombergii*. IUCN Red List of Threatened Species 2016:e.T44580012A44580021. http://dx.doi.org/10.2305/IUCN.UK.2016-3.RLTS.T44580012A44580021.en.

Cochran, D.M. 1941. The herpetology of Hispaniola. Bulletin of the United States National Museum 177.

Colston, T.J., F.G. Grazziotin, D.B. Shepard, L.J. Vitt, G.R. Colli, R.W. Henderson, S.B. Hedges, S. Bonatto, H. Zaher, and F.T. Burbrink. 2013. Molecular systematics and historical biogeography of tree boas (genus *Corallus* spp.). Molecular Phylogenetics and Evolution 66:953–959.

Comas González, A., and V. Berovides [Álvarez]. 1990. Densidad de la jutía conga (*Capromys pilorides*) en cayos del grupo insular Jardines de la Reina, Cuba. Revista Biología 4:15–20.

Comas González, A., R. González Brito, G. Cepero La Rosa, and V. Berovides Álvarez. 1989. Densidad de la jutía conga (*Capromys pilorides*) (Rodentia: Capromyidae) en el área protegida Sierra del Chorrillo, Camagüey. Ciencias Biológicas 21/22:115–129.

Comas González, A., F. Rosales Zequeira, R. González Brito, and U. Peláez Martínez. 1994. Ecología trófica de la jutía conga *Capromys pilorides* (Rodentia: Capromyidae), en el área protegida Sierra del Chorrillo; Camagüey, Cuba. Ciencias Biológicas 8:75–81.

Cooke, S.B., L.M. Dávalos, A.M. Mychajliw, S.T. Turvey, and N.S. Upham. 2017. Anthropogenic extinction dominates Holocene declines of West Indian mammals. Annual Review of Ecology, Evolution, and Systematics 48:301–327.

Cope, E.D. 1862. Synopsis of the species of *Holcosus* and *Ameiva*, with diagnoses of new West Indian and South American Colubridae. Proceedings of the Academy of Natural Sciences of Philadelphia 14:60–82.

Cope, E.D. 1871. Eighth contribution to the herpetology of tropical America. Proceedings of the American Philosophical Society [1869] [1870] 11:553–599.

Cope, E.D. 1893. Notes on some snakes from tropical America lately living in the collection of the Zoological Society of Philadelphia. Proceedings of the Academy of Natural Sciences of Philadelphia 1893:429–435.

Cope, E.D. 1895. The classification of the Ophidia. Transactions of the American Philosophical Society 18:186–219.

Coy Otero, A. 1999. Parasites. *In* L. Rodríguez Schettino, ed., The iguanid lizards of Cuba, pp. 77–85. Gainesville: University Press of Florida.

Coy Otero, A., and N. Lorenzo Hernández. 1982. Lista de los helmintos parásitos de los vertebrados silvestres cubanos. Poeyana 235:1–57.

Craton, M., and G. Saunders. 1992. Islanders in the stream: a history of the Bahamian people, vol 1. Athens: University of Georgia Press.

Cruz, A., and S. Gruber. 1981. The distribution, ecology and breeding biology of Jamaican Amazon parrots. *In* R.F. Pasquier, ed. Conservation of New World parrots, pp. 103–132. International Council for Bird Preservation Technical Publication 1. Washington, DC: Smithsonian Institution Press.

Cruz, J. de la. 1992. Bioecología de las grutas de calor. Mundos Subterráneos 3:7–21.

Cuarón, A.D., M.A. Martínez-Morales, K.W. McFadden, D. Valenzuela, and M.E. Gompper. 2004. The status of dwarf carnivores on Cozumel Island, Mexico. Biodiversity and Conservation 13:317–331.

Cuarón, A.D., D. Valenzuela-Galván, D. García-Vasco, M.E. Copa, S. Bautista, H. Mena, D. Martínez-Godínes, et al. 2009. Conservation of the endemic dwarf carnivores of Cozumel Island, Mexico. Small Carnivore Conservation 41:15–21.

Cundall, D., and H.W. Greene 2002. Feeding in snakes. *In* K. Schwenk, ed., Feeding: form, function, and evolution in tetrapod vertebrates, pp. 293–333. San Diego: Academic Press.

da Costa Silva, P., and R.W. Henderson. 2013. Barn Owl (*Tyto alba*) predation on *Corallus hortulanus* (Squamata: Boidae). Herpetology Notes 6:35.

Daltry, J.C. 2009. The status and management of Saint Lucia's forest reptiles and amphibians. Technical Report 2 to the National Forest Demarcation and Bio-Physical Resource Inventory Project, FCG International, Helsinki, Finland.

Daltry, J.C. 2018. *Boa orophias*. IUCN Red List of Threatened Species 2018:e.T74866530A75171346. https://dx.doi.org/10.2305/IUCN.UK.2018-2.RLTS.T74866530A75171346.en.

Daltry, J.C., R. Powell, and R.W. Henderson. 2018. *Boa nebulosa*. IUCN Red List of Threatened Species 2018: e.T74863215A75171341. https://dx.doi.org/10.2305/IUCN.UK.2018-2.RLTS.T74863215A75171341.en.

Da Silva, M.-A.O., J.T. Gade, C. Damsgaard, T. Wang, S. Heegaard, and M.F. Bertelsen. 2019. Morphology and evolution of the snake cornea. Journal of Morphology 2019:1–10.

Dávalos, L.M., and R. Eriksson. 2004. *Epicrates subflavus*. Foraging behavior. Herpetological Review 35:66.

Day, M., and P.[J.] Tolson. 1996. *Epicrates angulifer.* IUCN Red List of Threatened Species 1996:e.T7815A12852846.http://dx.doi.org/10.2305/IUCN.UK.1996.RLTS.T7815A12852846.en.

de Jesús, V. 2007. Encuentran majá de dos cabezas. Diario Granma 4:2. http://www.granma.cu/granmad/2007/01/04/nacional/artic07.html.

del Risco Rodríguez, E. 1995. Los bosques de Cuba: su historia y características. Havana: Editorial Científico-Técnica.

Denis, D. 2006. Humedales en Cuba. Capítulo 1. *In* L. Mugica, D. Denis, M. Acosta, A. Jiménez, and A. Rodríguez, eds., Aves acuáticas en los humedales de Cuba, pp. 8–25. Havana: Editorial Científico-Técnica.

Departamento de Recursos Naturales y Ambientales de Puerto Rico. 2004. Reglamento 6766. Reglamento para regir las especies vulnerables y en peligro de extinción en el estado libre asociado de Puerto Rico. http://drna.pr.gov.

Dewynter, M., J.-C. De Massary, C. Bochaton, R. Bour, I. Ineich, N. Vidal, and J. Lescure. 2019. Liste taxinomique de l'herpétofaune dans l'outre-mer français: 3. Collectivé territorial de Martinique. Bulletin de la Société Herpétologique de France 169:53–82.

Díaz, L.M. 2006. Anfibios y reptiles. *In* L.M. Díaz, W.S. Alverson, A. Barreto, and T. Wachter, eds., Cuba: Camagüey, Sierra de Cubitas, pp. 48–50. Field Museum Rapid Biological Inventories Report 8.

Díaz, L.M., and A. Cádiz. 2020. A new species of *Tropidophis* (Squamata: Tropidophiidae) and molecular phylogeny of the Cuban radiation of the genus. Novitates Caribaea 16:1–19.

Dinets, V. 2017. Coordinated hunting by Cuban boas. Animal Behavior and Cognition 4:24–29.

Dodd, C.K. 1986. Importation of live snakes and snake products into the United States, 1977–1983. Herpetological Review 17:76–79.

Domínguez Díaz, M., and L.V. Moreno García. 2006. *Alsophis cantherigerus* (Jubo, Jubo de Sabana, Jubo Sabanero). Size record. Herpetological Review 37:349.

Donihue, C.M., A.M. Kowaleski, J.B. Losos, A.C. Algar, S. Baeckens, R.W. Buchkowski, A.-C. Fabre, et al. 2020. Hurricane effects on Neotropical lizards span geographic and phylogenetic scales. Proceedings of the National Academy of Science of the USA117: 10429–10434.

Dowling, H.G., and J.M. Savage. 1960. A guide to the snake hemipenis: a survey of basic structure and systematic characteristics. Zoologica 45:17–28.

Drake, F., and M. Pierpont. 1996. Histoire naturelle des Indes: the Drake Manuscript in the Pierpont Morgan Library. New York: W.W. Norton.

Duméril, A.M.C., and G. Bibron. 1844. Erpétologie générale ou histoire naturelle complète des reptiles. 6. Paris: Librairie Encyclopédique de Roret.

Dunn, E.R. 1949. Relative abundance of some Panamanian snakes. Ecology 30:39–57.

Espeut, W.B. 1882. On the acclimatization of the Indian mongoose in Jamaica. Proceedings of the Zoological Society of London 1882:712–714.

Estrada, A.R. 1994. Herpetofauna de la Cuenca Banao-Higuanojo, Sancti Spíritus, Cuba. Revista de la Academia Colombiana de Ciencias 19:353–360.

Estrada, A.R. 2012. The Cuban archipelago. *In* R. Powell and R.W. Henderson, eds., Island lists of West Indian amphibians and reptiles, pp. 113–125. Bulletin of the Florida Museum of Natural History 51:85–166.

FAO (Food and Agriculture Organization of the United Nations). 2015. Global forest resources assessment 2015. Rome: FAO.

FAO (Food and Agriculture Organization of the United Nations). 2019. Study on the state of agriculture in the Caribbean. Rome: FAO and the Caribbean Development Bank.

Fernández de Oviedo y Valdés, G. (1535) 1851a. De las serpientes ó culebras de la isla de Cuba ó Fernandina, cap. 5. *In* J. Amador de Los Ríos, ed., Historia general y natural de Las Indias, islas y tierra-firme del mar océano, primera parte, tomo 1, pp. 500–501. Madrid: Imprenta de la Real Academia de la Historia.

Fernández de Oviedo y Valdés, G. (1535) 1851b. De las serpientes ó culebras é lagartos desta Isla Española y otras partes, cap. 8. *In* J. Amador de Los Ríos, ed., Historia general y natural de Las Indias, islas y tierra-firme del mar océano, primera parte, tomo 1, pp. 396–399. Madrid: Imprenta de la Real Academia de la Historia.

Fiorillo, B.F. and D. S. Batista. 2019. Predation on eggs of the Gray Tinamou (*Tinamus tao*, Tinamiformes: Tinamidae) by the rainbow boa (*Epicrates cenchria*, Serpentes: Boidae). Herpetology Notes 12:79–81.

Fischer, J. G. 1856. Neue Schlangen der Hamburgischen Naturhistorischen Museums. Abhandlungen aus dem Gebiete der Naturwissenschaften 3:79–116.

Fischer, J.G. 1888. Über eine kollektion reptilien und amphibien von Hayti. Jahrbuch der Hamburgischen Wissenschaftlichen Anstalten 5:23–45.

Fiske, R. S., and H. Sigurdsson. 1982. Soufriere Volcano, St. Vincent: observations of its 1979 eruption from the ground, aircraft, and satellites. Science 216:1105–1126.

Fitch, H.S., and R.O. Bare. 1978. A field study of the Red-tailed Hawk in eastern Texas. Transactions of the Kansas Academy of Science 81:1–13.

Fitch, H.S., F. Swenson, and D.F. Tillotson. 1946. Behavior and food habits of the Red-tailed Hawk. Condor 48:205–237.

Fitzpatrick, S.M. 2015. The pre-Columbian Caribbean: colonization, population dispersal, and island adaptations. PaleoAmerica 1:305–331.

Fitzpatrick, S.M., and W.F. Keegan. 2007. Human impacts and adaptations in the Caribbean islands: an historical ecology approach. Earth and Environmental Science Transactions of the Royal Society of Edinburgh 98:29–45.

Fitzpatrick, S.M., M. Kappers, Q. Kaye, C.M. Giovas, M.J. Lefebvre, M.H. Harris, S. Burnett, J.A. Pavia, K. Marsaglia, and J. Feathers. 2009. Precolumbian settlements on Carriacou, West Indies. Journal of Field Archaeology 34:247–266.

Fong, A. 2021. Chilabothrus angulifer. IUCN Red List of Threatened Species 2021:e.T7815A18979599. https://dx.doi.org/10.2305/IUCN.UK.2021-2.RLTS.T7815A18979599.en.

Forcart, L. 1951. Nomenclature remarks on some generic names of the snake family Boidae. Herpetologica 7:197–199.

Franz, R., C.K. Dodd Jr., and D.W. Buden. 1993. Distributional records of amphibians and reptiles from the Exuma Islands, including the first reports of a freshwater turtle and an introduced gecko. Caribbean Journal of Science 29:165–173.

Freze, V.I., and B. Rysavy. 1976. Cestodes of the suborder Proteocephalata Spassky, 1957 (Cestoda: Pseudophyllidea) from Cuba, and description of a new species *Ophiotaenia habanensis* sp. n. Folia Parasitologica 23:97–104.

Frynta, D., T. Vejvodová, and O. Šimková. 2016. Sex allocation and secondary sex ratio in Cuban boa (Chilabothrus angulifer): mother's body size affects the ratio between sons and daughters. Science of Nature 103:48.

Fuenmayor, G.R., G. Ugueto, R. Rivero, and A. Miralles. 2005. The herpetofauna of Isla de Margarita, Venezuela: new records and comments. Caribbean Journal of Science 41:346–351.

Funes Monzote, R. 2004. "El asiento de su riqueza." Los bosques y la ocupación del este de Cuba por el azúcar, 1898–1926. Anuario Instituto de Estudio Histórico Sociales 19:231–253.

Gaillard, S. 2020. Forestry division disturbed St Lucians killing and eating boas. Loop, September 15, 2020. https://stlucia.loopnews.com/content/forestry-division-disturbed-over-consumption-boas.

Gallardo, J.C., F.J. Vilella, and M.E. Colvin. 2019. A seasonal population matrix model of the Caribbean Red-tailed Hawk *Buteo jamaicensis jamaicensis* in eastern Puerto Rico. Ibis 161:459–466.

Gannon, M.R., A. Kurta, A. Rodríguez-Duran, and M.R. Willig. 2005. Bats of Puerto Rico: an island focus and a Caribbean perspective. Lubbock: Texas Tech University Press.

Garcia, M. 1992. Current status and distribution of *Epicrates monensis* in Puerto Rico. Department of Natural and Environmental Resources of Puerto Rico Project ES-I-8 Final Report (1991–1992), San Juan.

García, M., C.E. Diez, and A.O. Alvarez. 2001. The impact of feral cats on Mona Island Wildlife and recommendations for their control. Caribbean Journal of Science 37:107–108.

Garman, S. 1887. On West Indian reptiles in the Museum of Comparative Zoölogy, at Cambridge, Mass. Proceedings of the American Philosophical Society 24:278–286.

Garrido, O.H., and A. Schwartz. 1969. Anfibios, reptiles y aves de Cayo Cantiles. Poeyana, ser. A, 67:1–44.

Gerhardt, R.P., P.M. Harris, and M.A. Vásquez Marroquín. 1993. Food habits of nesting Great Black Hawks in Tikal National Park, Guatemala. Biotropica 25:349–352.

Gibbons, J.W., D.E. Scott, T.J. Ryan, K.A. Buhlmann, T.D. Tuberville, B.S. Metts, J.L. Greene, et al. 2000. The global decline of reptiles, déjà vu amphibians. BioScience 50:653–666.

Gibson, R. 1996a. *Chilabothrus subflavus*. IUCN Red List of Threatened Species 1996:e.T7826A12853495. http://dx.doi.org/10.2305/IUCN.UK.1996.RLTS.T7826A12853495.en.

Gibson, R.C. 1996b. The distribution and status of the Jamaican boa *Epicrates subflavus* Stejneger, 1901. Dodo, Journal of the Jersey Wildlife Preservation Trust 32:143–155.

Gibson, R., S.B. Hedges, and B.S. Wilson. 2021. *Chilabothrus subflavus*. IUCN Red List of Threatened Species 2021:e.T7826A18979286. https://dx.doi.org/10.2305/IUCN.UK.2021-2.RLTS.T7826A18979286.en.

Gillingham, J.C. 1987. Social behavior. *In* R.A. Seigel, J.T. Collins, and S.S. Novak, eds., Snakes: ecology and evolutionary biology, pp. 184–209. New York: Macmillan.

Giovas, C.M., M.J. LeFebvre, and S.M. Fitzpatrick. 2012. New records for prehistoric introduction of Neotropical mammals to the West Indies: evidence from Carriacou, Lesser Antilles. Journal of Biogeography 39:476–487.

Glenn, M.E. 1998. Population density of *Cercopithecus mona* on the Caribbean island of Grenada. Folia Primatologica 69:167–171.

Godínez, E., M. Gómez, J.A. Puentes, and S. Vargas. 1987. Características reproductivas de *Columba leucocephala* en la Península de Guanahacabibes, Cuba. Poeyana 340:1–8.

Golden, I. 2017. Origin of the invasive *Boa constrictor* population in St. Croix, U.S. Virgin Islands. University of North Carolina Asheville Journal of Undergraduate Research 2017: 283–288.

Gonzalez, R.C., T.B. Guedes, and P. Passos. 2021. Back in time to unlock the future: tracing the type-specimens of *Corallus hortulana* (Linnaeus, 1758) (Serpentes: Boidae), with designation of a lectotype for the Amazon tree boa. Zootaxa 4941:259–270.

González-Baca, C.A. 2006. Ecología de forrajeo de *Boa constrictor*. Un depredator introducido a la Isla Cozumel. MS thesis, Universidad Nacional Autonoma de México, Mexico City.

Gosse, P.H. 1851. A naturalist's sojourn in Jamaica. London: Longman, Brown, Green, and Longmans.

Gould, W.A., C. Alarcón, B. Fevold, M.E. Jimenez, S. Martinuzzi, G. Potts, M. Quiñones, M. Solórzano and E. Ventosa. 2008. The Puerto Rico gap analysis project. Vol. 1, Land cover, vertebrate species distributions, and land stewardship. U.S. Department of Agriculture, Forest Service, International Institute of Tropical Forestry, Río Piedras, Puerto Rico, General Technical Report IITF-39.

Grant, C. 1932a. Herpetology of Tortola; notes on Anegada and Virgin Gorda, British Virgin Islands. Journal of the Department of Agriculture of Puerto Rico 16:339–346.

Grant, C. 1932b. Notes on the boas of Puerto Rico and Mona. Journal of the Department of Agriculture of Puerto Rico 16:327–329.

Grant, C. 1933. Notes on *Epicrates inornatus*. Copeia 1933:224–225.

Grant, C. 1940. The reptiles. *In* The herpetology of Jamaica. Bulletin of the Institute of Jamaica 1:61–148.

Grant, P.R., and Grant, B.R. 2011. How and why species multiply: the radiation of Darwin's finches. Princeton, NJ: Princeton University Press.

Gray, J.E. 1842. Synopsis of prehensile-tailed snakes, or family Boidae. Zoological Miscellany 41-46.

Green, S. 2011. Ecology, conservation and commercial exploitation of the 'Hog Island' *Boa constrictor* in the Cayos Cochinos, Honduras. PhD dissertation, University of Kent, United Kingdom.

Groome, J.R. 1970. A natural history of the island of Grenada. Arima, Trinidad: Caribbean Printers.

Guglielmone, A.A., A. Estrada-Peña, J.E. Keirans, and R.G. Robbins. 2003. Ticks (Acari: Ixodida) of the Neotropical zoogeographic region. International Consortium on Ticks and Tick-borne Diseases, Atalanta, Hounten, Netherlands.

Gunderson, A.R., and M. Leal. 2012. Geographic variation in vulnerability to climate warming in a tropical Caribbean lizard. Functional Ecology 26:783–793.

Gundlach, J. 1883. Ornitología Cubana. Havana: La Moderna.

Gundlach, J. 1894. Erpetologia Cubana. Havana: Imprenta de G. Montiel y Ca.

Gundlach, J.C. 1880. Contribución a la erpetología Cubana. Havana: Imprenta de G. Montiel y Ca.

Günther, A. 1861. On a new species of the family Boidae. Proceedings of the Zoological Society of London 1861:142.

Günther, A. 1888. Notes on reptiles and frogs from Dominica, West Indies. Annals and Magazine of Natural History; Zoology, Botany, and Geology 6:362–366.

Hakluyt, R. ed. 1904. The voyage of captaine René Laudonniere to Florida in 1564. *In* The principal navigations voyages traffiques & discoveries of the English nation, vol. 9. Glasgow: James MacLehose and Sons.

Hanlon, R.W. 1964. Reproductive activity of the Bahaman boa (*Epicrates striatus*). Herpetologica 20:143–144.

Hanslowe, E.B., J.G. Duquesnel, R.W. Snow, B.G. Falk, A.A.Y. Adams, E.F. Metzger III, M.A.M. Collier, and R.N. Reed. 2018. Exotic predators may threaten another island ecosystem: a comprehensive assessment of python and boa reports from the Florida Keys. Management of Biological Invasions 9:369–377.

Hardy, J.D., Jr. 1957. Bat predation by the Cuban boa (*Epicrates angulifer*). Copeia 2:151–152.

Hardy, L.M., and R.W. McDiarmid. 1969. The amphibians and reptiles of Sinaloa, México. University of Kansas Publications, Museum of Natural History 18:39–252.

Harvey, D.S., and R.J. Platenberg. 2009. Predicting habitat use from opportunistic observations: a case study of the Virgin Islands tree boa (*Epicrates granti*). Herpetological Journal 19:111–118.

Haviser, J. 1997. Settlement strategies in the Early Ceramic Age. *In* S.M. Wilson, The indigenous people of the Caribbean, pp. 59–69. Gainesville: University of Florida Press.

Hedges, S.B. 2006. An overview of the evolution and conservation of West Indian amphibians and reptiles. Applied Herpetology 3:281–292.

Hedges, S.B., and C.E. Conn. 2012. A new skink fauna from Caribbean islands (Squamata, Mabuyidae, Mabuyinae). Zootaxa 3288:1–244.

Hedges, S.B., C.A. Hass, and T.K. Maugel. 1989. Physiological color change in snakes. Journal of Herpetology 23:450–455.

Hedges, S.B., R. Powell, R.W. Henderson, S. Hanson, and J.C. Murphy. 2019. An expanded definition of the Caribbean for biodiversity and conservation, with a checklist and recommendations for standardized common names of amphibians and reptiles. Caribbean Herpetology 67:1–53.

Helmer, E.H., T.A. Kennaway, D.H. Pedreros, M.L. Clark, H. Marcano-Vega, L.L. Tieszen, T.R. Ruzycki, S.R. Schill, and C.M.S. Carrington. 2008. Land cover and forest formation distributions for St. Kitts, Nevis, St. Eustatius, Grenada and Barbados from decision tree classification of cloud-cleared satellite imagery. Caribbean Journal of Science 44:175–198.

Henderson, R.W. 1990. Correlation of environmental variables and dorsal color in *Corallus enydris* (Serpentes: Boidae) on Grenada: some preliminary results. Caribbean Journal of Science 26:166–170.

Henderson, R.W. 1992. Consequences of predator introductions and habitat destruction on amphibians and reptiles in the post-Columbus West Indies. Caribbean Journal of Science 28:1–10.

Henderson, R.W. 1993. Foraging and diet in West Indian *Corallus enydris* (Serpentes: Boidae). Journal of Herpetology 27:24–28.

Henderson, R.W. 1997. A taxonomic review of the *Corallus hortulanus* complex of Neotropical tree boas. Caribbean Journal of Science 33:198–221.

Henderson, R.W. 2002. Neotropical treeboas: natural history of the *Corallus hortulanus* complex. Malabar, FL: Krieger Publishing.

Henderson, R.W. 2004. Correlation among dorsal body, iris, and tongue color in a local population of treeboas (*Corallus grenadensis*) on Grenada, Lesser Antilles. Caribbean Journal of Science 40:270–273.

Henderson, R.W. 2015. Natural history of Neotropical treeboas (genus *Corallus*). Frankfurt am Main, Germany: Edition Chimaira.

Henderson, R.W. 2019. The eyes have it: watching treeboas on the Grenada Bank. *In* H. B. Lillywhite and M. Martins, eds., Islands and snakes: isolation and adaptive evolution, pp. 156–180. New York: Oxford University Press.

Henderson, R.W., and A. Arias B[arreto]. 2001. *Epicrates angulifer.* Catalogue of American Amphibians and Reptiles 734:1–4.

Henderson, R.W., and C.S. Berg. 2005. A post-Hurricane Ivan assessment of frog and reptile populations on Grenada, West Indies. Herpetological Bulletin 91:4–9.

Henderson, R.W., and C.S. Berg. 2012. *Ameiva ameiva* (Squamata: Teiidae) and *Corallus grenadensis* (Squamata: Boidae) on Petite Martinique (Grenada Grenadines), West Indies. Herpetological Notes 5:439–440.

Henderson, R.W., and M. Breuil. 2012. Lesser Antilles. *In* R. Powell and R.W. Henderson, eds., Island lists of West Indian amphibians and reptiles, pp. 148–159. Bulletin of the Florida Museum of Natural History 51:85–166.

Henderson, R.W., and K.F. Henderson. 1995. Altitudinal

variation in body temperature in foraging tree boas (*Corallus enydris*) on Grenada. Caribbean Journal of Science 31:73–76.

Henderson, R.W., and M.J. Pauers. 2012. On the diets of Neotropical treeboas (Squamata: Boidae: *Corallus*). South American Journal of Herpetology 7:172–180.

Henderson, R.W., and R. Powell. 2001. Responses by the West Indian herpetofauna to human-influenced resources. Caribbean Journal of Science 37:41–54.

Henderson, R.W., and R. Powell. 2004. *Epicrates fordii*. Catalogue of American Amphibians and Reptiles 800:1–3.

Henderson, R.W., and R. Powell. 2009. Natural history of West Indian reptiles and amphibians. Gainesville: University Press of Florida.

Henderson, R.W., and R. Powell. 2018. Amphibians and reptiles of the St. Vincent and Grenada Banks, West Indies. Frankfurt am Main, Germany: Edition Chimaira.

Henderson, R.W., and R. Powell. 2021a. *Corallus cookii*. IUCN Red List of Threatened Species 2021:e. T203209A2762187. https://dx.doi.org/10.2305/IUCN.UK.2021-2.RLTS.T203209A2762187.en.

Henderson, R.W., and R. Powell. 2021b. *Corallus grenadensis*. IUCN Red List of Threatened Species 2021:e. T44580027A44580035. https://dx.doi.org/10.2305/IUCN.UK.2021-2.RLTS.T44580027A44580035.en.

Henderson, R.W., and R.A. Sajdak. 1983. Notes on reptiles from Isla Saona, República Dominicana. Florida Scientist 46:59–61.

Henderson, R.W., and R.A. Winstel. 1995. Aspects of habitat selection by an arboreal boa (*Corallus enydris*) in an area of mixed agriculture on Grenada. Journal of Herpetology 29:272–275.

Henderson, R.W., and R.A. Winstel. 1997. Daily activity in tree boas (*Corallus grenadensis*) on Grenada. Herpetological Natural History [1998] 5:175–180.

Henderson, R.W., T.A. Noeske-Hallin, J.A. Ottenwalder, and A. Schwartz. 1987. On the diet of the boa *Epicrates striatus* on Hispaniola, with notes on *E. fordi* and *E. gracilis*. Amphibia–Reptilia 8:251–258.

Henderson, R.W., R.A. Winstel, and J. Friesch. 1996. *Corallus enydris* (Serpentes: Boidae) in the post-Columbian West Indies: new habitats, new prey species, and new predators. *In* R. Powell and R.W. Henderson, eds., Contributions to West Indian herpetology: a tribute to Albert Schwartz. Contributions to Herpetology 12:417–423. Ithaca, NY: Society for the Study of Amphibians and Reptiles.

Henderson, R.W., R.A. Sajdak, and R.A. Winstel. 1998. Habitat utilization by the arboreal boa *Corallus grenadensis* in two ecologically disparate habitats on Grenada. Amphibia–Reptilia 19:203–214.

Henderson, R.W., M.L. Treglia, and S.D. Powell. 2007. *Corallus cookii* (St. Vincent treeboa). Foraging. Herpetological Review 38:466.

Henderson, R.W., C.S. Berg, B. Harrison, and D.T. Yorks. 2009. Notes on an unexpected decline of a population of *Corallus grenadensis* (Squamata: Boidae) in Grenada, West Indies. South American Journal of Herpetology 4:186–192.

Henderson, R.W., C.S. Berg, B. Harrison, R.A. Sajdak, and P. Felski. 2013a. Fifty! Reptiles & Amphibians 20:184–189.

Henderson, R.W., M.J. Pauers, and T.J. Colston. 2013b. On the congruence of morphology, trophic ecology, and phylogeny in Neotropical treeboas (Squamata: Boidae: *Corallus*). Biological Journal of the Linnean Society 109:466–475.

Henderson, R.W., B. Harrison, C.S. Berg, and E.M. Rush. 2015. Glimpses of social behavior in Grenada Bank treeboas (*Corallus grenadensis*). IRCF Amphibians & Reptiles 22:98–101.

Henderson, R.W., E.T. Hileman, R.A. Sajdak, B.C. Harrison, R. Powell, and D. R. Bradke. 2021a. Effects of body size, diet, and transience on the demography of the arboreal boid snake *Corallus grenadensis* on Carriacou (Grenada Grenadines, West Indies). Population Ecology 63:177–188.

Henderson, R.W., S. Incháustegui, and M.A. Landestoy T. 2021b. *Chilabothrus gracilis*. IUCN Red List of Threatened Species 2021:e.T15155194A15155225. https://dx.doi.org/10.2305/IUCN.UK.2021-2.RLTS.T15155194A15155225.en.

Henderson, R.W., R.A. Sajdak, B.C. Harrison, and R. Powell. 2021c. *Corallus grenadensis*. Foraging and post-predation behavior. Herpetological Review 52:662.

Hernandez Martínez, F.R., and O. Pimentel . 2005. Enfermedades, parásitos y depredadores de la jutía conga (*Capromys pilorides* Say) en el macizo forestal central de la cordillera de Guaniguanico. Revista Electrónica de Veterinaria 6.

Herrmann, N.C., S. Yates, J.R. Fredette, M.K. Leavens, R. Moretti, and R.G. Reynolds. 2018. Lizards on islands within islands: microhabitat use, movement, and cannibalism in *Anolis sagrei* and *Anolis smaragdinus*. Caribbean Naturalist 50:1–17.

Hileman, E.T., R. Powell, G. Perry, K. Mougey, R. Thomas, and R.W. Henderson. 2017. Demography of the Puerto Rican racer, *Borikenophis portoricensis* (Squamata: Dipsadidae), on Guana Island, British Virgin Islands. Journal of Herpetology 51:454–460.

Hoagland, D.B., G.R. Horst, and C.W. Kilpatrick. 1989. Biogeography and population biology of the mongoose in the West Indies. *In* C.A. Woods, ed., Biogeography of the West Indies: past, present, and future, pp. 611–633. Gainesville, FL: Sandhill Crane Press.

Hoefer, S., N.J. Robinson, and T. Pinou. 2021. Size matters: sexual dimorphism in the pelvic spurs of the Bahamian boa (*Chilabothrus strigilatus strigilatus*). Herpetology Notes 14:201–203.

Holanova, V., and J. Hribal. 2004. Dos anolis cubanos: *Anolis bartschi* y *Anolis lucius*. Reptilia 47:62–68.

Horst, G.R., D.B. Hoagland, and C.W. Kilpatrick. 2001. The mongoose in the West Indies: the biogeography and population biology of an introduced species. *In* C.A. Woods and F.E. Sergile, eds., Biogeography of the West Indies: patterns and perspectives, 2nd edition, pp. 409–424. Boca Raton, FL: CRC Press.

Horlbeck, G. 1988. Successful breeding of the Cuban boa—*Epicrates angulifer*. Snake Keeper 2:6–8.

Howard, R.A. 1952. The vegetation of the Grenadines, Windward Islands, British West Indies. Contribution of the Gray Herbarium, Harvard University 174:1–129

Hübener, D.R. 1898. Die Inseln Mona und Monito. Globus 74:368–372.

Huey, R.B., C.A. Deutsch, J.J. Tewksbury, L.J. Vitt, P.E. Hertz, H.J. Álvarez-Pérez, and T. Garland. 2009. Why tropical forest lizards are vulnerable to climate warming. Proceedings of the Royal Society B 276: 1939–1948.

Huey, R.B., M.R. Kearney, A. Krockenberger, J.A.M. Holtum, M. Jess, and S.E. Williams. 2012. Predicting organismal vulnerability to climate warming: roles of behaviour, physiology and adaptation. Philosophical Transactions of the Royal Society B 367:1665–1679.

Huff, T.A. 1976. Breeding the Cuban boa *Epicrates angulifer* at the Reptile Breeding Foundation. International Zoo Yearbook 16:81–82.

Huff, T.A. 1978. Breeding the Puerto Rican boa at the Reptile Breeding Foundation. International Zoo Yearbook 18:96–97.

Huff, T.A. 1979. Breeding the Jamaican boa (*Epicrates subflavus*) in captivity. Proceedings of the American Association of Zoological Parks and Aquariums, Regular Workshop, Wheeling, West Virginia, pp. 339–345.

Huff, T.A. 1980. Captive propagation of the subfamily Boidae with emphasis on the genus *Epicrates*. *In* Murphy, J.B. and J.T. Collins, eds., Reproductive biology and diseases of captive reptiles, pp. 125–134. Society for the Study of Amphibians and Reptiles Contributions to Herpetology 1.

Isaacs, M. 2014. Report on the smuggling of Bahamian rock iguanas. Sixty-fifth Meeting of the Standing Committee of the Convention on International Trade in Endangered Species of Wild Fauna and Flora, Geneva, Switzerland, SC65 Information Document 4:1–6.

Island Conservation. 2018. Post-hurricane assessments on the Puerto Rican Bank focusing on habitat suitability for the highly endangered VI Boa (*Chilabothrus granti*) and other Caribbean priority species. Island Conservation report to U.S. Fish and Wildlife Service, 1614F17AC01191, CFDA 15.630.

Iturralde-Vinent, M.A., R.D.E. MacPhee, S. Díaz-Franco, R. Rojas-Consuegra, W. Suárez, and A. Lomba. 2000. Las Breas de San Felipe, a Quaternary fossiliferous asphalt seep near Martí (Matanzas Province, Cuba). Caribbean Journal of Science 36:300–313.

Island Resources Foundation [and Jost van Dyke Preservation Society]. 2009. An environmental profile of the Island of Jost van Dyke, British Virgin Islands, including Little Jost van Dyke, Sandy Cay, Green Cay and Sandy Spit. Jost van Dyke, British Virgin Islands: Jost van Dyke Preservation Society. Available at: http://www.jvdps.org.

Iverson, J.B. 1978. The impact of feral cats and dogs on populations of the West Indian rock iguana, *Cyclura carinata*. Biological Conservation 14:63–73.

Jaimez Salgado, E., D. Gutierrez Calvache, R.D.E. MacPhee, and G.C. Gould. 1992. The monkey caves of Cuba. Cave Science 19:25–28.

Jan, G. 1863. Elenco sistematico degli ofidi descriti e disegnati per l'iconografia generale. Milano: A. Lombardi.

Jan, G. 1864. Iconographie générale des ophidiens. Paris: J.-B. Baillière et fils.

Jiménez, A., D. Denis, M. Acosta, L. Mugica, O. Torres, and A. Rodríguez. 2002. Algunos aspectos de la ecología reproductiva de la Cachiporra (*Himantopus mexicanus*) en una colonia de nidificación en la ciénaga de Birama, Cuba. El Pitirre 15:34–37.

Jiménez Vázquez, O., M.M. Condis, and E. García Cancio. 2005. Vertebrados post-glaciales en un residuario fósil de *Tyto alba scopoli* (Aves: Tytonidae), en el occidente de Cuba. Revista Mexicana de Mastozoología 9:85–112.

Joglar, R. L., A.O. Álvarez, T.M. Aide, D. Barber, P.A. Burrowes, M.A. García, A. León-Cardona, et al. 2011. Conserving the Puerto Rican herpetofauna. *In* A. Hailey, B. Wilson, and J. Horrocks, eds., Conservation of Caribbean island herpetofaunas. Vol. 2, Regional accounts of the West Indies, pp. 339–358. Leiden, Netherlands: Brill.

Joseph, C. 2019. Snake alert! Boa constrictor captured near Horsford Hill. Antigua Observer, December 19, 2019. https://antiguaobserver.com/snake-alert-boa-constrictor-captured-at-horsford-hill/.

Juventud Revelde. 2007. Encuentran majá vivo con dos cabezas. El ejemplar de Santa María rompió la armonía en un batey en la provincia cubana de Matanzas. Diario Juventud Revelde, January 4, 2007. http://www.juventudrebelde.cu/cuba/2007-01-04/encuentran-maja-vivo-con-dos-cabezas.

Keegan, W.F. 1994. West Indian archaeology. 1. Overview and foragers. Journal of Archaeological Research 2:255–284.

Keegan, W.F. 1996. West Indian archaeology. 2. After Columbus. Journal of Archaeological Research 4:265–294.

Kemp, M.E., A.M. Mychajliw, J. Wadman, and A. Goldberg. 2020.7000 years of turnover: historical contingency and human niche construction shape the Caribbean's Anthropocene biota. Proceedings of the Royal Society B 287:20200447.

Kennerley, R.J., M.A. Nicoll, R.P. Young, S.T. Turvey, J.M. Nuñez-Miño, J.L. Brocca, and S.J. Butler. 2019. The impact of habitat quality inside protected areas on distribution of the Dominican Republic's last endemic non-volant land mammals. Journal of Mammalogy 100:45–54.

Kephart, R. 2019. Carriacou animal stories. Anthropology News, May 16, 2019. https://doi.org/10.1111/AN.1169.

King, W., and T. Krakauer. 1966. The exotic herpetofauna of southeast Florida. Quarterly Journal of the Florida Academy of Sciences 29:144–154.

Kluge, A.G. 1988a. Parsimony in vicariance biogeography: a quantitative method and a Greater Antillean example. Systematic Zoology 37:315–328.

Kluge, A.G. 1988b. Relationships of the Cenozoic boine snakes *Paraepicrates* and *Pseudoepicrates*. Journal of Vertebrate Paleontology 8:229–230.

Kluge, A.G. 1989. A concern for evidence and a phylogenetic hypothesis of relationships among *Epicrates* (Boidae: Serpentes). Systematic Zoology 38:7–25.

Kluge, A.G. 1991. Boine snake phylogeny and research cycles. University of Michigan Museum of Zoology Miscellaneous Publications 178.

Knapp, C.R., and A.K. Owens. 2004. Diurnal refugia and novel ecological attributes of the Bahamian boa, *Epicrates striatus fowleri* (Boidae). Caribbean Journal of Science 40:265–270.

Knapp, C.R., M. Greenaway, A. James, and L. Prince. 2009. *Boa constrictor*. Diet. Herpetological Review 40:229.

Knight, R.L., and A.W. Erickson. 1976. High incidence of snakes in the diet of nesting Red-tailed Hawks. Journal of Raptor Research 10:108–111.

Koenig, N. 1953. A comprehensive agricultural program for Puerto Rico. Washington, DC: U.S. Department of Agriculture and Commonwealth of Puerto Rico.

Koenig, S.E. 2001. The breeding biology of Black-billed Parrot *Amazona agilis* and Yellow-billed Parrot *A. collaria* in Cockpit Country, Jamaica. Bird Conservation International 11:205–225.

Koenig, S.E. 2019. Jamaican boa home range, attraction–avoidance behaviour, and habitat preferences in Cockpit Country, Jamaica. Windsor Research Centre, 2016–17. http://www.cockpitcountry.com/JamaicanBoaTelemetry.html.

Koenig, S.E., and M. Schwartz. 2003. *Epicrates subflavus* (yellow boa). Diet. Herpetological Review 34:374–375.

Koenig, S.E., J.W. Wunderle Jr., and E.C. Enkerlin-Hoeflich. 2007. Vines and canopy contact: a route for snake predation on parrot nests. Bird Conservation International 17:79–91.

Koenig, S., C. Levy, and V. Turland. 2015. Jamaica's endemic parakeet *Eupsittula* [formerly *Aratinga*] *nana* (Vigors, 1830): Natural history and conservation status. Broadsheet 100:6–18.

Koopman, K., and R. Ruibal. 1955. Cave-fossil vertebrates from Camaguey, Cuba. Breviora 46:1–8.

Krysko, K.L., J.P. Burgess, M.R. Rochford, C.R. Gillette, D. Cueva, K.M. Enge, L.A. Somma, et al. 2011. Verified non-indigenous amphibians and reptiles in Florida from 1863 through 2010: outlining the invasion process and identifying invasion pathways and stages. Zootaxa 3028:1–64.

Krysko, K.L., D.W. Steadman, J.I. Mead, NA. Albury, C.A. MacKenzie-Krysko, and S.L. Swift. 2013. New island records for amphibians and reptiles on the Little Bahama Bank, Commonwealth of the Bahamas. Amphibians and Reptiles: Conservation and Natural History 20:152–154.

Krysko, K.L., L.A. Somma, D.C. Smith, C.R. Gillette, D. Cueva, J.A. Wasilewski, K.M. Enge, et al. 2016. New verified nonindigenous amphibians and reptiles in Florida through 2015, with a summary of over 152 years of introductions. Reptiles & Amphibians 23:110–143.

Labat, J.-B. 1724. Nouveau voyage aux isles de l'Amérique. Paris: Guillaume Cavelier.

Landestoy T., M.A., R.W. Henderson, and S. Incháustegui.

2018. *Chilabothrus striatus*. IUCN Red List of Threatened Species 2018:e.T62231A18979323. http://dx.doi.org/10.2305/IUCN.UK.2018-2.RLTS.T62231A18979323.en.

Landestoy T., M.A., S. Incháustegui, and R.W. Henderson. 2021a. *Chilabothrus fordii*. IUCN Red List of Threatened Species 2021:e.T15155091A15155181. https://dx.doi.org/10.2305/IUCN.UK.2021-2.RLTS.T15155091A15155181.en.

Landestoy T., M.A., R.G. Reynolds, and R.W. Henderson. 2021b. A small new arboreal species of West Indian boa (Boidae; *Chilabothrus*) from southern Hispaniola. Breviora 571:1–20.

Lando, R.V., and E.E. Williams. 1969. Notes on the herpetology of the U.S. Naval Base at Guantanamo Bay, Cuba. Studies on the Fauna Curaçao and other Caribbean Islands 31:159–201.

Laurenti, J.N. 1768. Specimen medicum, exhibens synopsis reptilium emendatam cum experimentis circa venena et antidota reptilium austriacorum. Vienna: J.T. Nobilis de Trattnern.

Lazell, J.D. 1980. Report: British Virgin Islands, 1980. Conservation Agency, Jamestown, RI. *In* J.D. Lazell. 2005. Island: fact and theory in nature, p. 382. Berkeley: University of California Press.

Lazell, J.D. 1983. Biogeography of the herpetofauna of the British Virgin Islands, with description of a new anole (Sauria: Iguanidae). *In* A.G.J. Rhodin and K. Miyata, eds., Advances in herpetology and evolutionary biology, pp 99–117. Cambridge, MA: Museum of Comparative Zoology.

Lazell, J.D. 2005. Island: fact and theory in nature. Berkeley: University of California Press.

Lazell, J.D., Jr. 1964. The Lesser Antillean representatives of *Bothrops* and *Constrictor*. Bulletin of the Museum of Comparative Zoology 132:245–273.

LeFebvre, M.J. 2007. Zooarchaeological analysis of prehistoric vertebrate exploitation at the Grand Bay site, Carriacou, West Indies. Coral Reefs 26:931–944.

Leite, G.A., and T.F. Dorado-Rodrigues. 2017. *Epicrates cenchria* (Salamanta; rainbow boa) Diet. Herpetological Review 48:449–450.

Lemke, T.O. 1978. Predation upon bats by *Epicrates cenchris* in Colombia. Herpetological Review 9:47.

Lescure, J. 1979. Étude taxinomique et éco-éthologique d'un amphibian des petites Antilles: *Leptodactylus fallax* Muller, 1926 (Leptodactylidae). Bulletin du Muséum National d'Histoire Naturelle, Paris 4, ser. 1, sect. A, no. 3:757–774.

Lescure, J. 2001. Caractéristiques biogéographiques des Petites Antilles et herpétofaune. *In* J.-L. d'Hondt and J. Lorenz, éds., L'exploration naturaliste des Antilles et de la Guyane, pp. 95–106. Actes du 123e Congrès national des Sociétés historiques et scientifiques, Antilles-Guyane, 6–10 avril 1998. Paris: Comité des Travaux Historiques et Scientifiques.

Lescure, J., C. Bochaton, M. Breuil, I. Ineich, J-C De Massary, and N. Vidal. 2020. Liste taxinomique des serpentes des Petites Antilles. Bulletin de la Société Herpétologique de France 174:59–92.

Letsch, W. 1986. Haltung und vermehrung von *Epicrates angulifer* (Cocteau and Bibron, 1840). Elaphe 8(3): 41–44; 8(4): 74.

Lillywhite, H.W., and R.W. Henderson. 1993. Behavioral and functional ecology of arboreal snakes. *In* R.A. Siegel and J.T. Collins, eds., Snakes: ecology and behavior, pp. 1–48. New York: McGraw-Hill.

Linares Rodríguez, J.L., V. Berovides Álvarez, J.A. Camejo Lamas, L. Márquez Llauger, A. Rojas Valdez, and O. Borrego Fernández. 2011. Estudio de las características del hábitat, distribución geográfica y uso de la especie majá de Santa María (*Epicrates angulifer*) en la Reserva de Biosfera Península de Guanahacabibes. Cubazoo 23:27–32.

Linnaeus, C. 1758. Systema naturæ per regna tria naturæ, secundum classes, ordines, genera, species, cum characteribus, differentiis, synonymis, locis. Tomus I. Editio decima, reformata. Holmiæ [Stockholm]: Laurentii Salvii.

Lippold, L.K. 1991. Animal resource utilization by Saladoid peoples at Pearls, Grenada, West Indies. *In* E.N. Ayubi and J.B. Haviser, eds., Proceedings of the 13th International Congress for Caribbean Archeology. Reports of the Archaeological-Anthropological Institute of the Netherlands Antilles, Willemstad, Curaçao 9:264–268.

Long, E.G. 1974. The serpent's tale: reptiles and amphibians of St. Lucia. Iouanaloa, series 2. Morne, St. Lucia: University of the West Indies, Department of Extra-Mural Studies.

Longueira, A.R. 2006. Composición, distribución y conservación de la fauna exclusiva de las cuevas de calor de Cuba. MS thesis, Instituto de Geografía Tropical, Ministerio de Ciencia, Tecnología y Medio Ambiente, Havana.

Lorch, J.M., S. Knowles, J.S. Lankton, K. Michell, J.L. Edwards, J.M. Kapfer, R.A. Staffen, et al. 2016. Snake fungal disease: an emerging threat to wild snakes. Philosophical Transactions of the Royal Society B 37:20150457.

Lorvelec, O., M. Pascal, C. Pavis, and P. Feldmann. 2007. Amphibians and reptiles of the French West Indies: inventories, threats and conservation. Applied Herpetology 4:131–161.

Louppe, V., B. Leroy, A. Herrel, and G. Veron. 2020. The globally invasive small Indian mongoose *Urva auropunctata* is likely to spread with climate change. Scientific Reports. https://doi.org/10.1038/s41598-020-64502-6.

Lourdais, O., X. Bonnet, R. Shine, D. DeNardo, G. Naulleau, and M. Guillon. 2002. Capital-breeding and reproductive effort in a variable environment: a longitudinal study of viviparous snakes. Journal of Animal Ecology 71:470–479.

Lourdais, O., F. Brischoux, D. DeNardo, and R. Shine. 2004. Protein catabolism in pregnant females (*Epicrates cenchria maurus*, Boidae) compromises musculature and performance after parturition. Journal of Comparative Physiology B 174:383–391.

Lourdais, O., F. Brischoux, R. Shine, and X. Bonnet. 2005. Adaptive maternal cannibalism in snakes (*Epicrates cenchria maurus*, Boidae). Biological Journal of the Linnean Society 84:767–774.

Lynch, J.D. 2013. El context de las serpients de Colombia con un análisis de las amenazas en contra de su conservación. Revista de la Academia Colombiana de Ciencias Exactas, Físicas y Naturales 36:435–449.

Lynch, T.E. 1856. The wonders of the West Indies. London: Seeley, Jackson, and Halliday.

Machado-Filho, P.R., M.R. Duarte, L.F. do Carmo, and F.L. Franco. 2011. New record of *Corallus cropanii* (Boidae, Boinae): a rare snake from the Vale do Ribeira, State of São Paulo, Brazil. Salamandra 47:112–115.

MacLean, W.P. 1982. Reptiles and amphibians of the Virgin Islands. London: Macmillan.

Madsen, T., B. Stille, and R. Shine. 1996. Inbreeding depression in an isolated population of adders *Vipera berus*. Biological Conservation 75:113–118.

Malfait, B.T. and M.G. Dinkelman. 1972. Circum-Caribbean tectonic and igneous activity and the evolution of the Caribbean Plate. Geological Society of America Bulletin 83:251–271.

Malhotra, A., and R.S. Thorpe. 1999. Reptiles & amphibians of the eastern Caribbean. London: Macmillan Education.

Malhotra, A., R.S. Thorpe, E. Hypolite, and A. James. 2011. A report on the status of the herpetofaunas of the Commonwealth of Dominica, West Indies. *In* A. Hailey, B.S. Wilson, and J.A. Horrocks, eds., Conservation of Caribbean island herpetofaunas. Vol. 2, Regional accounts of the West Indies, pp. 149–166. Leiden, Netherlands: Brill.

Mallery, C.S., Jr., M.A. Marcum, R. Powell, J.S. Parmerlee Jr., and R.W. Henderson. 2007. Herpetofaunal communities of the leeward slopes and coasts of St. Vincent: a comparison of sites variously altered by human activity. Applied Herpetology 4:313–325.

Mancina, C.A. 2004. Bat community structure in an evergreen forest in western Cuba. Poeyana 491:8–12.

Mancina, C.A. 2011. Introducción a los murciélagos. *In* R. Borroto-Páez and C.A. Mancina, eds., Mamíferos en Cuba, pp. 123–133. Vaasa, Finland: UPC Print.

Mancina, C.A., and A. Llanes Sosa. 1997. Indicios de depredación de huevos de *Hirundo fulva* (Passeriformes: Hirundinidae) por *Epicrates angulifer* (Serpentes: Boidae). El Pitirre 10:95–96.

Mancina, C.A., L. García-Rivera, and B.W. Miller. 2012. Wing morphology, echolocation, and resource partitioning in syntopic Cuban mormoopid bats. Journal of Mammalogy 93:1308–1317.

Mann, P., 1999. Caribbean sedimentary basins; classification and tectonic setting from Jurassic to present. Sedimentary Basins of the World 4:3–31.

Marcano Vega, H. 2019. Los bosques de Puerto Rico, 2014. U.S. Department of Agriculture, Southern Forest Experimental Station, Asheville, North Carolina, Resource Bulletin SRS–224.

Marichal Arbona, E. 2016. El majá de Santa María, *Chilabothrus angulifer* (Squamata: Boidae), en el Archipiélago Jardines de la Reina: nuevo registro de distribución. Poeyana 503:64–65.

Martin, J.A. 2007. A–Z of Grenada heritage. Oxford: Macmillan Caribbean.

Martin-Kaye, P.H.A., 1969. A summary of the geology of the Lesser Antilles. Overseas Geology and Mineral Resources 10:172–206.

Martin-Solano, S., T. Toulkeridis, A. Addison, and W.E. Pozo-Rivera. 2016. Predation of *Desmodus rotundus* Geoffroy, 1818 (Phyllostomidae, Chiroptera) by *Epicrates cenchria* (Linnaeus, 1758) (Boidae, Reptilia) in an Ecuadorian cave. Subterranean Biology 19: 41–50.

Martínez-Morales, M.A., and A.D. Cuarón. 1999. *Boa constrictor*, an introduced predator threatening the endemic fauna on Cozumel Island, Mexico. Biodiversity and Conservation 8:957–963.

Martínez Reyes, M., and A. Arias Barreto. 2014. Composición y distribución de los reptiles terrestres diurnos. *In* D. Rodríguez Batista, A. Arias Barreto, and E. Ruiz Rojas, eds., Fauna terrestre del archipiélago de Sabana-Camagüey, Cuba, pp. 188–204. Havana: Editorial Academia.

Martins, M., and M.E. Oliveira. 1998. Natural history of snakes in forests of the Manaus Region, Central Amazonia, Brazil. Herpetological Natural History [1999] 6:78–150.

Martins, R.A., F. Vélez-García, C.V. Mira-Mendes, and Y. Le Pendu. 2018. A case of predation of *Hylaeamys laticeps* (Lund 1840) by *Epicrates cenchria* (Linnaeus, 1758) (Serpentes: Boidae) in the Atlantic Forest of southern Bahia, Brazil. Herpetology Notes 11:513–514.

Mayer, G.C., 2012. Puerto Rico and the Virgin Islands. *In* R. Powell, and R.W. Henderson, eds., Island lists of West Indian amphibians and reptiles. Bulletin of the Florida Museum of Natural History 51:136–147.

Mayer, G.C., and J.D. Lazell Jr. 1988. Distributional records for reptiles and amphibians from the Puerto Rico Bank. Herpetological Review 19:23–24.

McDiarmid, R.W., T'S.A. Touré, and J.M. Savage. 1996. The proper name of the Neotropical tree boa often referred to as *Corallus enydris* (Serpentes: Boidae). Journal of Herpetology 30:320–326.

Mead, J.I., and D.W. Steadman. 2017. Late Pleistocene snakes (Squamata: Serpentes) from Abaco, the Bahamas. Geobios 50:431–440.

Medeiros de Pinho, G., D. Oliveira de Lima, P. Nogueira Da Costa, and F.A. Dos Santos Fernandez. 2009. *Epicrates cenchria* (Brazilian rainbow boa). Diet. Herpetological Review 40:354–355.

Meerwarth, H. 1901. Die Westindischen reptilien und batracher des Naturhistorischen Museums im Hamburg. Mitteilungen aus dem Naturhistorischen Museum in Hamburg 18:1–41.

Mercado, J.E., E. Terranova, and J.M. Wunderle Jr. 2002. Avian mobbing of the Puerto Rican boa (*Epicrates inornatus*). Caribbean Journal of Science 38:125–126.

Mertens, R. 1939. Herpetologische ergebnisse einer reise nach der insel Hispaniola, Westindien. Abhandlungen der Senckenberg Naturforschenden Gesellschaft 449:1–84.

Micucci, P.A., and T. Waller. 2007. The management of the yellow anaconda (*Eunectes notaeus*) in Argentina: from historical misuse to resource appreciation. Iguana 14:160–171.

Miersma, E.E. 2010. Movements, activity range, habitat use, and conservation of the Jamaican (yellow) boa, *Epicrates subflavus*. MS thesis, University of Montana, Missoula.

Miller, G.S., Jr. 1904. Notes on the bats collected by William Palmer in Cuba. Proceedings of the United States National Museum 27:337–348.

Ministerio de Justicia. 2011. Resolución No. 160. Regulaciones para el control y la protección de especies de especial significación para la diversidad biológica en el país. Gaceta Oficial de la República de Cuba 26:723–745.

Miranda, E.B.P., R.P. Ribeiro Jr, B.F. Camera, M. Barros, J. Draque, P. Micucci, T. Waller, and C. Strussmann. 2017. Penny and penny laid up will be many: large yellow anacondas do not disregard small prey. Journal of Zoology 301:301–309.

Mistretta, B.A. 2019. Grenada's extinct rice rats (Oryzomyini): zooarchaeological evidence for taxonomic diversity. Journal of Archaeological Science: Reports 24:71–79.

Mitchell, N., R. Haeffner, V. Veer, M. Fulfordgardner, W. Clerveaux, C.R. Veitch, and G. Mitchell. 2002. Cat eradication and the restoration of endangered iguanas (*Cyclura carinata*) on Long Cay, Caicos Bank, Turks and Caicos Islands, British West Indies. *In* C.R. Veitch and M.N. Clout, eds., Turning the tide: the eradication of invasive species, pp. 206–212. Gland, Switzerland: IUCN.

Mittermeier, M.G. 2011. Consumption of Bahamian racers (*Cubophis vudii*) by a boa (*Epicrates striatus strigilatus*) in captivity. Reptiles & Amphibians 18:214–215.

Montgomery, C.E., and O. da Cunha. 2018. *Boa imperator.* IUCN Red List of Threatened Species 2018:e. T203879A2771951. http://dx.doi.org/10.2305/IUCN.UK.2018-2.RLTS. T203879A2771951.en.

Montgomery, C.E., S.M. Boback, R.N. Reed, and J.A. Frazier. 2015. An assessment of the impact of the pet trade on five CITES-Appendix II case studies—*Boa constrictor imperator.* Twenty-eighth meeting of the Animals Committee, Conference Report, Convention of International Trade in Endangered Species of Wild Fauna and Flora, AC28 Information Document 7.

Morell Savall, E. 2009. Desarrollo y conducta del majá de Santa María en cautiverio, durante el primer año de vida. Cubazoo 20:65–67.

Morell Savall, E., R. Díaz Aguiar, and O. Alfonso Álvarez. 1998. El majá de Santa María (*Epicrates angulifer*): la boa de la mayor de las Antillas. Flora y Fauna 1:40–42.

Morgan, G.S., and C.A. Woods. 1986. Extinction and zoogeography of West Indian land mammals. *In* L.R. Heaney and B.D. Patterson, eds., Island biogeography of mammals, pp. 167–203. London: Linnaean Society.

Morgan, G.S., C.E. Ray, and O. Arredondo. 1980. A giant extinct insectivore from Cuba (Mammalia: Insectivora: Solenodontidae). Proceedings of the Biological Society of Washington 93:597–608.

Mugica, L., M. Acosta, and A. Sanz. 1989. Nidificación de la Gallareta Azul (*Gallinula martinica*). Miscelánea Zoológica 43:1–2.

Mulero Oliveras, E. 2019. Population ecology, spatial ecology, and habitat selection of the Puerto Rican boa (*Chilabothrus inornatus*) in an urban fragmented landscape. MS thesis, University of Puerto Rico–Mayagüez.

Muñoz, A., L. Gonzales, D. Embert, J. Aparicio, and R. Aguayo. 2016. *Eunectes beniensis*. IUCN Red List of Threatened Species 2016:e.T174126A18978378. http://dx.doi.org/102305/IUCN.UK2010-4.RLTS.T174126A18978378.en.

Murphy, J.B., and R.K. Guese. 1977. Reproduction in the Hispaniolan boa *Epicrates fordii fordii* at Dallas Zoo. International Zoological Yearbook 17:132–133.

Murphy, J.B., D.G. Barker, and B.W. Tryon. 1978. Miscellaneous notes on the reproductive biology of reptiles. 2. Eleven species of the family Boidae, genera *Candoia*, *Corallus*, *Epicrates* and *Python*. Journal of Herpetology 12:385–390.

Murphy, J.C. 1997. Amphibians and reptiles of Trinidad and Tobago. Malabar: Krieger.

Murray-Wallace, C., and C. Woodroffe. 2014. Quaternary sea-level changes: a global perspective. Cambridge: Cambridge University Press.

Myers, N., R.A. Mittermeier, C.G. Mittermeier, G.A.B. Fonseca, and J. Kent. 2000. Biodiversity hotspots for conservation priorities. Nature 403:853–858.

National Environment and Planning Agency. 2007. A recovery action plan for the Jamaican boa *Epicrates subflavus*: a look at the conservation needs of Jamaica's largest native terrestrial predator. Draft. Ministry of Housing and Environment, Kingston.

Nellis, D.W., R.L. Norton, and W.P. MacLean. 1983. On the biogeography of the Virgin Islands boa, *Epicrates monensis granti*. Journal of Herpetology 17:413–417.

Netting, M.G., and C.J. Goin. 1944. Another new boa of the genus *Epicrates* from the Bahamas. Annals of the Carnegie Museum 306:71–76.

Newman, B.C., S.E. Henke, S.E. Koenig, and R.L. Powell. 2016. Distribution and general habitat use analysis of the Jamaican boa (*Chilabothrus subflavus*). South American Journal of Herpetology 11:228–234.

Newman, B.C., S.E. Henke, D.B. Wester, T.M. Shedd, H.L. Perotto-Baldivieso, and D.C. Rudolph. 2019. Determining the suitability of the Jamaican boa (*Chilabothrus subflavus*) for short-distance translocation in Cockpit Country, Jamaica. Caribbean Journal of Science, 49:222–238.

Newman, B.C., S.E. Henke, R.L. Powell, and D.C. Rudolph. 2020. Review of the Jamaican boa (*Chilabothrus subflavus*): biology, ecology, and conservation management of a vulnerable species. Herpetological Conservation and Biology 15:390–408.

Newsom, L.A. and E.S. Wing. 2004. On land and sea: Native American uses of biological resources in the West Indies. Tuscaloosa: University of Alabama Press.

Nichols, H.A.A. 1891. Unpublished diary of a trip through the Grenadines. Cambridge, MA: Gray Herbarium Library, Harvard University.

Noonan, B.P., and P.T. Chippendale. 2006. Dispersal and vicariance: the complex evolutionary history of boid snakes. Molecular Phylogenetics and Evolution 40:347–358.

Noonan, B.P., and J.W. Sites Jr. 2010. Tracing the origins of iguanid lizards and boine snakes of the Pacific. American Naturalist 175:61–72.

Norberg, U.M., and J.M.V. Rayner. 1987. Ecological morphology and flight in bats (Mammalia; Chiroptera): wing adaptations, flight performance, foraging strategy and echolocation. Philosophical Transactions of the Royal Society B 316:335–427.

Nowinski, B. 1977. Voortplanting van de Cubaanse boa, *Epicrates angulifer*, in het terrarium. Lacerta 35:144–147.

Ober, F.A. 1899. Camps in the Caribees: the adventures of a naturalist in the Lesser Antilles. Boston: Lee and Shepard.

Oliver, W.L.R. 1982. The coney and the yellow snake: the distribution and status of the Jamaican hutia *Geocapromys brownii* and the Jamaica boa *Epicrates subflavus*. Dodo, Journal of the Jersey Wildlife Preservation Trust 19:6–33.

Onary, S., and A.S. Hsiou. 2018. Systematic revision of the early Miocene fossil *Pseudoepicrates* (Serpentes: Boidae): implications for the evolution and historical biogeography of the West Indian boid snakes (*Chilabothrus*). Zoological Journal of the Linnean Society 184:453–470.

Oswald, J.A., J.M. Allen, M.J. LeFebvre, B.J. Stucky, R.A. Folk, N.A. Albury, G.S. Morgan, R.P. Guralnick, and D.W. Steadman. 2020. Ancient DNA and high-resolution chronometry reveal a long-term human role in the historical diversity and biogeography of the Bahamian hutia. Scientific Reports 10:1–10.

Ottenwalder, J.A. 1980. *Epicrates striatus* como predador de aves. Naturalista Postal 21:1–2.

Ottenwalder, J.A. 1985. The distribution and habitat of *Solenodon* in the Dominican Republic. MS thesis, University of Florida, Gainesville.

Outhwaite, W. 2019. Proving legality: the trade in

endemic Caribbean reptiles. TRAFFIC and Flora & Fauna International. https://www.traffic.org/site/assests/files/12516/trade_in_endemic_reptiles.pdf.

Pendlebury, G.B. 1974. Stomach and intestine contents of *Corallus enydris*; a comparison of island and mainland specimens. Journal of Herpetology 8:241-244.

Penman, T.D., D.A. Pike, J.K. Webb, and R. Shine. 2010. Predicting the impact of climate change on Australia's most endangered snake, *Hoplocephalus bungaroides*. Diversity and Distributions 16:109–118.

Pérez Vigueras, I. 1934. On the ticks of Cuba, with description of a new species, *Amblyoma torrei*, from *Cyclura macleayi* Gray. Psyche 41:13–18.

Peters, J.A., and B. Orejas-Miranda. 1970. Catalogue of the Neotropical Squamata. Part I. Snakes. Washington, DC: Smithsonian Institution Press.

Petersen, C., F. Burns, and P.[J.] Tolson. 2007. Toledo Zoo and Navy partner to study Cuban boa: researchers use GIS and other technology to collect biological data. Currents (Winter) 2007:46–49.

Petersen, C., P.[J.] Tolson, and J. Jackson. 2015. Cuban boa helps to maintain ecosystem balance at Guantánamo Bay. Currents (Summer) 2015:38–41.

Pizzatto, L., and O.A.V. Marques. 2007. Reproductive ecology of boine snakes with emphasis on Brazilian species and a comparison to pythons. South American Journal of Herpetology 2:107–122.

Pizzatto, L., O. Marques, and K. Facure. 2009. Food habits of Brazilian boid snakes: overview and new data, with special reference to *Corallus hortulanus*. Amphibia-Reptilia 30:533–544.

Platenberg, R. 2021. *Chilabothrus granti*. IUCN Red List of Threatened Species 2021:e.T7829A18979910. https://dx.doi.org/10.2305/IUCN.UK.2021-2.RLTS.T7829A18979910.en.

Platenberg, R.J., and D.S. Harvey. 2010. Endangered species and land use conflicts: a case study of the Virgin Islands boa (*Epicrates granti*). Herpetological Conservation and Biology 5:548–554.

Poey, F. 1866. Repertorio físico-natural de la Isla de Cuba. Tomo 2. Havana: Imprenta de la viuda de Barcina y Comp.

Polo Leal, J.L., and L.V. Moreno. 2007. Reproducción y cría del majá de Santa María la boa de Cuba *Epicrates angulifer* (Bibron, 1843) en el Parque Zoológico Nacional de Cuba. Cubazoo 17:33–38.

Polo Leal, J.L., and T.M. Rodríguez[-]Cabrera. 2012. *Epicrates angulifer* Cocteau y Bibron, 1843. *In* H. González Alonso, L. Rodríguez Schettino, A. Rodríguez, C.A. Mancina, and I. Ramos García, eds., Libro rojo de los vertebrados de Cuba, pp. 160–164. Havana: Editorial Academia.

Potter, R.B., D. Barker, D. Conway, and T. Klak. 2004. The contemporary Caribbean. London: Pearson Education.

Powell, R., and R.W. Henderson. 2007. The St. Vincent (Lesser Antilles) herpetofauna: conservation concerns. Applied Herpetology 4:295–312.

Powell, R., and R.W. Henderson. 2008. Urban herpetology in the West Indies. *In* J.C. Mitchell, R.E. Jung Brown, and B. Bartholomew, eds., Urban herpetology. Herpetological Conservation 3, pp. 389–404. Salt Lake City: Society for the Study of Amphibians and Reptiles.

Powell, R., and R.W. Henderson, eds. 2012. Island lists of West Indian amphibians and reptiles. Bulletin of the Florida Museum of Natural History 51:85-166.

Powell, R., J.A. Ottenwalder, and S.J. Incháustegui. 1999. The Hispaniolan herpetofauna: diversity, endemism, and historical perspectives, with comments on Navassa Island. *In* B.I. Crother, Caribbean amphibians and reptiles, pp. 93-168. San Diego: Academic Press.

Powell, R., R.W. Henderson, M.C. Farmer, M. Breuil, A.C. Echternacht, G. van Buurt, C.M. Romagosa, and G. Perry. 2011. Introduced amphibians and reptiles in the Greater Caribbean: patterns and conservation implications. *In* A. Hailey, B.S. Wilson, and J.A. Horrocks, eds., Conservation of Caribbean island herpetofaunas. Vol. 1, Conservation biology and the wider Caribbean, pp. 63–143. Leiden, Netherlands: Brill.

Powell, R., R.W. Henderson, and J.S. Parmerlee Jr. 2015. The reptiles and amphibians of the Dutch Caribbean: Saba, St. Eustatius, and St. Maarten, 2nd edition. Nature Guide Series 004. Kralendijk, Bonaire: Dutch Caribbean Nature Alliance.

Powell, S.B., M.L. Treglia, R.W. Henderson, and R. Powell. 2007. Treeboas in the West Indies: responses of *Corallus cookii* and *C. grenadensis* to disturbed habitats. *In* R.W. Henderson and R. Powell, eds., Biology of the boas and pythons, pp. 375–386. Eagle Mountain, Utah: Eagle Mountain Publishing.

Pregill, G.K. 1981. An appraisal of the vicariance hypothesis of Caribbean biogeography and its application to West Indian terrestrial vertebrates. Systematic Zoology 27:1–16.

Pregill, G.K., D.W. Steadman, S.L. Olson, and F.V. Grady. 1988. Late Holocene fossil vertebrates from Burma Quarry, Antigua, Lesser Antilles. Smithsonian Contributions to Zoology 463:1–25.

Pregill, G.K., D.W. Steadman, and D.R. Watters. 1994. Late Quaternary vertebrate faunas of the Lesser Antilles: historical components of Caribbean biogeography. Bulletin of Carnegie Museum of Natural History 30.

Prior, K.A., and R.C. Gibson. 1997. Observations on the foraging behavior of the Jamaican boa, *Epicrates subflavus*. Herpetological Review 28:72–73.

Progscha, K.-H. 1972. Außergewöhnliche Heilung einer Verletzung bei einem jungen *Corallus enydris* (Serpentes, Boidae). Salamandra 8: 176.

Puente-Rolón, A.R. 1999. Foraging behavior, home range, movements, activity patterns and habitat characterization of the Puerto Rican boa (*Epicrates inornatus*) at Mata de Plátano Reserve in Arecibo, Puerto Rico. MS thesis, University of Puerto Rico–Mayagüez.

Puente-Rolón, A.R. 2001. Current status and distribution of the Virgin Islands boa (*Epicrates monensis granti*) in Puerto Rico. Final report. Department of Natural and Environmental Resources of Puerto Rico, San Juan.

Puente-Rolón, A. 2012. Reproductive ecology, fitness and management of the Puerto Rican boa (*Epicrates inornatus*, Boidae). PhD dissertation, University of Puerto Rico, Río Piedras.

Puente-Rolón, A.R., and F.J. Bird-Picó. 2004. Foraging behavior, home range, movements and activity patterns of *Epicrates inornatus* (Boidae) at Mata de Plátano Reserve in Arecibo, Puerto Rico. Caribbean Journal of Science 40:343–352.

Puente-Rolón, A.R., R.G. Reynolds, and L. Revell. 2013. Preliminary genetic analysis supports cave populations as targets for conservation in the endemic endangered Puerto Rican Boa (Boidae: *Epicrates inornatus*). PLOS ONE 8(5): e63899.

Puente-Rolón, A.R., S.I. Vega-Castillo, L. Stemberg, and E. Cuevas. 2016. Diet comparison of free-ranging and cave-associated Puerto Rican boas, *Chilabothrus inornatus* (Reinhardt, 1843) (Reptilia: Boidae), using stable carbon and nitrogen isotopes. Life: The Excitement of Biology 4:88–99.

Pyron, R.A., F.T. Burbrink, and J.J. Wiens. 2013. A phylogeny and revised classification of Squamata, including 4161 species of lizards and snakes. BMC Evolutionary Biology 13:93.

Pyron, R.A., R.G. Reynolds, and F.T. Burbrink. 2014. A taxonomic revision of boas (Serpentes: Boidae). Zootaxa 3846:249–260.

Quick, J.S., H.K. Reinert, E.R. de Cuba, and R.A. Odum, 2005. Recent occurrence and dietary habits of *Boa constrictor* on Aruba, Dutch West Indies. Journal of Herpetology 39:304–307.

Quinn, D.P., A.L. McTaggart, R.W. Henderson, and R. Powell. 2011. *Corallus grenadensis* (Grenada Bank treeboa, Congo snake). Habitat and abundance. Herpetological Review 42:438.

Rabe, A.M., N.C. Herrmann, K.A. Culbertson, C.M. Donihue, and S.R. Prado-Irwin. 2020. Post-hurricane shifts in the morphology of island lizards. Biological Journal of the Linnean Society 130:156–165.

Ramos Donato, C., M.A. Trindade Dantas, and P.A. da Rocha. 2012. *Epicrates cenchria* (rainbow boa). Diet and foraging behavior. Herpetological Review 43:343–344.

Reagan, D.P. 1984. Ecology of the Puerto Rican boa (*Epicrates inornatus*) in the Luquillo Mountains of Puerto Rico. Caribbean Journal of Science 20:119–127.

Reagan, D.P., and C.P. Zucca. 1982. Inventory of the Puerto Rican boa (*Epicrates inornatus*) in the Caribbean National Forest. Final Report to U.S. Forest Service, U.S. Department of Agriculture, Atlanta, GA.

Reed, R.N., and G.H. Rodda. 2009. Giant constrictors: biological and management profiles and an establishment risk assessment for nine large species of pythons, anacondas, and the boa constrictor. U.S. Geological Survey Open-File Report 2009–1202, Reston, VA.

Reed, R.N., S.M. Boback, C.E. Montgomery, S. Green, Z. Stevens, and D. Watson. 2007. Ecology and conservation of an exploited insular population of *Boa constrictor* (Squamata: Boidae) on the Cayos Cochinos, Honduras. *In* R.W. Henderson and R. Powell, eds., Biology of the boas and pythons, pp. 388–403. Eagle Mountain, Utah: Eagle Mountain Publishing.

Rehák, I. 1987. Color change in the snake *Tropidophis feicki* (Reptilia: Squamata: Tropidophiidae). Véstník Ceskoslovenské Spolecnosti Zoologické 51:300–303.

Rehák, I. 1996. International studbook of Cuban boas *Epicrates angulifer*. Prague: Zoological Garden.

Reinhart, I. Th. 1843. Beskrivelse ad nogle nye Slangerater. Danske Vid. Selsk Afhandl. 10:233–279.

Reynolds, R.G. 2011a. *Eleutherodactylus planirostris planirostris* (Cuban flathead frog), distribution. Caribbean Herpetology 27:1.

Reynolds, R.G. 2011b. *Hemidactylus mabouia* (tropical house gecko), distribution. Caribbean Herpetology 28:1.

Reynolds, R.G. 2011c. Islands, metapopulations, and archipelagos: genetic equilibrium and non-equilibrium dynamics of structured populations in the context of conservation. PhD dissertation, University of Tennessee, Knoxville.

Reynolds, R.G. 2011d. Status, conservation, and introduction of amphibians and reptiles in the Turks and Caicos Islands, British West Indies. *In* A. Hailey, B.S. Wilson, and J.A. Horrocks, eds., Conservation of Caribbean island herpetofaunas. Vol. 2, Regional accounts of the West Indies, pp. 377–406. Leiden, Netherlands: Brill.

Reynolds, R.G. 2012. *Epicrates chrysogaster.* Catalogue of American Amphibians and Reptiles 898:1–5.

Reynolds, R.G. 2017. *Chilabothrus argentum.* IUCN Red List of Threatened Species 2017:e.T118470875A118470877. https://dx.doi.org/10.2305/IUCN.UK.2017-3.RLTS.T118470875A118470877.en.

Reynolds, R.G., and S. Buckner. 2016. *Chilabothrus exsul.* IUCN Red List of Threatened Species 2016:e.T15155078A15155082. https://dx.doi.org/10.2305/IUCN.UK.2016-3.RLTS.T15155078A15155082.en.

Reynolds, R.G., and S. Buckner. 2019. *Chilabothrus strigilatus.* IUCN Red List of Threatened Species 2019:e.T74872197A74874898. https://dx.doi.org/10.2305/IUCN.UK.2019-2.RLTS.T74872197A74874898.en.

Reynolds, R.G., and S. Buckner.2021. *Chilabothrus chrysogaster.* IUCN Red List of Threatened Species 2021:e.T15154880A15154888. https://dx.doi.org/10.2305/IUCN.UK.2021-2.RLTS.T15154880A15154888.en.

Reynolds, R.G, and C. Deal. 2010. Where do all the babies go? Understanding the biology of juvenile rainbow boas. Green Pages, Times of the Islands Magazine (Winter) 93:34–36.

Reynolds, R.G., and G.P. Gerber. 2012. Ecology and conservation of the Turks Island boa (*Epicrates chrysogaster chrysogaster*: Squamata: Boidae) on Big Ambergris Cay. Journal of Herpetology 46:578–586.

Reynolds, R.G., and R.W. Henderson. 2018. Boas of the world (superfamily Booidae): a checklist with systematic, taxonomic, and conservation assessments. Bulletin of the Museum of Comparative Zoology 162:1–58.

Reynolds, R.G., and B.N. Manco. 2013. *Epicrates chrysogaster chrysogaster* (Turks Island boa). Climbing Behavior. Herpetological Review 44:690–691.

Reynolds, R.G., and M.L. Niemiller. 2009. Expedition report and recommendations for the Department of Environment and Coastal Resources. Unpublished technical report for the Ministry of Natural Resources, Turks and Caicos Islands.

Reynolds, R.G., and M.L. Niemiller. 2010. *Epicrates chrysogaster* (Southern Bahamas boa), distribution. Caribbean Herpetology 14:1.

Reynolds, R.G., and M.L. Niemiller. 2011. New scales: reptiles invade the Turks and Caicos. Green Pages, Times of the Islands Magazine (Spring) 94:30–33.

Reynolds, R.G., and A.R. Puente-Rolón. 2016. *Chilabothrus strigilatus* (Bahamas boa). Distribution. Herpetological Review 47:425.

Reynolds, R.G., and B.M. Riggs. 2011a. *Iguana iguana* (green iguana). Distribution. Herpetological Review 42:241.

Reynolds, R.G., and B.M. Riggs. 2011b. *Pantherophis guttatus* (corn snake). Distribution. Herpetological Review 42:243.

Reynolds, R.G., J. Batchasingh, and B.M. Riggs. 2011a. *Ramphotyphlops braminus* (Brahminy blind snake). Distribution. Herpetological Review 42:244.

Reynolds, R.G., G.P. Gerber, and B.M. Fitzpatrick. 2011b. Unexpected shallow genetic divergence in Turks Island boas (*Epicrates c. chrysogaster*) reveals single evolutionarily significant unit for conservation. Herpetologica 67:477–486.

Reynolds, R.G., M.L. Niemiller, and B.N. Manco. 2011c. *Epicrates chrysogaster chrysogaster* (Turks Island boa), maximum size. Herpetological Review 42:239.

Reynolds, R.G., B.M. Riggs, M. Hibbert, and E. Salamanca. 2011d. *Anolis equestris* (Knight anole). Distribution. Herpetological Review 42:239.

Reynolds, R.G., B.M. Riggs, M. Hibbert, and E. Salamanca. 2011e. *Trachemys scripta elegans* (red-eared slider). Distribution. Herpetological Review 42:239.

Reynolds, R.G., M.L. Niemiller, S.B. Hedges, A. Dornburg, A.R. Puente–Rolón, and L.J. Revell. 2013a. Molecular phylogeny and historical biogeography of West Indian boid snakes (*Chilabothrus*). Molecular Phylogenetics and Evolution 68:461–470.

Reynolds, R.G., A.R. Puente-Rolón, R.N. Reed, and L.J. Revell. 2013b. Genetic analysis of a novel invasion of Puerto Rico by an exotic constricting snake. Biological Invasions 15: 953–959.

Reynolds, R.G., M.L. Niemiller, and L.J. Revell. 2014a. Toward a tree-of-life for the boas and pythons: multilocus species-level phylogeny with unprecedented taxon sampling. Molecular Phylogenetics and Evolution 71:201–213.

Reynolds, R.G., A.R. Puente-Rolón, M. Barandiaran, and L.J. Revell. 2014b. Hispaniolan boa (*Chilabothrus striatus*) on Vieques Island, Puerto Rico. Herpetology Notes 7:121–122.

Reynolds, R.G., A.R. Puente-Rolón, R. Platenberg, R.K. Tyler, P.J. Tolson, and L.J. Revell. 2015. Large divergence and low diversity suggest genetically informed conservation strategies for the endangered Virgin Islands boa (*Chilabothrus monensis*). Global Ecology and Conservation 3:487–502.

Reynolds, R.G., D.C. Collar, S.A. Pasachnik, M.L. Niemiller, A.R. Puente-Rolón, and L.J. Revell. 2016a. Ecological specialization and morphological diversification in Greater Antillean boas. Evolution 70:1882–1895.

Reynolds, R.G., S.T. Giery, W. Jesse, and Q. Quach. 2016b. Preliminary assessment of road mortality in the northern Bahamas boa, *Chilabothrus exsul*. Caribbean Naturalist 34:1–10.

Reynolds, R.G., A.R. Puente-Rolón, K.J. Aviles-Rodriguez, A.J. Geneva, and N.C. Herrmann. 2016c. Discovery of a remarkable new boa from the Conception Island Bank, Bahamas. Breviora 549:1–19.

Reynolds, R.G., J. Burgess, G. Waters, and G.P. Gerber. 2017. *Chilabothrus chrysogaster chrysogaster* (Turks Island boa). Diet. Herpetological Review 48:857.

Reynolds, R.G., A.R. Puente-Rolón, J.P. Burgess, and B.O. Baker. 2018. Rediscovery and a redescription of the Crooked-Acklins boa, *Chilabothrus schwartzi* (Buden, 1975), comb. nov. Breviora 558:1–16.

Reynolds, R.G., J.P. Burgess, G. Waters, B.N. Manco, and G.P. Gerber. 2020a. Remarkable color pattern polymorphism and color variation in Turks Island boas, *Chilabothrus c. chrysogaster.* Journal of Herpetology 54: 337–346.

Reynolds, R.G., G. Colosimo, K. Peek, A. Vanerelli, K. Bradley, and G.P. Gerber. 2020b. *Chilabothrus chrysogaster chrysogaster* (Turks Island boa). Diet. Herpetological Review 51:610–611.

Reynolds, R.G., J.J. Kolbe, R.E. Glor, M. López-Darias, C.V. Gómez Pourroy, A.S. Harrison, K. de Queiroz, L.J. Revell, and J.B. Losos. 2020c. Phylogeographic and phenotypic outcomes of brown anole colonization across the Caribbean provide insight into the beginning stages of an adaptive radiation. Journal of Evolutionary Biology 33:468–494.

Reynolds, R.G., A.H. Miller, and A. Puente-Rolón, A. 2021. *Chilabothrus schwartzi.* IUCN Red List of Threatened Species 2021:e.T162923732A162923901. https://dx.doi.org/10.2305/IUCN.UK.2021-2.RLTS.T162923732A162923901.en.

Rick, T.C., P.V. Kirch, J.M. Erlandson, and S.M. Fitzpatrick. 2013. Archaeology, deep history, and the human transformation of island ecosystems. Anthropocene 4:33–45.

Ríos-Franceschi, A., J.G. García-Cancel, F.J. Bird-Píco, and L.D. Carrasquillo. 2016. Spatiotemporal changes of the herpetofaunal community in Mount Resaca and Luis Pena Cay, Culebra National Wildlife Refuge, Culebra, Puerto Rico. Life: The Excitement of Biology 3:254–289.

Ríos-López, N., and T.M. Aide. 2007. Herpetofaunal dynamics during secondary succession. Herpetologica 63:35–50.

Rivalta González, V., A. Chamizo Lara, L.V. Moreno García, A. Sampedro Marín, and A. Torres Barboza. 2003. Mitos, creencias populares y usos. *In* L. Rodríguez Schettino, ed., Anfibios y reptiles de Cuba, pp. 144–155. Vaasa, Finland: UPC Print.

Rivas, J.A. 2020. Anaconda: the secret life of the world's largest snake. New York: Oxford University Press.

Rivera, P.C., V. Di Cola, J.J. Martínez, C.N. Gardenal, and M. Chiaraviglio. 2011. Species delimitation in the continental forms of the genus *Epicrates* (Serpentes, Boidae) integrating phylogenetics and environmental niche models. PLOS ONE 6(9): e22199.

Rivero, J.A. 1978. Los anfibios y reptiles de Puerto Rico. Rio Piedras: Editorial de la Universidad de Puerto Rico.

Rivero, J.A. 1998. Los anfibios y reptiles de Puerto Rico. San Juan: Editorial de la Universidad de Puerto Rico.

Rivero, J. 2006. Los anfibios y reptiles de Puerto Rico, 2nd ed. San Juan: Editorial de la Universidad de Puerto Rico.

Rivero, J.A., R. Joglar, and I. Vázquez. 1982. Cinco nuevos ejemplares del culebrón de la Mona, *Epicrates m. monensis* (Ophidia: Boidae). Caribbean Journal of Science 17:7–13.

Robiou Lamarche, S. 2004. La gran serpiente en la mitología taína. Boletín del Gabinete de Arqueología 3:51–58.

Rodríguez, C., G.C. Mayer, and P.J. Tolson. 2018. *Chilabothrus inornatus.* IUCN Red List of Threatened Species 2018:e.T7821A74870228. https://dx.doi.org/10.2305/IUCN.UK.2018-2.RLTS.T7821A74870228.en.

Rodríguez, C., G.C. Mayer, and P.J. Tolson. 2021. *Chilabothrus monensis.* IUCN Red List of Threatened Species 2021:e.T7823A18979328. https://dx.doi.org/10.2305/IUCN.UK.2021-2.RLTS.T7823A18979328.en.

Rodríguez, G., and D.P. Reagan.1984. Bat predation by the Puerto Rican boa (*Epicrates inornatus*). Copeia 1984:219–220.

Rodríguez Batista, D., A. Arias Barreto, E. Ruiz Rojas, R. Rodríguez-León Merino, I. Fernández García, M. Martínez [Reyes], M. Hernández Quinta, and O. Barrio Valdés. 2014. Conservación. *In* D. Rodríguez Batista, A. Arias Barreto, and E. Ruiz Rojas, eds., Fauna terrestre del Archipiélago de Sabana-Camagüey, Cuba, pp. 405–438. Havana: Editorial Academia.

Rodríguez-Cabrera, T.M., J. Torres [López], and R. Marrero. 2015. At the lower size limit of snake preying on bats in the West Indies: the Cuban boa, *Chilabothrus angulifer* (Boidae). Reptiles & Amphibians 22:8–15.

Rodríguez-Cabrera, T.M., R. Marrero, and J. Torres [López]. 2016a. An overview of the past, present, and future of the Cuban boa, *Chilabothrus angulifer* (Squamata: Boidae): a top terrestrial predator on an oceanic island. Reptiles & Amphibians 23:152–168.

Rodríguez-Cabrera, T.M., J. Torres López, R. Marrero, E. Morell Savall, and A. Sanz Ochotorena. 2016b. Sexual maturation in free-ranging *Chilabothrus angulifer* (Serpentes: Boidae). Phyllomedusa 15:163–174.

Rodríguez-Cabrera, T.M., A. Fong G., and J. Torres [López]. 2020a. New dietary records for three Cuban snakes in the genus *Tropidophis* (Tropidophiidae), with comments on possible niche partitioning by Cuban tropes. Reptiles and Amphibians 27:201–208.

Rodríguez-Cabrera, T.M., E.M. Savall, S. Rodríguez-Machado, and J. Torres [López]. 2020b. Trophic ecology of the Cuban boa, *Chilabothrus angulifer* (Boidae). Reptiles & Amphibians 27(2): 169–200.

Rodríguez-Cabrera, T.M., A. Hernández Gómez, and J. Torres [López]. 2022. First recorded birth of suspected non-identical twins in the Cuban boa *Chilabothrus angulifer.* The Herpetological Bulletin 160:31–34.

Rodríguez-Durán, A. 1996. Foraging ecology of the Puerto Rican boa (*Epicrates inornatus*): bat predation, carrion feeding, and piracy. Journal of Herpetology 30:533–536.

Rodríguez-Durán, A., J. Pérez, M.A. Montalbán, and J.M. Sandoval. 2010. Predation by free-roaming cats on an insular population of bats. Acta Chiropterologica 12:359–362.

Rodríguez Ferrer, M. 1876. Naturaleza y civilización de la grandiosa isla de Cuba, parte primera: naturaleza. Madrid: Imprenta de J. Noguera a cargo de M. Martínez.

Rodríguez-Robles, J.A., and H.W. Greene. 1996. Ecological patterns in Greater Antillean macrostomatan snake assemblages, with comments on body-size evolution in *Epicrates* (Boidae). *In* R. Powell and R.W. Henderson, eds., Contributions to West Indian herpetology: a tribute to Albert Schwartz. Contributions to Herpetology 12:339–357. Ithaca, NY: Society for the Study of Amphibians and Reptiles.

Rodríguez-Robles, J.A., and M. Leal. 1993. Natural history notes: *Alsophis portoricensis* (Puerto Rican Racer). Diet. Herpetological Review 24:150–151.

Rodríguez-Robles J.A., T. Jezkova, M.K. Fujita, P.J. Tolson, and M.A. García. 2015. Genetic divergence and diversity in the Mona and Virgin Islands boas, *Chilabothrus monensis* (*Epicrates monensis*) (Serpentes: Boidae), West Indian snakes of special conservation concern. Molecular Phylogenetics and Evolution 88:144–153.

Rodríguez Schettino, L., V. Rivalta González, and E. Pérez Rodríguez. 2010. Distribución regional y altitudinal de los reptiles de Cuba. Poeyana 498:11–20.

Rodríguez Schettino, L., C.A. Mancina, and V. Rivalta González. 2013. Reptiles of Cuba: checklist and geographic distribution. Smithsonian Herpetological Information Service 144:1–96.

Rojas[-]Consuegra, R., O. Jiménez Vásquez, M.M. Condis Fernández, and S. Díaz[-]Franco. 2012. Tafonomía y paleoecología de un yacimiento paleontológico del Cuaternario en la Cueva del Indio, Havana, Cuba. Espelunc@digital 12:1–15.

Romero-Nájera, I., A.D. Cuarón, and C. González-Baca. 2006. Distribution, abundance, and habitat use of introduced *Boa constrictor* threatening the native biota of Cozumel Island, Mexico. Biodiversity and Conservation 16:1183–1195.

Rovatsos, M., J. Vukić, P. Lymberakis, and L. Kratochvíl. 2015. Evolutionary stability of sex chromosomes in snakes. Proceedings of the Royal Society B 282: 20151992.

Roze, J.A. 1966. La taxonomia y zoogeografia de los ofidios en Venezuela. Caracas: Universidad Central de Venezuela, Ediciones de la Biblioteca.

Rush, E.M., and R.W. Henderson. 2014. *Corallus grenadensis* (Grenada Bank treeboa). Diet. Herpetological Review 45:143.

Rush, E.M., R.W. Henderson, M. Drake, and N.M. Lonce. 2013. Predation on the introduced frog *Eleutherodactylus johnstonei* by the arboreal boid *Corallus grenadensis.* Herpetology Notes 6:353.

Rush, E.M., V.A. Amadi, R. Johnson, N. Lonce, and H. Hariharan. 2020. *Salmonella* serovars associated with Grenadian tree boa (*Corallus grenadensis*) and their antimicrobial susceptibility. Veterinary Medicine and Science 6(3): 565–569.

Sajdak, R.A., and R.W. Henderson. 2017. *Corallus grenadensis.* Diet. Herpetological Review 48:206–207.

Sajdak, R.A., and R.W. Henderson. 2019. *Corallus grenadensis.* Diet. Herpetological Review 50:388–389.

Sampedro Marín, A. 1998. Adaptaciones morfométricas y conductuales de *Trachemys decussata decussata* (Chelonia: Emydidae). PhD dissertation, Universidad de La Habana, Havana.

Sampedro Marín, A., and L. Montañez Huguez. 1989. Estrategia reproductiva de la jicotea cubana (*Pseudemys decussata*) en la Ciénaga de Zapata. Havana: Editorial Academia.

Santana, E.C., and S.A. Temple. 1988. Breeding biology and diet of Red-tailed Hawks in Puerto Rico. Biotropica 20:151–160.

Savidge, J.A., M.W. Hopken, G.W. Witmer, S.M. Jojola,

J.J. Pierce, P.W. Burke, and A.J. Piaggio. 2012. Genetic evaluation of an attempted *Rattus rattus* eradication on Congo Cay, US Virgin Islands, identifies importance of eradication units. Biological Invasions 14:2343–2354.

Schaefer, R.R., S.E. Koenig, G.R. Graves, and D.C. Rudolph. 2019. Observations of Jamaican Crows (*Corvus jamaicensis*) mobbing Jamaican boas (*Chilabothrus subflavus*). Journal of Caribbean Ornithology 32:73–76.

Schlaepfer, M.A., C. Hoover, and C.K. Dodd Jr. 2005. Challenges in evaluating the impact of the trade in amphibians and reptiles on wild populations. BioScience 55:256–264.

Schmidt, K.P. 1926. The amphibians and reptiles of Mona Island, West Indies. Field Museum of Natural History Publications 12(12): 149–163.

Schmidt, K.P. 1928. Amphibians and land reptiles of Porto Rico. Vol. 10, part 1 of Scientific Survey of Porto Rico and the Virgin Islands. New York: New York Academy of Science.

Schnell, J.H. 1994. Common Black-hawk (*Buteogallus anthracinus*). *In* A. Poole and F. Gill, eds., The birds of North America, 122. Philadelphia: Academy of Natural Sciences; Washington, DC: American Ornithologists' Union.

Schulte, R. 1988. Observaciones sobre la boa verde, *Corallus caninus*, en el Departamento San Martín-Perú. Boletin Lima 55:21–26.

Schwartz, A. 1979. The herpetofauna of Île à Cabrit, Haiti, with the description of two new subspecies. Herpetologica 35:248–255.

Schwartz, A., and R.W. Henderson. 1991. Amphibians and reptiles of the West Indies: descriptions, distributions, and natural history. Gainesville: University of Florida Press.

Schwartz, A., and L.H. Ogren. 1956. A collection of reptiles and amphibians from Cuba, with the descriptions of two new forms. Herpetologica 12:91–110.

Segovia Vega, Y., A.E. Reyes Vázquez, and A. Fong G. 2013. *Epicrates angulifer.* Diet. Herpetological Review 44:153–154.

Seixas, F., F. Morinha, C. Luis, N. Alvura, and M.A. Pires. 2020. DNA-validated parthenogenesis: first case in a captive Cuban boa (*Chilabothrus angulifer*). Salamandra 56:83–86.

Ševčík, J., and J. Ševčík. 1988. Hroznýšovec kubánský ve volné přírodě *Epicrates angulifer.* Akvárium Terárium 31 (1):31–32.

Shaw, C.E. 1957. Longevity of snakes in captivity in the United States as of January 1, 1957. Copeia 1957:310.

Sheplan, B.R., and A. Schwartz. 1974. Hispaniolan boas of the genus *Epicrates* (Serpentes: Boidae) and their Antillean relationships. Annals of the Carnegie Museum 45:57–143.

Sherrod, S.K. 1978. Diets of North American Falconiformes. Raptor Research 12:49–121.

Shine, R., and M. Wall. 2007. Why is intraspecific niche partitioning more common in snakes than in lizards? *In* S.M. Reilly, L.D. McBrayer, and D.B. Miles, eds, Lizard ecology: the evolutionary consequences of foraging mode, pp. 173–208. Cambridge: Cambridge University Press.

Silva[-]Taboada, G. 1979. Los murciélagos de Cuba. Havana: Editorial Academia.

Silva[-]Taboada, G. 1988. Sinopsis de la espeleofauna cubana. Havana: Editorial Científico-Técnica.

Silva-Taboada, G., and K.F. Koopman. 1964. Notes on the occurrence and ecology of *Tadarida laticaudata yucatanica* in Eastern Cuba. American Museum Novitates 2174:1–6.

Šimková, O. 2007. Vznik a vývin velikostního pohlavního dimorfismu vybraných druhů hroznýšovitých hadů (Boidae). Diplomová práce. Přírodovědecká fakulta univerzity Karlovy, Katedra zoologie.

Slavens, F., and K. Slavens. 2003. Longevity–snake index. Reptiles and Amphibians in Captivity. Available at: http://www.pondturtle.com/lsnake.html.

Snider, A.T., and J.K. Bowler. 1992. Longevity of reptiles and amphibians in North American collections, 2nd ed. Society for the Study of Amphibians and Reptiles, Herpetological Circular 21:1–40.

Snow, R.W., K.L. Krysko, K.M. Enge, L. Oberhofer, A. Warren-Bradley, and L. Wilkins. 2007. Introduced populations of *Boa constrictor* (Boidae) and *Python molurus bivitattus* (Pythonidae) in southern Florida. *In* R.W. Henderson and R. Powell, eds., Biology of the boas and pythons, pp. 416–438. Eagle Mountain, Utah: Eagle Mountain Publishing.

Starostová, Z., I. Hynková, and D. Frynta. 2006. Progress in project: perspectives of captive populations of endangered reptiles: genetic variation in two model species (*Epicrates angulifer* and *Cyclura nubila*). Herpetologické informace 5:13–14.

Steadman, D.W., G.K. Pregill, and S.L. Olson. 1984. Fossil vertebrates from Antigua, Lesser Antilles: evidence for late Holocene human-caused extinctions in the West Indies. Proceedings of the National Academy of Sciences of the USA 81: 4448–4451.

Steadman, D.W., R. Franz, G.S. Morgan, N.A. Albury, B. Kakuk, K. Broad, S.E. Franz, et al. 2007. Exceptionally well preserved late Quaternary plant and vertebrate fossils from a blue hole on Abaco, the Bahamas. Proceedings of the National Academy of Sciences of the USA 104:19897–19902.

Stejneger, L. 1901. A new systematic name for the yellow boa of Jamaica. Proceedings of the United States National Museum 23:469–470.

Stejneger, L. 1904. The herpetology of Porto Rico. Annual Report of the United States National Museum for 1902:549–734.

Stouvenot, C., S. Grouard, S. Bailon, D. Bonnissent, A. Lenoble, N. Serrand, and V. Sierpe. 2014. L'abri sous roche Cadet 3 (Marie-Galante): un gisement á accumulations de faune et á vestiges archéologiques. *In* B. Bérard, and C. Losier, eds., Archéologie Caraïbe, Taboui 2, pp. 77–102. Leiden, Netherlands: Sidestone.

Straker, L. 2008. Concerns expressed about the use of live snakes during carnival jump up. eTurboNews, http://www.eturbonews.com/print/4317/concerns-expressed-about-the-use-live-snakes-during-carnival-jump-up.

Stull, O.G. 1933. Two new subspecies of the family Boidae. Occasional Papers of the Museum of Zoology, University of Michigan 267:1–4.

Stull, O.G. 1935. A check list of the family Boidae. Proceedings of the Boston Society of Natural History 40:387–408.

Suárez, W. 2000. Fossil evidence for the occurrence of Cuban Poorwill *Siphonorhis daiquiri* in western Cuba. Cotinga 14:66–68.

Suárez-Atilano, M., A.D. Cuarón, and E. Vázquez-Domínguez. 2019. Deciphering geographical affinity and reconstructing invasion scenarios of *Boa imperator* on the Caribbean island of Cozumel. Copeia 107:606–621.

Tavárez, H., and L. Elbakidze. 2019. Valuing recreational enhancements in the San Patricio Urban Forest of Puerto Rico: a choice experiment approach. Forest Policy and Economics 109:102004.

Taylor, K., H.P. Nelson, and A. Lawrence. 2011. Population density of the Cook's tree boa (*Corallus ruschenbergerii*) in the Caroni Swamp. *In* A. Lawrence and H.P. Nelson, eds., Proceedings of the 1st Research Symposium on Biodiversity in Trinidad and Tobago, pp. 8–18. St. Augustine: University of the West Indies.

Teubner, V.A. 1986. Endocrine and behavioral aspects of the reproductive cycle of the Haitian boa (*Epicrates striatus*). PhD dissertation, University of Toledo, Toledo, Ohio.

Teubner, V.A., and P.J. Tolson. 1984. Behavioral and endocrine aspects of the reproductive cycle in the Haitian boa *Epicrates striatus*. American Zoologist 24:114A.

Thomas, O., and S.J.R. Allain. 2021. A review of prey taken by anacondas (Squamata, Boidae, *Eunectes*). Reptiles & Amphibians 28(2): 329–334.

Thompson, N.E., E.W. Lankau, and G.M. Rogall. 2018. Snake fungal disease in North America. U.S. Geological Survey Fact Sheet 2017–3064. https://doi.org/10.3133/fs20173064.

Tilman, D., R.M. May, C.L. Lehman, and M.A. Nowak. 2002. Habitat destruction and the extinction debt. Nature 371:65–66.

Tolson, P.J. 1980. Comparative reproductive behavior of four species of snakes in the boid genus *Epicrates*. *In* Proceedings of the 4th Reptile Symposium on Captive Propagation and Husbandry, pp. 87–97. Thurmont, Maryland: Zoological Consortium.

Tolson, P.J. 1983. Captive propagation and husbandry of the Cuban boa, *Epicrates angulifer*. *In* Proceedings of the 6th Reptile Symposium on Captive Propagation and Husbandry, pp. 248–292. Thurmont, Maryland: Zoological Consortium.

Tolson, P.J. 1985. A field report on the status of the Virgin Islands boa, *Epicrates monensis granti*, on Cayo Diablo, Puerto Rico.

Tolson, P.J. 1986. A field report on the status of the Virgin Islands boa, *Epicrates monensis granti*, on Cayo Diablo, Puerto Rico, part II. The 1985–1986 field season. Final Report to the Puerto Rico Department of Natural Resources.

Tolson, P.J. 1987. Phylogenetics of the boid snake genus *Epicrates* and Caribbean vicariance theory. Occasional Papers of the Museum of Zoology, University of Michigan 715:1–59.

Tolson, P.J. 1988. Critical habitat, predator pressures, and the management of *Epicrates monensis* (Serpentes: Boidae) on the Puerto Rico Bank: a multivariate analysis. *In* R.C. Szaro, K.E. Severson, and D.R. Patton, eds., Management of amphibians, reptiles, and small mammals in North America, pp. 228–238. U.S. Department of Agriculture Forest Service, Fort Collins, Colorado, General Technical Report RM-166.

Tolson, P.J. 1989. Breeding the Virgin Islands boa, *Epicrates monensis granti*, at the Toledo Zoological Gardens. International Zoo Yearbook 28:163–167.

Tolson, P.J. 1991a. Captive breeding and reintroduction: recovery efforts for the Virgin Islands boa, *Epicrates monensis granti*. Endangered Species UPDATE 8(1): 52–53.

Tolson, P.J. 1991b. *Epicrates* (West Indian boa). Reproductive longevity. Herpetological Review 22:100.

Tolson, P.J. 1991c. The conservation status of *Epicrates monensis* (Serpentes Boidae) on the Puerto Rico

Bank. *In* J.A. Moreno, ed., Status y distribución de los anfibios y reptiles de Puerto Rico, pp. 11–63. Departamento de Recursos Naturales de Puerto Rico Publicaciones Científicas Misceláneas 1.

Tolson, P.J. 1992a. A report on the presence of the Virgin Islands boa, *Epicrates monensis granti*, on Punta Soldado, Isla Culebra, Puerto Rico. Report to the Puerto Rico Department of Natural Resources.

Tolson, P.J. 1992b. The reproductive biology of the Neotropical boid genus *Epicrates* (Serpentes: Boidae). *In* W.C. Hamlett, ed., Reproductive biology of South American vertebrates, pp 165–178. New York: Springer-Verlag.

Tolson, P.J. 1994. The reproductive management of the insular species of *Epicrates* (Serpentes: Boidae) in captivity. *In* J. B. Murphy, K. Adler, and J. T. Collins, eds., Captive management and conservation of amphibians and reptiles. Contributions to Herpetology 11:353–357. Ithaca, NY: Society for the Study of Amphibians and Reptiles.

Tolson, P. 1996a. *Chilabothrus granti*. IUCN Red List of Threatened Species 1996:e.T7829A12853577. http://dx.doi.org/10.2305/IUCN.UK.1996.RLTS.T7829A12853577.en.

Tolson, P. 1996b. *Chilabothrus monensis*. IUCN Red List of Threatened Species 1996:e.T7823A12853323. https://dx.doi.org/10.2305/IUCN.UK.1996.RLTS.T7823A12853323.en.

Tolson, P.J. 1996c. Conservation of *Epicrates monensis* on the satellite islands of Puerto Rico. *In* R. Powell and R.W. Henderson, eds., Contributions to West Indian herpetology: a tribute to Albert Schwartz. Contributions to Herpetology 12:407–416. Ithaca, NY: Society for the Study of Amphibians and Reptiles.

Tolson, P.J. 1997. Population census and habitat assessment for the Puerto Rican boa, *Epicrates inornatus*, at the U.S. Naval Security Group Activity, Sabana Seca, Puerto Rico. Final Report. Puerta de Tierra, Puerto Rico: Departamento de Recursos Naturales y Ambientales.

Tolson, P.J. 2000. Ecological studies of the Mona Island boa, *Epicrates monensis monensis*, on Isla Mona, Puerto Rico. Final Report to the U.S. Fish and Wildlife Service as per purchase order 40181–5–6048.

Tolson, P.J. 2004. Surveys for the Puerto Rican boa, *Epicrates inornatus*, and the Virgin Islands boa, *Epicrates monensis granti*, at the U.S. Naval Base, Roosevelt Roads, Puerto Rico. Prepared for Geo-Marine, Plano, Texas.

Tolson, P.J. 2005. Reintroduction evaluation and habitat assessment of the Virgin Islands boa, *Epicrates monensis granti*, to the U. S. Virgin Islands. Final Report to the U.S. Fish and Wildlife Service.

Tolson, P. J. 2012. Battle to the death in the Graffiti Hill arroyo: Cuban anole and boa fight to the end. Anole Annals. http://www.anoleannals.org/2012/11/22/battle-to-the-death-in-the-graffiti-hill-arroyo/.

Tolson, P.J., and R.W. Henderson. 1993. The natural history of West Indian boas. Taunton, England: R&A Publishing.

Tolson, P.J., and R.W. Henderson. 2006. An overview of snake conservation in the West Indies. Applied Herpetology 3:345–356.

Tolson, P.J., and R.W. Henderson. 2011. An overview of snake conservation in the West Indies. *In* A. Hailey, B.S. Wilson, and J.A. Horrocks, eds., Conservation of Caribbean island herpetofaunas. Vol. 1, Conservation biology and the wider Caribbean, pp. 49–62. Leiden, Netherlands: Brill.

Tolson, P.J., and C. Petersen. 2008. Homing in on hutias at Gtmo: the Navy & the Toledo Zoo partner to study a little-known rodent. Currents (Summer) 2008:8–15.

Tolson, P.J., and J.L. Piñero. 1985. A field report on the status of the Virgin Islands boa, *Epicrates monensis granti*, on Cayo Diablo, Puerto Rico. Report to the Departmento de Recursos Naturales, Puerto Rico.

Tolson, P.J., and V.A. Teubner. 1987. The role of social manipulation and environmental cycling in propagation of the boid genus *Epicrates*: lessons from the field and laboratory. American Association of Zoological Parks and Aquariums Regional Conference Proceedings 1987:606–613.

Tolson, P.J., M.A. García, and C.L. Ellsworth. 2007. Habitat use by the Mona boa (*Epicrates m. monensis*) on Isla Mona, West Indies. *In* R.W. Henderson and R. Powell, eds., Biology of the boas and pythons, pp 118–126. Eagle Mountain, Utah: Eagle Mountain Publishing.

Tolson, P.J., M.A. García, and J.J. Pierce. 2008. Reintroduction of the Virgin Islands boa to the Puerto Rico Bank. Special issue of Re-introduction News 27(2008): 76.

Tucker, A.M., C.P. McGowan, E.S. Mulero Oliveras, N.F. Angeli, and J.P. Zegarra. 2020. A demographic projection model to support conservation decision making for an endangered snake with limited monitoring data. Animal Conservation 24:291–301.

Turvey, S.T., and K. Helgen. 2017. *Oryzomys antillarum*. IUCN Red List of Threatened Species 2017:e.T136540A22388029. http://dx.doi.org?10.2305/IUCN.UK.2017-3.RLTS.T136540A22388029.en.

Turvey, S.T., F.V. Grady, and P. Rye. 2006. A new genus and species of 'giant hutia' (*Tainotherium valei*) from the Quaternary of Puerto Rico: an extinct arboreal quadruped? Journal of Zoology 270:585–594.

Turvey, S.T., J.R. Oliver, Y.M. Narganes Storde, and P. Rye. 2007. Late Holocene extinction of Puerto Rican native land mammals. Biology Letters 22:193–196.

Turvey, S.T., M. Weksler, E.L. Morris, and M. Nokkert. 2010. Taxonomy, phylogeny, and diversity of the extinct Lesser Antillean rice rats (Sigmodontinae: Oryzomyini), with description of a new genus and species. Zoological Journal of the Linnaean Society 160: 748-772.

Turvey, S.T., R.J. Kennerley, J.M. Nuñez-Miño, and R.P. Young. 2017. The last survivors: current status and conservation of the non-volant land mammals of the insular Caribbean. Journal of Mammalogy 98:918-936.

Tyler, R.E. 1850. Notes on the serpents of St. Lucia. Annals of the Magazine of Natural History 6(2): 130–134.

Tyler, R.K., K.M. Winchell, and L.J. Revell. 2016. Tails of the city: caudal autotomy in the tropical lizard, *Anolis cristatellus*, in urban and natural areas of Puerto Rico. Journal of Herpetology 50:435–441.

Tzika, A.C., S. Koenig, R. Miller, G. Garcia, C. Remy, and M.C. Milinkovitch. 2008. Population structure of an endemic vulnerable species, the Jamaican boa (*Epicrates subflavus*). Molecular Ecology 17:533–544.

Tzika, A.C., C. Remy, R. Gibson, and M.C. Milinkovitch. 2009. Molecular genetic analysis of a captive-breeding program: the vulnerable endemic Jamaica yellow boa. Conservation Genetics 10:69–77.

Uetz, P., and A. Stylianou. 2018. The original descriptions of reptiles and their subspecies. Zootaxa 4375: 257-264.

Ugueto, G.N., and G.A. Rivas. 2010. Amphibians and reptiles of Margarita, Coche and Cubagua. Frankfurt am Main: Edition Chimaira.

USFWS (U.S. Fish and Wildlife Service). 1978. Final determination of threatened status and critical habitat for the Mona boa and Mona ground iguana. Federal Register 43(24): 4618.

USFWS. 1980. Status of Virgin Islands boa clarified. Endangered Species Technical Bulletin 5:12.

USFWS. 1984. Mona boa recovery plan. USFWS, Atlanta, GA.

USFWS. 1986a. Recovery plan for the Puerto Rican boa (*Epicrates inornatus*). USFWS, Atlanta, GA.

USFWS. 1986b. Virgin Islands tree boa recovery plan. USFWS, Atlanta, GA.

USFWS. 1991. Endangered and threatened wildlife and plants: 5-year review of listed species. Federal Register 56:56882–56900.

USFWS. 2009. Virgin Islands tree boa (*Epicrates monensis granti*) 5-year review: summary and evaluation. USFWS, Boquerón, Cabo Rojo, Puerto Rico.

USFWS. 2011. 5-Year review of the Puerto Rican boa (*Epicrates inornatus*): summary and evaluation. USFWS, Boquerón, Cabo Rojo, Puerto Rico.

USFWS. 2014. 5-Year review of Mona boa (*Epicrates monensis monensis*). USFWS, Atlanta, GA.

USFWS. 2018. Virgin Islands tree boa (*Chilabothrus granti*) species status assessment. Version 1.0. USFWS, Atlanta, GA.

Valencia, J.H., E. Arbeláez, K. Garzón, and P. Picerno-Toala. 2008. Notes on *Corallus blombergi* (Rendahl and Vestergren, 1941) from Ecuador. Herpetozoa 21:91–94.

Vandeventer, T.L. 1992. In search of the Tete'chein [sic]: observations on the natural history of *Boa constrictor nebulosus*. 16th International Herpetological Symposium on Captive Propagation and Husbandry, pp. 1–9. International Herpetological Symposium.

Vanzolini, P.E. 1952. Fossil snakes and lizards from the lower Miocene of Florida. Journal of Paleontology 3:452–457.

Vareschi, E., and W. Janetzky. 1998. Bat predation by the yellow snake or Jamaican boa, *Epicrates subflavus*. Jamaica Naturalist 5:34–35.

Varona, L.S., and O. Arredondo. 1979. Nuevos fósiles de Capromyidae (Rodentia: Caviomorpha). Poeyana 195:1–51.

Vázquez-Domínguez, M. Suárez-Atilano, W. Booth, C. González-Baca, and A.D. Cuarón. 2012. Genetic evidence of a recent successful colonization of introduced species on islands: *Boa constrictor imperator* on Cozumel Island. Biological Invasions 14:2101–2116.

Vázquez Milian, A., and E. Nieves Lorenzo. 1980. Aspectos de la reproducción de la torcaza cabeciblanca (*Columba leucocephala* Linneo) en la EFI "Playa Larga." BS dissertation, Centro Universitario de Pinar del Río, Pinar del Río, Cuba.

Vega-Ross, M., 2018. Movement, habitat use and diet of an invasive snake, *Boa constrictor* (Boidae), in Puerto Rico. PhD dissertation, University of Puerto Rico–Mayagüez.

Vergner, J. 1978. Tři druhy rodu *Epicrates* z Antilské oblasti. Akvárium Terárium 21(5): 140–142.

Vergner, I. 1989. K ekologii, etologii a rozmnožování hroznýšovce kubánského, *Epicrates angulifer*, v teráriu. Akvárium Terárium 32(1): 26–31.

Viña Dávila, N., and L.F. de Armas. 1989. Depredación de

*Tropidophis melanurus* (Serpentes: Tropidophiidae) por *Epicrates angulifer* (Serpentes: Boidae). Miscelánea Zoológica 41:2–3.

Vinnikov, K.Y., A. Robock, R.J. Stouffer, J.E. Wals, C.L. Parkinson, D.J. Cavalieri, J.F.B. Mitchell, D. Garrett, and V.F. Zakharov.1999. Global warming and northern hemisphere sea ice extent. Science 286:1934–1937.

Viñola[-]López, L.W., O.H. Garrido, and A. Nermúdez. 2018. Notes on *Mesocapromys sanfelipensis* (Rodentia: Capromyidae) from Cuba. Zootaxa 4410:164–176.

Vitt, L.J., and L.D. Vangilder. 1983. Ecology of a snake community in northeastern Brazil. Amphibia-Reptilia 4:273–296.

Voss, R.S., and S.A. Jansa. 2012. Snake-venom resistance as a mammalian trophic adaptation: lessons from didelphid marsupials. Biological Reviews 87:822–837.

Wadsworth, F.W. 1949. The development of forest land resources of the Luquillo Mountains, Puerto Rico. PhD dissertation, University of Michigan, Ann Arbor.

Waller, T., E. Buongermini, and P.A. Micucci. 2001. *Eunectes notaeus* (yellow anaconda). Diet. Herpetological Review 32:47.

Watts, D. 1987. The West Indies: patterns of development, culture and environmental change since 1492. Cambridge: Cambridge University Press.

Wetherbee, D.K. 1989. The patronym and type-locality of *Pelophilus* (*Epicrates*) *fordii* Günther 1861 of Hispaniola. Herpetological Review 20:8.

Wiewandt, T.A. 1977. Ecology, behavior and management of the Mona Island ground iguana. PhD dissertation, Cornell University, Ithaca, NY.

Wiley, J.W. 2003. Habitat association, size, stomach contents, and reproductive condition of Puerto Rican boas (*Epicrates inornatus*). Caribbean Journal of Science 39:189–194.

Wiley, J.W. 2010. Food habits of the endemic Ashy-faced Owl (*Tyto glaucops*) and recently arrived Barn Owl (*Tyto alba*) in Hispaniola. Journal of Raptor Research 44:87–100.

Wiley, J.W., and B.N. Wiley. 1981. Breeding season ecology and behavior of Ridgway's Hawk (*Buteo ridgwayi*). Condor 83:132–151.

Wilson, B.S., J.A. Horrocks, and A. Hailey. 2006. Conservation of insular herpetofaunas in the West Indies. Applied Herpetology 3:181–195.

Wilson, B.S., S.E. Koenig, R. van Veen, E. Miersma, and D.C. Rudolph. 2011. Cane toads a threat to West Indian wildlife: mortality of Jamaican boas attributable to toad ingestion. Biological Invasions 13:55–60.

Wing, E.S. 2001. Native American use of animals in the Caribbean. *In* C.A. Woods and F.E. Sergile, eds., Biogeography of the West Indies: patterns and perspectives, 2nd ed., pp. 481–518. Boca Raton: CRC Press.

Woods, C.A. 1996. The land mammals of Puerto Rico and the Virgin Islands. Annals of the New York Academy of Science 776:131–149.

World Bank Group. 2018. Population density. The World Bank. https://data.worldbank.org/indicator/en.pop.dnst.

Wunderle, J.M., Jr., J.E. Mercado, B. Parresol, and E. Terranova. 2004. Spatial ecology of Puerto Rican boas (*Epicrates inornatus*) in a hurricane impacted forest. Biotropica 36:555–571.

Yorks, D.T., K.E. Williamson, R.W. Henderson, R. Powell and J.S. Parmerlee Jr. 2003. Foraging behavior in the arboreal boid *Corallus grenadensis*. Studies on the Neotropical Fauna and Environment 38:167–172.

Zajicek, D., and M. Mauri Méndez. 1969. Hemoparásitos de algunos animals de Cuba. Poeyana 66:1–10.

Zenneck, I. 1898. Die Ziechnung der Boiden. Zeitschrift für wissenschaftliche Zoologie 64:1–384.

# INDEX

Numbers in boldface indicate photos; numbers followed by *fig* indicate figures and by *t* indicate tables.

# AUTHOR BIOGRAPHIES

**R. Graham Reynolds** is an associate professor of biology at the University of North Carolina–Asheville, an associate at the Museum of Comparative Zoology at Harvard University, and a National Geographic Explorer. He earned a BA in biology from Duke University, followed by a PhD in ecology and evolutionary biology at the University of Tennessee–Knoxville. He has coedited two books on reptiles and amphibians and has authored more than fifty scientific papers and book chapters—many on the West Indian boas. His research focuses on the biology and conservation of reptiles and amphibians in the Caribbean and in the southeastern United States, with special focus on evolutionary genetics of West Indian reptiles.

**Robert W. Henderson** is curator emeritus of herpetology at the Milwaukee Public Museum. For the past 38 years the geographic focus of his research has been the West Indies with, for about 30 years, a taxonomic emphasis on the boid genus *Corallus* (especially *Co. grenadensis* on the Grenada Bank). He has written, cowritten, or coedited fifteen books (ten of which have a strong West Indian focus) and over 250 papers and book chapters. His current research continues to focus on *Co. grenadensis* with an emphasis on demography and the impact of human activity on its natural history.

*Left*: R. Graham Reynolds holding *Chilabothrus s. strigilatus* on Long Island, Bahamas; photo by Alberto R. Puente-Rolón. *Middle*: Robert W. Henderson holding *Corallus grenadensis*; photo by R. Sajdak. *Right*: Luis M. Díaz holding *Chilabothrus angulifer*; photo by Leodan Roque.

*Top*: Alberto R. Puente-Rolón holding *Chilabothrus granti*. *Bottom*: Tomás M. Rodríguez-Cabrera holding *Chilabothrus angulifer*. Photos courtesy of ARPR and Raimundo López-Silvero.

**Luis M. Díaz** is curator of herpetology at the National Museum of Natural History in Cuba since 1998. He earned a PhD in biological sciences from the University of Havana. He has published about one hundred papers on the taxonomy, bioacoustics, natural history, evolution, and conservation of Cuban amphibians and reptiles, with a few contributions in Hispaniola as well. In 2008 he published the book *Guía Taxonómica de los Anfibios de Cuba* and is now working on several projects focused on systematics, conservation, and new field guides of Cuban and Caribbean herpetofauna.

**Tomás M. Rodríguez-Cabrera** is an autodidact researcher with over 20 years of experience studying wildlife in Cuba who currently works at the Instituto de Ecología y Sistemática in Havana. His research interests include ecology, systematics, and conservation—mostly of reptiles (Squamata) and arthropods (arachnids, crustaceans). The Cuban boa (*Chilabothrus angulifer*) has been the focus of his investigations for about 14 years. He worked as a wildlife conservation specialist at the Cienfuegos Botanic Garden (formerly Atkins Garden and Research Laboratory of the Harvard University Arnold Arboretum at Soledad, Cuba) for 5 years and has been working as a scientific consultant and writer for nature documentary films since 2017. He has authored and coauthored over eighty scientific papers and book chapters. Currently his research focuses on the ecology and reproductive biology of Cuban snakes.

**Alberto R. Puente-Rolón** is an associate professor of biology at the University of Puerto Rico–Mayagüez. He earned an MS in biology at the University of Puerto Rico Mayagüez and a PhD in Biology at the University of Puerto Rico Río Piedras. A broadly trained naturalist, he has studied wildlife in Puerto Rico and the Greater Antilles for more than 25 years, both as an academic and with the Department of Natural Resources in San Juan. His work has largely focused on herpetology, and he has published over thirty scientific papers and book chapters and serves as a consultant for nature documentary films.